Frontiers in Mathematics

Advisory Editors

William Y. C. Chen, Nankai University, Tianjin, China

Laurent Saloff-Coste, Cornell University, Ithaca, NY, USA

Igor Shparlinski, The University of New South Wales, Sydney, NSW, Australia

Wolfgang Sprößig, TU Bergakademie Freiberg, Freiberg, Germany

This series is designed to be a repository for up-to-date research results which have been prepared for a wider audience. Graduates and postgraduates as well as scientists will benefit from the latest developments at the research frontiers in mathematics and at the "frontiers" between mathematics and other fields like computer science, physics, biology, economics, finance, etc. All volumes are online available at SpringerLink.

Mohammad Sal Moslehian · Hiroyuki Osaka

Advanced Techniques with Block Matrices of Operators

Mohammad Sal Moslehian
Department of Pure Mathematics
Ferdowsi University of Mashhad
Mashhad, Iran

Hiroyuki Osaka
Department of Mathematical Sciences
Ritsumeikan University
Kyoto, Japan

ISSN 1660-8046 ISSN 1660-8054 (electronic)
Frontiers in Mathematics
ISBN 978-3-031-64545-7 ISBN 978-3-031-64546-4 (eBook)
https://doi.org/10.1007/978-3-031-64546-4

Mathematics Subject Classification: 15A60, 47B10, 47B15, 15B48, 46L05, 46L10, 46L30, 47A05, 47A08, 47A30, 47A50, 47A63, 47A64, 47C15

© The Editor(s) (if applicable) and The Author(s), under exclusive license to Springer Nature Switzerland AG 2024

This work is subject to copyright. All rights are solely and exclusively licensed by the Publisher, whether the whole or part of the material is concerned, specifically the rights of translation, reprinting, reuse of illustrations, recitation, broadcasting, reproduction on microfilms or in any other physical way, and transmission or information storage and retrieval, electronic adaptation, computer software, or by similar or dissimilar methodology now known or hereafter developed.

The use of general descriptive names, registered names, trademarks, service marks, etc. in this publication does not imply, even in the absence of a specific statement, that such names are exempt from the relevant protective laws and regulations and therefore free for general use.

The publisher, the authors and the editors are safe to assume that the advice and information in this book are believed to be true and accurate at the date of publication. Neither the publisher nor the authors or the editors give a warranty, expressed or implied, with respect to the material contained herein or for any errors or omissions that may have been made. The publisher remains neutral with regard to jurisdictional claims in published maps and institutional affiliations.

This book is published under the imprint Birkhäuser, www.birkhauser-science.com by the registered company Springer Nature Switzerland AG
The registered company address is: Gewerbestrasse 11, 6330 Cham, Switzerland

If disposing of this product, please recycle the paper.

Dedicated to our dear math teachers, who introduced us to the mysterious world of mathematics:

Ali Abaei, Mohammad Ashraf Hesari, Mahmoud Elhami, Ebrahim Rezazadeh, and Hassan Taghizadeh (in high school)

Michi Nishizuka (in junior high school) with respect and affection

Preface

The main goal of this modern book is to introduce several powerful techniques and fundamental ideas involving block matrices of operators (matrices) as well as matrices with entries in a C^*-algebra $\mathscr{A}$. These techniques and ideas can be used to solve problems that are difficult to address in $\mathscr{A}$ itself. In particular, 2×2 operator matrices provide important mathematical inequalities in many areas of operator theory and matrix analysis. We employ these matrices to simplify problems. For example, the classical proof of the Putnam–Fuglede theorem is based on 2×2 matrix techniques. Moreover, such methods are applied to investigate n-positive maps, completely positive maps, operator means, nonlinear positive maps, and various operator and norm inequalities. In recent decades, operator matrices in quantum information and computing theories have received significant attention.

This book is suitable as a textbook for an advanced undergraduate or graduate course or as a supplement for researchers and students in mathematics and physics who have a basic knowledge of linear algebra, functional analysis, and operator theory. The book provides detailed arguments and relevant technical material for most results. Some portions are drawn from various sources and presented in a self-contained, unified, and logically consistent manner.

By addressing existing literature, we ensure that readers can effectively understand our aim and derive essential techniques for working with block matrices in a clear, coherent, and integrated manner. This approach allows us to enrich the book with a deep contextual background and established methods.

This book is divided into five chapters.

Chapter 1 introduces the reader to basic concepts and theorems from functional analysis, operator theory, and matrix analysis. These concepts and results serve as essential tools for the subsequent chapters.

Chapter 2 is the heart of the book. It introduces the concept of block matrices of operators through the isomorphism $\mathbb{M}_n(\mathbb{B}(\mathscr{H})) \simeq \mathbb{B}(\mathscr{H}^{\oplus n})$. This chapter provides a comprehensive exposition of dilation theory, presenting numerous characterizations of the positivity of 2×2 operator matrices of the form $\begin{bmatrix} A & X \\ X^* & B \end{bmatrix}$. It also investigates the properties of 2×2 matrices with entries in a C^*-algebra. In addition, operator matrices are used to derive several inequalities related to the eigenvalues and unitarily invariant norms of matrices.

Chapter 3 is devoted to the study of operator monotone and operator convex functions, along with a thorough investigation of their foundational characteristics. In this chapter, we establish connections between operator monotone functions and operator means using the Kubo–Ando theory. We also address the verification of positive, n-positive, weakly n-positive, and completely positive maps.

Chapter 4 extends the concepts of variance and covariance beyond classical probability theory to a noncommutative framework. In this context, we provide upper bounds for unitarily invariant norms of the covariance of bounded linear operators and matrices.

Chapter 5 is concerned with the topic of nonlinear positive maps. We examine Lieb maps and their essential properties, explore the concept of 3-positivity in nonlinear maps, and investigate the continuity of 3-positive maps. Throughout this chapter, we extensively utilize block techniques to facilitate our analysis.

At the end of each chapter, readers can expect a variety of exercises and problems with references to the relevant literature. Some of these problems involve open questions, while others are challenging and provide suggestions for future research. The book also includes an extensive bibliography with about 230 references. It is worth noting that several results are due to esteemed mathematicians in the field such as Tsuyoshi Ando, Rajendra Bhatia, Jean-Christophe Bourin, Man-Duen Choi, Fuad Kittaneh, Vern I. Paulsen, and others. In addition, the book includes our favorite strategies involving block matrices. However, readers may find techniques in the literature other than those covered in this book.

The authors would like to express their gratitude to Jean-Christophe Bourin, Ali Dadkhah, Masatoshi Fujii, Fumio Hiai, Minghua Lin, Pei Yuan Wu, and Qingxiang Xu for their valuable comments and suggestions.

Mashhad, Iran	Mohammad Sal Moslehian
Kyoto, Japan	Hiroyuki Osaka
Spring 2024	

Contents

1 **Matrices and Hilbert Space Operators** 1
   1.1 Fundamental Information 1
   1.2 Some Decompositions of Operators 16
   1.3 Unitarily Invariant Norms 23
   1.4 Moore–Penrose Inverse 26
   1.5 Differences Between Real and Complex Hilbert Spaces 28
   1.6 Exercises and Problems 29

2 **Block Matrices of Operators** 33
   2.1 Operator Matrices of Size $n \times n$ 33
   2.2 Dilation 43
   2.3 2×2 Matrices with Entries in a C^*-Algebra 50
   2.4 Positivity of 2×2 Matrices of Operators 55
   2.5 Diagonal Blocks and Unitary Orbits 69
   2.6 Inequalities Follow from the Positivity of 2×2 Block Matrices 73
   2.7 Inequalities Related to Unitarily Invariant Norms and Numerical
       Radius 77
   2.8 Some Inequalities Involving Eigenvalues and Singular Values 86
   2.9 Exercises and Problems 94

3 **Operator Monotone Functions and Positive Maps** 101
   3.1 Operator Monotone and Operator Convex Functions 101
   3.2 Operator Means 112
   3.3 Positive and Completely Positive Linear Maps 129
   3.4 Weakly 2-Positive Maps 147
   3.5 Positive Linear Maps on Matrix Algebras 151
   3.6 Exercises and Problems 164

4 **Operator Variance and Covariance** 167
   4.1 Bounds for the Operator Norm of the Covariance 167
   4.2 Bounds for Unitarily Invariant Norms of the Covariance 174

	4.3 Concluding Remarks	180
	4.4 Exercises and Problems	182
5	**Nonlinear Positive Maps**	185
	5.1 Lieb Maps and Their Fundamental Properties	186
	5.2 3-Positivity of Nonlinear Maps	189
	5.3 Continuity of 3-Positive Maps	193
	5.4 Exercises and Problems	201
Bibliography		205
Index		215

Matrices and Hilbert Space Operators

We introduce basic concepts and key results from functional analysis, operator theory, and matrix analysis. These concepts and results will play a central role as indispensable tools in the upcoming chapters.

1.1 Fundamental Information

Throughout the book, all vector spaces are considered to be complex vector spaces, and all operators are assumed to be linear and bounded unless explicitly stated otherwise. A capital letter stands for an operator, a matrix, or an element of a C^*-algebra. We use the following notations:

A *normed space* is a vector space $\mathcal{X}$ equipped with a so-called *norm* $\|\cdot\| : \mathcal{X} \to [0, \infty)$ satisfying the following properties: (i) $\|x\| = 0$ if and only if $x = 0$; (ii) $\|\lambda x\| = |\lambda| \, \|x\|$; (iii) $\|x + y\| \leq \|x\| + \|y\|$ for all $x, y \in \mathcal{X}$ and all scalars λ in the field $\mathbb{C}$ of complex numbers. If (i) replaced by $\|x\| = 0$ if $x = 0$, then $\|\cdot\|$ is called a *seminorm*.

If a normed space $(\mathcal{X}, \|\cdot\|)$, endowed with the metric $d(x, y) := \|x - y\|$, is a complete metric space, in the sense that every Cauchy sequence in $\mathcal{X}$ converges to some vector in $\mathcal{X}$, then it is called a *Banach space*. The *dual space* $\mathcal{X}'$ of a normed space $\mathcal{X}$ is defined as the Banach space consisting of all continuous linear functionals $f : \mathcal{X} \to \mathbb{C}$.

Every finite-dimensional normed space $\mathcal{X}$ is a Banach space and any two norms $\|\cdot\|_1$ and $\|\cdot\|_2$ on it are equivalent. This means that there exist positive numbers α and β such that $\alpha \|x\|_2 \leq \|x\|_1 \leq \beta \|x\|_2$ for all $x \in \mathcal{X}$. The space $\mathbb{M}_n$ of $n \times n$ matrices with complex entries together with any one of the norms $\|[a_{ij}]\|_\sigma = \sum_{i,j=1}^{n} |a_{ij}|$ and $\|[a_{ij}]\|_{\max} = \max_{1 \leq i,j \leq n} |a_{ij}|$ is a Banach space. Hilbert spaces are significant examples of Banach spaces as defined below.

Suppose that $\mathcal{H}$ is a vector space. A function $\langle \cdot, \cdot \rangle : \mathcal{H} \times \mathcal{H} \to \mathbb{C}$ is called a *semi-inner product* if

(i) $\langle x + \lambda y, z \rangle = \langle x, z \rangle + \lambda \langle y, z \rangle$,
(ii) $\langle y, x \rangle = \overline{\langle x, y \rangle}$,
(iii) $\langle x, x \rangle \geq 0$

for all $x, y, z \in \mathcal{H}$ and all scalars $\lambda \in \mathbb{C}$. Then, the pair $(\mathcal{H}, \langle \cdot, \cdot \rangle)$ is called a *semi-inner product space* and $\|x\| = \sqrt{<x,x>}$ gives a seminorm. In such a space, it holds that

$$\langle x, y \rangle = \frac{1}{4} \sum_{k=0}^{3} i^k \|x + i^k y\|^2 \quad (x, y \in \mathcal{H}) \tag{1.1.1}$$

which is called the *polarization identity*.

If $x = 0$ whenever $\langle x, x \rangle = 0$, then $\mathcal{H}$ is called an *inner product space* and $\|x\| = \langle x, x \rangle^{\frac{1}{2}}$ can be shown to be a norm on $\mathcal{H}$ by using the following *Cauchy–Schwarz inequality*:

Theorem 1.1.1 (Cauchy–Schwarz inequality) *In a semi-inner product space $\mathcal{H}$, it holds that*

$$|\langle x, y \rangle| \leq \langle x, x \rangle^{1/2} \langle y, y \rangle^{1/2} \quad (x, y \in \mathcal{H}).$$

If the norm $\|\cdot\|$ is complete, then $\mathcal{H}$ is called a *Hilbert space*. It is known that every Hilbert space is of the form $L_2(\Omega, \mu)$, which is the *space of square integrable functions* (by identifying functions that are equal almost everywhere) in a measure space (Ω, μ). In particular, the space ℓ_2 consists of square summable sequences (x_n) equipped with the inner product

$$\langle (x_n), (y_n) \rangle = \sum_{n=1}^{\infty} x_n \overline{y_n}.$$

Also, the closed subspace $\mathbb{C}^n$ of ℓ^2 is known as *n-dimensional Euclidean space*.

A vector $x \in \mathcal{H}$ is called *orthogonal* to $y \in \mathcal{H}$ if $\langle x, y \rangle = 0$. The set $\mathcal{M}^\perp$ consists of all $x \in \mathcal{H}$ that are orthogonal to every element in a subset $\mathcal{M}$ of $\mathcal{H}$, and it is called the *orthogonal complement* of $\mathcal{M}$. Two vectors are called *orthonormal* if they are unit vectors (that is, of norm 1) and orthogonal to each other. An orthonormal basis for $\mathcal{H}$ is an orthonormal set $(e_j)_{j \in J}$ of vectors such that

$$x = \sum_j \langle x, e_j \rangle e_j$$

for every $x \in \mathcal{H}$. Also,

$$\|x\|^2 = \sum_j |\langle x, e_j \rangle|^2$$

1.1 Fundamental Information

is called the *Parseval identity*. The Gram–Schmidt process is a method used to create orthogonal vectors from a finite number of linearly independent vectors in a Hilbert space. Employing Zorn's lemma along with the Gram–Schmidt process, we can show that every Hilbert space admits an orthonormal basis. The standard basis for ℓ_2 is composed of the vectors $e_n = (\delta_{1n}, \delta_{2n}, \delta_{3n}, \ldots)$ for $n = 1, 2, \ldots$, where δ represents the Kronecker delta defined as

$$\delta_{ij} = \begin{cases} 1 & \text{if } i = j \\ 0 & \text{if } i \neq j. \end{cases}$$

A linear map $A : \mathcal{H} \to \mathcal{H}$ is called a *bounded operator* if

$$\sup\{\|Ax\| : x \in \mathcal{H}, \|x\| = 1\} < \infty.$$

In this case, its *operator norm* is defined by

$$\|A\| = \sup\{\|Ax\| : x \in \mathcal{H}, \|x\| = 1\}.$$

The space $\mathbb{B}(\mathcal{H})$ of all bounded linear operators on a complex Hilbert space $\mathcal{H}$ equipped with the operator norm $\|\cdot\|$ is a Banach space. The identity operator in $\mathbb{B}(\mathcal{H})$ is denoted by I. We can turn the space $\mathbb{B}(\mathcal{H})$ into an algebra by defining the multiplication of $A, B \in \mathbb{B}(\mathcal{H})$ as the composition $AB := A \circ B$.

On $\mathbb{B}(\mathcal{H})$, we can define the *weak operator topology* by the convergence of nets as $A_\alpha \xrightarrow{\text{WOT}} A$ (or $w-\lim A_\alpha = A$) whenever $\langle A_\alpha x, y \rangle \to \langle Ax, y \rangle$ for all $x, y \in \mathcal{H}$. We can also define the *strong operator topology* by $A_\alpha \xrightarrow{\text{SOT}} A$ (or $s-\lim A_\alpha = A$) whenever $A_\alpha x \to Ax$ for all $x \in \mathcal{H}$.

We denote the range, kernel, and rank of an operator A by ran A, ker A, and $\text{rank}(A) = \dim(\text{ran } A)$. The closure of a set $\mathcal{D}$ in the norm topology is denoted by $\overline{\mathcal{D}}$. The restriction of an operator A to a set $\mathcal{D}$ is presented as $A|_\mathcal{D}$. A linear map between Hilbert spaces is called *bounded below* if there is a positive number c such that $\|Ax\| \geq c\|x\|$ for all $x \in \mathcal{H}$. In general, a bounded below linear map may not be bounded. For example, the map $A : \ell_2 \to \ell_2$ defined by $A(x_1, x_2, x_3, \ldots) = (x_1, 2x_2, 3x_3, \ldots)$ is bounded below since

$$\|A(x_1, x_2, x_3, \ldots)\|_2^2 = \sum_{n=1}^\infty n^2 |x_n|^2 \geq \sum_{n=1}^\infty |x_n|^2 = \|(x_1, x_2, x_3, \ldots)\|.$$

However, it is not bounded since $\|A(e_n)\| = n$.

If $\dim \mathcal{H} = n$, we can identify $\mathbb{B}(\mathcal{H})$ with the matrix algebra $\mathbb{M}_n$. We denote the identity of $\mathbb{M}_n$ as I_n, or simply I if there is no ambiguity. Therefore, a matrix can be viewed as an operator. The standard *system of matrix units* for $\mathbb{M}_n$ is the family $(E_{ij})_{1 \leq i, j \leq n}$. The (i, j)-entry of E_{ij} is one, while all other entries are zero. In fact, we can express $E_{ij} = e_i^* e_j$, where e_i is the ith vector in the standard basis of $\mathbb{C}^n$. A matrix obtained by deleting certain rows or columns from a given matrix is called a *submatrix* of the original matrix. A *principal* *submatrix* is a specific type of submatrix where the remaining row indices are identical to the remaining column indices.

Let us recall the tensor product of matrices $A = [a_{ij}] \in \mathbb{M}_m$ and $B = [b_{pq}] \in \mathbb{M}_n$. Let $(e_1, e_2, \ldots, e_m)$ and $(f_1, f_2, \ldots, f_n)$ be the standard orthonormal bases for $\mathbb{C}^m$ and $\mathbb{C}^n$, respectively. The *tensor product of matrices* A and B is the matrix $A \otimes B$ represented relative to the basis elements $e_i \otimes f_p$ of the tensor product $\mathbb{C}^m \otimes \mathbb{C}^n$ via

$$\langle (A \otimes B)(e_j \otimes f_q), e_i \otimes f_p \rangle = \langle Ae_j \otimes Bf_q, e_i \otimes f_p \rangle = \langle Ae_j, e_i \rangle \langle Bf_q, f_p \rangle = a_{ij} b_{pq}.$$

It is easy to see that the tensor product of positive semidefinite matrices is positive semidefinite.

The following result is commonly used. To prove that two operators $A, B \in \mathbb{B}(\mathscr{H})$ are equal it is sufficient to use it and show that $\langle Ax, x \rangle = \langle Bx, x \rangle$ for all $x \in \mathscr{H}$.

Proposition 1.1.2 *An operator $A \in \mathbb{B}(\mathscr{H})$ is the zero operator 0 if and only if $\langle Ax, x \rangle = 0$ for all $x \in \mathscr{H}$.*

Proof Suppose that $\langle Ax, x \rangle = 0$ for all $x \in \mathscr{H}$. It follows from (1.1.1) that

$$\langle Ax, y \rangle = \frac{1}{4} \sum_{k=0}^{3} i^k \langle A(x + i^k y), (x + i^k y) \rangle$$

that $\langle Ax, y \rangle = 0$ for all $x, y \in \mathscr{H}$. Set $y = Ax$ to get $\|Ax\|^2 = \langle Ax, Ax \rangle = 0$. Hence, $A = 0$. The reverse assertion is clear. □

The following theorem characterizes the dual of a Hilbert space.

Theorem 1.1.3 (Riesz representation theorem) *Let $\mathscr{H}$ be a Hilbert space and $f : \mathscr{H} \to \mathbb{C}$ be a bounded linear functional. Then there exists a unique element $z \in \mathscr{H}$ such that $\|z\| = \|f\|$ and $f(x) = \langle x, z \rangle$ for all $x \in \mathscr{H}$.*

Proof If $f = 0$, then taking $z = 0$ is sufficient. Let's assume that $f \neq 0$. Thus, $\ker f$ is a proper closed subspace. This implies that $\mathscr{H} = \mathbb{C} z' \oplus \ker f$ for some $z' \in \mathscr{H}$ with $z' \perp \ker f$. We can assume that $f(z') = 1$. Then, $x - f(x)z' \in \ker f$ and

$$\langle x, z' \rangle = \langle x - f(x)z', z' \rangle + \langle f(x)z', z' \rangle = f(x)\|z'\|^2.$$

By choosing $z = \frac{z'}{\|z'\|^2}$, we obtain $f(x) = \langle x, z \rangle$. If there exist two vectors z_1 and z_2 such that $f(x) = \langle x, z_1 \rangle = \langle x, z_2 \rangle$, then $\langle x, z_1 - z_2 \rangle = 0$ for all $x \in \mathscr{H}$. Setting $x = z_1 - z_2$, we deduce that $\langle z_1 - z_2, z_1 - z_2 \rangle = 0$, which ensures that $z_1 = z_2$. This confirms the uniqueness assertion. □

1.1 Fundamental Information

For each $A \in \mathbb{B}(\mathscr{H})$ and $y \in \mathscr{H}$, we can use the Riesz representation Theorem 1.1.3 to the bounded linear functional $x \mapsto \langle Ax, y \rangle$ to find a unique vector denoted by A^*y satisfying $\|A^*y\| \leq \|A\| \|y\|$ and $\langle Ax, y \rangle = \langle x, A^*y \rangle$ for all $x \in \mathscr{H}$. Therefore, we get a bounded linear operator $y \mapsto A^*y$ on $\mathscr{H}$. Thus, the so-called *adjoint operation* $A \mapsto A^*$ can be defined. The algebra $\mathbb{B}(\mathscr{H})$ endowed with the adjoint operation can be considered as a *-*algebra*, which means an algebra endowed with an involution $*$ that satisfies the following properties:

- $(A + \lambda B)^* = A^* + \bar{\lambda} B^*$;
- $(AB)^* = B^* A^*$;
- $(A^*)^* = A$.

For matrices, if $A = [a_{ij}]$, then $A^* = [\overline{a_{ji}}]$. The matrix $A^t = [a_{ji}]$ is said to be the *transpose* of A.

A *C^*-algebra* is a complex $*$-algebra $\mathscr{A}$, which is at the same time a Banach space and satisfies the submultiplicative property $\|AB\| \leq \|A\| \|B\|$ and the C^*-condition

$$\|A^*A\| = \|A\|^2 \qquad (1.1.2)$$

for all $A, B \in \mathscr{A}$. The identity element of a C^*-algebra $\mathscr{A}$ is denoted by I. It is easy to verify that $\mathbb{B}(\mathscr{H})$ is a C^*-algebra. In addition, if Ω is any compact Hausdorff space, then the algebra $C(\Omega)$ of all continuous complex-valued functions on Ω equipped with the sup-norm $\|f\|_\infty = \sup_{x \in \Omega} |f(x)|$ and the $*$-operation $f \mapsto \bar{f}$ is a C^*-algebra. Here, $\bar{f}(x) := \overline{f(x)}$ for all $x \in \Omega$.

The C^*-subalgebra generated by a subset of $\mathscr{A}$ is defined as the intersection of all C^*-subalgebras of $\mathscr{A}$ containing the subset. Especially, the C^*-algebra generated by a self-adjoint element $A \in \mathscr{A}$ and the identity operator I, denoted by $C^*(A, I)$, is the closure of all polynomials in A, and so it is commutative. We use $\mathscr{A}$ and $\mathscr{B}$ to denote C^*-algebras and Φ and Ψ for arbitrary (linear or nonlinear) maps between C^*-algebras.

By a *-*homomorphism* $\pi : \mathscr{A} \to \mathscr{B}$, we mean a linear map that satisfies the multiplicative property $\pi(AB) = \pi(A)\pi(B)$ and the $*$-preserving property $\pi(A^*) = \pi(A)^*$. A $*$-homomorphism $\pi : \mathscr{A} \to \mathbb{B}(\mathscr{H})$ is called a *representation*. An *irreducible representation* is a representation $\pi : \mathscr{A} \to \mathbb{B}(\mathscr{H})$, where the algebra $\pi(\mathscr{A})$ has no invariant closed subspace other than 0 and itself. Recall that a closed subspace $\mathscr{K}$ of $\mathscr{H}$ is called an *invariant subspace* for a subset S of $\mathbb{B}(\mathscr{H})$ if for each operator $A \in S$ we have $A(\mathscr{K}) \subseteq \mathscr{K}$.

Every C^*-algebra can be regarded as a norm-closed $*$-subalgebra of $\mathbb{B}(\mathscr{H})$ for some Hilbert space $\mathscr{H}$. This representation is known as the *Gelfand–Naimark–Segal representation*; refer to [176, Theorem 3.4.1].

A one-to-one surjective $*$-homomorphism is called a $*$-*isomorphism*. If a C^*-algebra $\mathscr{A}$ is $*$-isometric to a C^*-algebra $\mathscr{B}$, then we write $\mathscr{A} \simeq \mathscr{B}$. It is well-known [176, Theorem 3.1.5] that every $*$-isomorphism $\pi : \mathscr{A} \to \mathscr{B}$ between C^*-algebras is isometric, that is, $\|\pi(A)\| = \|A\|$ for all $A \in \mathscr{A}$.

The Gelfand–Naimark–Segal representation enables us to naturally define various types of operators such as self-adjoint operators, normal operators, contractions, isometries, coisometries partial isometries, unitaries, idempotents, and projections for elements of a C^*-algebra. An operator A in a C^*-algebra $\mathscr{A}$ is said to be

- *self-adjoint* (*Hermitian matrix*) if $A^* = A$;
- *normal* if $A^*A = AA^*$;
- *contraction* if $\|A\| \leq 1$;
- *isometry* if $A^*A = I$;
- *coisometry* if A^* is an isometry;
- *partial isometry* if $AA^*A = A$;
- *unitary* if $A^*A = AA^* = I$;
- *idempotent* if $A^2 = A$;
- *nilpotent* if $A^n = 0$ for some n in the set of natural numbers $\mathbb{N}$;
- (*orthogonal*) *projection* if it is a self-adjoint idempotent.

For a closed subspace $\mathscr{M}$, it holds that $\mathscr{H} = \mathscr{M} \oplus \mathscr{M}^\perp$, which means that every element $x \in \mathscr{H}$ can be uniquely expressed as $x = y + z$ where $y \in \mathscr{M}$ and $z \in \mathscr{M}^\perp$. The map $x \mapsto y$ gives a projection, denoted by $P_\mathscr{M}$, on $\mathscr{H}$ whose range is $\mathscr{M}$. Conversely, for every projection $P \in \mathbb{B}(\mathscr{H})$, the range of P is a closed subspace of $\mathscr{H}$. In fact, there exists a one-to-one correspondence between the set of all closed subspaces of $\mathscr{H}$ and the set of projections in $\mathbb{B}(\mathscr{H})$. Given two projections P and Q, the projection onto $\mathrm{ran}(P) \cap \mathrm{ran}(Q)$ is denoted as $P \wedge Q$.

If A is a partial isometry, then A^*A (AA^*, respectively) is a projection and its range is called the *initial space* (*final space*, respectively) of A.

The Cartesian decomposition of an element $A \in \mathscr{A}$ is $A = B + iC$, where $B = \mathrm{Re}(A) := (A + A^*)/2$ and $C = \mathrm{Im}(A) := (A - A^*)/(2i)$ are self-adjoint and called the *real part* and *imaginary part* of A. The real linear space of all self-adjoint elements of $\mathscr{A}$ is denoted by $\mathscr{A}_{\mathrm{sa}}(\mathscr{H})$. In particular, the set of self-adjoint operators on a Hilbert space $\mathscr{H}$ is denoted by $\mathbb{B}_{\mathrm{sa}}(\mathscr{H})$.

The spectrum of an operator $A \in \mathbb{B}(\mathscr{H})$ refers to the set $\mathrm{sp}(A)$ of all complex numbers λ such that $A - \lambda I$ is not invertible in $\mathbb{B}(\mathscr{H})$. It is a compact and nonempty set. When A is a matrix in $\mathbb{M}_n$, the set of its eigenvalues is exactly $\mathrm{sp}(A)$. The *spectral radius* of an operator A is defined as

$$r(A) = \max\{|\lambda| : \lambda \in \mathrm{sp}(A)\}.$$

The *Gelfand–Beurling formula* states that

$$r(A) = \inf_n \|A^n\|^{1/n} = \lim_n \|A^n\|^{1/n}.$$

In particular, $r(A) \leq \|A\|$. For $A, B \in \mathbb{B}(\mathscr{H})$,

$$\mathrm{sp}(AB) \cup \{0\} = \mathrm{sp}(BA) \cup \{0\}, \qquad (1.1.3)$$

1.1 Fundamental Information

and so $r(AB) = r(BA)$. The equality (1.1.3) follows from the relation

$$\begin{bmatrix} I & A \\ 0 & I \end{bmatrix}^{-1} \begin{bmatrix} AB & 0 \\ B & 0 \end{bmatrix} \begin{bmatrix} I & A \\ 0 & I \end{bmatrix} = \begin{bmatrix} 0 & 0 \\ B & BA \end{bmatrix}.$$

Example 1.1.4 Suppose that $A = \begin{bmatrix} 0 & 0 \\ 1 & 0 \end{bmatrix}$ and $B = \begin{bmatrix} 0 & 1 \\ 0 & 0 \end{bmatrix}$. Then

(i) $A^2 = 0$, so $r(A) = \inf_n \|A^n\|^{\frac{1}{n}} = 0$, but $\|A\| = 1$. Hence, equality may not hold in $r(A) \le \|A\|$ ([134, Theorem 3.2.3]).
(ii) $r(AB) > r(A)r(B)$ and $r(A+B) > r(A) + r(B)$. Thus, the spectral radius is neither submultiplicative nor subadditive in general; see [134, Proposition 3.2.10].

Example 1.1.5 Let $A = \begin{bmatrix} 1 & -1 \\ -1 & 2 \end{bmatrix}$ and $B = \begin{bmatrix} 0 & 0 \\ 0 & 1 \end{bmatrix}$. Then $\mathrm{sp}(A+B)$ is not a subset of $\mathrm{sp}(A) + \mathrm{sp}(B)$. It is notable that if A and B commute, then $\mathrm{sp}(A+B) \subseteq \mathrm{sp}(A) + \mathrm{sp}(B)$ [134, Proposition 3.2.10].

Let $A \in \mathbb{B}(\mathcal{H})$. The set $W(A) = \{\langle Ax, x \rangle : \|x\| = 1\}$ is said to be the *numerical range* of A and $w(A) = \sup\{|\lambda| : \lambda \in W(A)\}$ is called the *numerical radius* of A.

It is known that $W(A)$ is convex, according to the Toeplitz-Hausdorff theorem. It is invariant under unitary similarity, which means that $W(A) = W(U^*AU)$ for all A and all unitaries U. In addition, thus the convex hull of $\mathrm{sp}(A)$ is contained in the closure of $W(A)$. However, the numerical range does not have similarity invariance. To illustrate this, consider the matrix

$$A_\lambda = \begin{bmatrix} 0 & \lambda \\ 0 & 0 \end{bmatrix}.$$

Then, $W(A_\lambda) = \lambda W(A_1)$, which is the closed disc centered at the origin with radius $|\lambda|$. Thus, $W(A_\lambda)$'s are distinct sets. However, for $\lambda \ne 0$, the matrix A_λ is similar to A_1. This can be seen by considering $B_\lambda = \begin{bmatrix} \lambda & 0 \\ 0 & 1 \end{bmatrix}$, which satisfies $A_\lambda = B_\lambda A_1 B_\lambda^{-1}$; see [207]. Interestingly, $w(\cdot)$ is a norm on $\mathbb{B}(\mathcal{H})$ that is equivalent to the operator norm. More precisely,

$$\frac{1}{2}\|A\| \le w(A) \le \|A\| \qquad (1.1.4)$$

holds for each $A \in \mathbb{B}(\mathcal{H})$.

Theorem 1.1.6 *If* $A \in \mathbb{B}_{\mathrm{sa}}(\mathcal{H})$, *then*

(i) $r(A) = \|A\|$;

(ii) sp(A) *is a subset of the set* $\mathbb{R}$ *of real numbers;*
(iii) *at least one of* $\|A\|$ *or* $-\|A\|$ *is in* sp(A) *and* sp(A) $\subseteq [-\|A\|, \|A\|]$;
(iv)
$$\|A\| = w(A).$$

Proof (i) $\|A^{2^n}\| = \|A\|^{2^n}$ for all n, so
$$r(A) = \lim_n \|A^n\|^{1/n} = \lim_n \|A^{2^n}\|^{1/2^n} = \lim_n \|A\| = \|A\|.$$

(ii) Let $\lambda = \alpha + i\beta \in \text{sp}(A)$, where α and β are real numbers. Assume that $\beta \neq 0$. For each n, put $A_n = A - (\alpha - in\beta)I$. Then $i(n+1)\beta \in \text{sp}(A_n)$ and
$$\begin{aligned}(n+1)^2\beta^2 &= |i(n+1)\beta|^2 \leq r(A_n)^2 \leq \|A_n\|^2 = \|A_n^* A_n\| \\ &= \|(A - (\alpha + in\beta)I)(A - (\alpha - in\beta)I)\| = \|(A - \alpha I)^2 + n^2\beta^2 I\| \\ &\leq \|A - \alpha I\|^2 + n^2\beta^2,\end{aligned}$$
whence $n \leq \frac{1}{2}\left(\|A - \alpha I\|^2/\beta^2 - 1\right)$ for all n, which is impossible. Hence, $\lambda = \alpha \in \mathbb{R}$.
(iii) It follows from (i) and compactness of sp(A) that $\|A\| \in \{|\lambda| : \lambda \in \text{sp}(A) \subseteq \mathbb{R}\}$. Thus, $\|A\|$ or $-\|A\|$ is in sp(A). In addition, sp(A) $\subseteq [-\|A\|, \|A\|]$, since $r(A) \leq \|A\|$.
(iv) Since $r(A) \leq w(A) \leq \|A\|$, it follows from (i) that
$$\|A\| = w(A).$$
□

There is a valuable formula in which the numerical radius of an operator is expressed in terms of the norm of specific operators as follows.

Corollary 1.1.7 ([233, p. 85]) *Let* $A \in \mathbb{B}(\mathcal{H})$. *Then*
$$w(A) = \sup_{\theta \in \mathbb{R}} \|\text{Re}(e^{i\theta}A)\| = \frac{1}{2} \sup_{\theta \in \mathbb{R}} \|A + e^{-2i\theta}A^*\|. \tag{1.1.5}$$

Proof Since $\text{Re}(e^{i\theta}A)$ is self-adjoint, we have
$$\begin{aligned}\sup_{\theta \in \mathbb{R}} \|\text{Re}(e^{i\theta}A)\| &= \sup_{\theta \in \mathbb{R}} w\left(\text{Re}(e^{i\theta}A)\right) = \sup_{\theta \in \mathbb{R}} \sup_{\|x\|=1} |\langle \text{Re}(e^{i\theta}A)x, x\rangle| \\ &= \sup_{\|x\|=1} \sup_{\theta \in \mathbb{R}} |\text{Re}\left(e^{i\theta}\langle Ax, x\rangle\right)| = \sup_{\|x\|=1} |\langle Ax, x\rangle| = w(A).\end{aligned}$$

The last equality is concluded from the definition of the real part of an operator. □

An element $A \in \mathscr{A}$ is called *positive* if it is self-adjoint and its spectrum is contained in the interval $[0, \infty)$. In this case, we write $A \geq 0$. We say A is strictly positive (positive definite in the setting of matrices) and write $A > 0$ if it is positive and invertible. For self-adjoint elements (Hermitian matrices, respectively) A and B, we say $B \geq A$ ($B > A$, respectively) if $B - A \geq 0$ ($B - A > 0$, respectively). This order is known as the *Löwner order*. The *Schur product theorem* states that the *Hadamard product* or *Schur product* $A \circ B = [a_{ij}b_{ij}]$ of two positive semidefinite matrices $A = [a_{ij}]$ and $B = [b_{ij}]$ is also positive semidefinite; see [116, Theorem 2.18].

The (continuous) functional calculus for a self-adjoint operator provides a powerful tool for establishing connections between continuous functions and bounded linear operators. Let us consider an operator $A \in \mathbb{B}_{sa}(\mathcal{H})$. If we take a function f in the space $C(sp(A))$ of all continuous functions defined on the spectrum of A, we can find a unique operator $f(A)$ in $\mathbb{B}(\mathcal{H})$. This operator possesses a special property: if we have a sequence of polynomials (p_n) such that $\lim_{n\to\infty} p_n = f$ in sup-norm on $C(sp(A))$, then it follows that $f(A) = \lim_{n\to\infty} p_n(A)$ in operator norm on $\mathbb{B}(\mathcal{H})$.

This function-to-operator map, denoted as $f \mapsto f(A)$, represents a unique isometric $*$-isomorphism from the space of continuous functions defined on the spectrum of A to the C^*-subalgebra of $\mathbb{B}(\mathcal{H})$ generated by A and the identity operator I. Notably, this map assigns the function $f(t) = t$ to the operator A and the constant function $f(t) = 1$ to the identity operator I. Consequently, if we have two functions f and g in $C(sp(A))$ such that $f(t) \leq g(t)$ for all $t \in sp(A)$, then it follows that $f(A) \leq g(A)$. In addition, $f(A)$ is self-adjoint if and only if $f(t) = \overline{f(t)}$ for all $t \in sp(A)$ or, equivalently, the range of f is a subset of $\mathbb{R}$.

The *spectral theorem* further asserts that $sp(f(A)) = f(sp(A))$ for any $A \in \mathcal{A}$.

Since A is self-adjoint, $sp(A) \subseteq \mathbb{R}$. By the compactness of the spectrum, there are real numbers m, M such that $sp(A) \subseteq [m, M]$. By functional calculus, this is equivalent to $mI \leq A \leq MI$ or simply $m \leq A \leq M$.

Let $A \in \mathbb{B}_{sa}(\mathcal{H})$. Consider the continuous functions $f_+(t) = \max\{t, 0\}$ and $f_-(t) = \max\{-t, 0\}$. These functions satisfy $f = f_+ - f_-$, $f_+ f_- = 0$, $|f_\pm| \leq |f|$. Using the functional calculus for A, we get two positive operators $A_+, A_- \in \mathbb{B}(\mathcal{H})$ such that $A = A_+ - A_-$, $A_+ A_- = 0 = A_- A_+$, $\|A_\pm\| \leq \|A\|$. The decomposition $A = A_+ - A_-$ is called the *Jordan decomposition*, and A_+ and A_- are called the *positive part* and the *negative part* of A, respectively.

Let us consider the case where $A \in \mathbb{B}(\mathcal{H})$ is a normal operator and $\mathcal{B}$ is the C^*-algebra generated by A and I. Then, there exists a $*$-isometrically isomorphism between $\mathcal{B}$ and $C(sp(A))$, which maps A to the inclusion map of $sp(A)$ in $\mathbb{C}$. Consequently, if $sp(A) \subseteq \mathbb{R}$, then $z = \bar{z}$ on $sp(A)$, implying that A is self-adjoint.

We can use the aforementioned fact to show that if P_1 and P_2 are projections such that $P_1 P_2$ is normal, then P_1 and P_2 commute. To see this, note that $sp(P_1 P_2) \cup \{0\} = sp(P_1 P_2 P_2) \cup \{0\} = sp(P_2 P_1 P_2) \cup \{0\} \subseteq \mathbb{R}$ since $P_2 P_1 P_2$ is self-adjoint. Hence, $P_1 P_2$ is self-adjoint, and thus $P_1 P_2 = (P_1 P_2)^* = P_2 P_1$.

A *von Neumann algebra* $\mathcal{A}$ acting on a Hilbert space $\mathcal{H}$ is a $*$-subalgebra of the algebra $\mathbb{B}(\mathcal{H})$ such that $\mathcal{A} = (\mathcal{A}^c)^c$. Here, the *commutant of a set* $\mathcal{D} \subseteq \mathbb{B}(\mathcal{H})$ is defined by

$\mathcal{D}^c = \{Y \in \mathbb{B}(\mathcal{H}): XY = YX \text{ for all } X \in \mathcal{D}\}$. Equivalently, a von Neumann algebra is a C^*-algebra $\mathcal{A}$ that is the dual of a Banach space $\mathcal{A}_*$. The latter space is indeed the space of all normal linear functionals on $\mathcal{A}$. A bounded linear functional is called a *normal linear functional* when for a bounded increasing net of self-adjoint operators (A_α), we have $f(\sup_\alpha A_\alpha) = \sup_\alpha f(A_\alpha)$. Here, $\sup_\alpha A_\alpha$ means the least upper bound of the self-adjoint operators A_α's with respect to the Löwner order.

Projections P and Q are called *Murray–von Neumann equivalent*, denoted as $P \sim Q$, if there is a partial isometry U such that $U^*U = P$ and $UU^* = Q$. A von Neumann algebra is said to be *properly infinite* if there exist projections P_1 and P_2 such that $P_1 \sim I$, $P_2 \sim I$, and $P_1 P_2 = 0$.

The *spectral representation* for a self-adjoint operator $A \in \mathbb{B}_{sa}(\mathcal{H})$ states that if $m = \min\{\lambda : \lambda \in \text{sp}(A)\}$ and let $M = \max\{\lambda : \lambda \in \text{sp}(A)\}$, then there exists a certain family of projections $\{E_\lambda\}_{\lambda \in \mathbb{R}}$ known as the *spectral family* of A, which has the following properties

(a) $E_\lambda \leq E_{\lambda'}$ for $\lambda \leq \lambda'$;
(b) $E_{m-0} = 0$, $E_M = I$ and $E_{\lambda+} = E_\lambda$ for all $\lambda \in \mathbb{R}$;
(c) $A = \int_{m-0}^{M} \lambda \, dE_\lambda$;
(d) for every continuous complex-valued function f defined on $\mathbb{R}$, the Riemann–Stieltjes operator-valued integral

$$f(A) = \int_{m-0}^{M} f(\lambda) \, dE_\lambda$$

holds. This integral means that for every $\varepsilon > 0$ there exists a $\delta > 0$ such that

$$\left\| f(A) - \sum_{k=1}^{n} f(\lambda'_k)(E_{\lambda_k} - E_{\lambda_{k-1}}) \right\| \leq \varepsilon$$

when $\lambda_0 < m = \lambda_1 < \cdots < \lambda_{n-1} < \lambda_n = M$, $\lambda_k - \lambda_{k-1} \leq \delta$ for $1 \leq k \leq n$, and $\lambda'_k \in (\lambda_{k-1}, \lambda_k)$ for $1 \leq k \leq n$.

If $\dim \mathcal{H} = n$, then A can be considered as a Hermitian matrix in $\mathbb{M}_n$. The Schur decomposition for matrices states that there exists a unitary matrix $U \in \mathbb{M}_n$ such that $A = U^*DU$, where $D = \text{diag}(\lambda_1, \ldots, \lambda_n)$ is a diagonal matrix whose diagonal entries are the eigenvalues λ_j ($1 \leq j \leq n$) of A. If $E_j = U^* \text{diag}(\underbrace{1, 1, \ldots, 1}_{j \text{ terms}}, 0, \ldots, 0) U$, then we get the spectral representation of A as follows:

$$A = U^*DU = \sum_{j=1}^{n} \lambda_j \Delta E_j,$$

where $\Delta E_j = E_j - E_{j-1}$ and $E_0 = 0$ are pairwise orthogonal projections; see also [116, p. 18]. Also, if f is a real-valued continuous function on an interval containing the eigenvalues of A, then

$$f(A) = \sum_{j=1}^{n} f(\lambda_j) \Delta E_j.$$

Here $f(A)$ is understood as $f(A)$ defined by functional calculus, or simply as $f(A) = U^*D'U$, where $D' = \text{diag}(f(\lambda_1), \ldots, f(\lambda_n))$.

Now, we present a fundamental theorem.

Theorem 1.1.8 *Let $A \in \mathbb{B}(\mathcal{H})$. The following assertions are equivalent:*
(a) A is positive;
(b) A is of the form B^2 for some positive operator $B \in \mathbb{B}(\mathcal{H})$;
(c) A is of the form B^*B for some $B \in \mathbb{B}(\mathcal{H})$;
(d) $\langle Ax, x \rangle \geq 0$ holds for every $x \in \mathcal{H}$.

Proof (a) $\Longrightarrow$ (b). Since $\text{sp}(A) \subseteq [0, \infty)$, the function $t^{1/2}$ is continuous on $\text{sp}(A)$. Using the functional calculus for A, we get the positive operator $A^{1/2}$ satisfying $A = (A^{1/2})A^{1/2}$. So we reach (b) with $B = A^{1/2}$.
(b) $\Longrightarrow$ (c). It is clear.
(c) $\Longrightarrow$ (d). We have

$$\langle Ax, x \rangle = \langle B^*Bx, x \rangle = \langle Bx, Bx \rangle = \|Bx\|^2 \geq 0.$$

(d) $\Longrightarrow$ (a). It follows from $\langle Ax, x \rangle \geq 0$ that

$$\langle Ax, x \rangle = \overline{\langle x, Ax \rangle} = \langle x, Ax \rangle = \langle A^*x, x \rangle.$$

Hence, $\langle (A - A^*)x, x \rangle = 0$ for all $x \in \mathcal{H}$. It follows from Proposition 1.1.2 that $A = A^*$. Let $A = A_+ - A_-$ be the Jordan decomposition of A. Since $\langle Ax, x \rangle \geq 0$, we have $\langle A_-x, x \rangle \leq \langle A_+x, x \rangle$. Replacing x with A_-x in the latter inequality, we get $0 \leq \langle A_-^3 x, x \rangle = \langle A_-^2 x, A_-x \rangle \leq \langle A_+A_-x, A_-x \rangle = 0$, because $A_+A_- = 0$. By Proposition 1.1.2, we have $A_-^3 = 0$. Applying functional calculus, we infer that $A_- = 0$. Therefore, $A = A_+ - A_- = A_+ \geq 0$. $\square$

The operator B in Part (b) of the above theorem is unique and called the *positive square root* of A and is denoted by $A^{1/2}$. In addition, for each operator $A \in \mathbb{B}(\mathcal{H})$, the positive square root of the positive operator A^*A is denoted by $|A|$ and is called *absolute value* of A.

Let $A \in \mathbb{B}_{\text{sa}}(\mathcal{H})$. Then A is positive if there is a nonnegative real number c such that $\|A - cI\| \leq c$, since by utilizing the functional calculus for A, this norm inequality is equivalent to $\sup_{t \in \text{sp}(A)} |t - c| \leq c$, which yields $t \geq 0$ and this, in turn, implies that $A \geq 0$. Conversely, in the same way, we observe that if $A \geq 0$, then $\|A - cI\| \leq c$ for all nonnegative real numbers $c \geq \|A\|/2$.

The following properties of the Löwner order and positive operators are frequently used without being referred to.

Theorem 1.1.9 *Let $A, B \in \mathbb{B}(\mathcal{H})$ be self-adjoint. Then,*
 (i) *if $A \geq 0$ and $t \in [0, \infty)$, then $tA \geq 0$;*
 (ii) *if A and B are positive, then so is $A + B$;*
 (iii) *If $-B \leq A \leq B$, then $\|A\| \leq \|B\|$;*
 (iv) *if $A \leq B$, then $X^*AX \leq X^*BX$ for all $X \in \mathbb{B}(\mathcal{H})$;*
 (v) *$0 < A$ if and only if there is $m > 0$ such that $0 < m \leq A$;*
 (vi) *if $0 < A \leq B$, then B is invertible and $B^{-1} \leq A^{-1}$;*
 (vii) *the set of positive operators is closed in $\mathbb{B}(\mathcal{H})$;*
 (viii) *if $A, B \in \mathbb{B}(\mathcal{H})$ are positive and $AB = BA$, then $AB \geq 0$;*
 (ix) *if $A, B \in \mathbb{B}(\mathcal{H})$ are positive, then $\mathrm{sp}(AB) \subseteq [0, \infty)$.*

Proof (i) It is deduced from $\mathrm{sp}(tA) = t\,\mathrm{sp}(A)$.
(ii) By the note preceding this theorem, we have $\|A - \|A\|I\| \leq \|A\|$ and $\|B - \|B\|I\| \leq \|B\|$. Hence

$$\|(A + B) - (\|A\| + \|B\|)\| \leq \|A - \|A\|I\| + \|B - \|B\|I\| \leq \|A\| + \|B\|,$$

from which we conclude, from the note preceding this theorem, that $A + B \geq 0$.
(iii) By making use of (ii), we infer that $\leq$ is a partial order on $\mathbb{B}_{\mathrm{sa}}(\mathcal{H})$. It follows from $t \leq \sup_{t \in \mathrm{sp}(A)} |t|$ and the functional calculus for C that $C \leq \|C\|I$ for any self-adjoint operator C. Thus, $-\|B\|I \leq A \leq B \leq \|B\|I$. By the functional calculus for B, $-\|B\| \leq t \leq \|B\|$ for all $t \in \mathrm{sp}(A)$. Thus, $\sup_{t \in \mathrm{sp}(A)} |t| \leq \|B\|$, whence $\|A\| \leq \|B\|$.
(iv) Since $0 \leq A \leq B$, we have $B - A \geq 0$, so

$$X^*(B - A)X = X^*(B - A)^{1/2}(B - A)^{1/2}X = ((B - A)^{1/2}X)^*((B - A)^{1/2}X) \geq 0,$$

from which we conclude that $X^*AX \leq X^*BX$.
(v) Let A be a positive invertible operator. Hence, $\mathrm{sp}(A) \subseteq (0, \infty)$. The spectrum of any operator is compact, so there is a real positive number m such that $0 < m \leq t$ for all $t \in \mathrm{sp}(A)$. Applying the functional calculus for A, we get $m \leq A$. The proof of the reverse assertion is similar.
(vi) Since $0 < m \leq A \leq B$ we infer that B is invertible. Using the functional calculus for B, we observe that the operator $B^{-1/2}$ corresponding to the continuous function $t^{-1/2}$ is well-defined. Employing (iv), we deduce from $A \leq B$ that $B^{-1/2}AB^{-1/2} \leq B^{-1/2}BB^{-1/2} = I$. Using the functional calculus for $B^{-1/2}AB^{-1/2}$ we see that $t \leq 1$ for all $t \in \mathrm{sp}(B^{-1/2}AB^{-1/2})$. Hence, $t^{-1} \geq 1$ for all $t \in \mathrm{sp}(B^{-1/2}AB^{-1/2})$. Another use of

functional calculus entails that $(B^{-1/2}AB^{-1/2})^{-1} \geq I$. Therefore, $B^{1/2}A^{-1}B^{1/2} \geq I$, from which we get $A^{-1} \geq B^{-1/2}IB^{-1/2} = B^{-1}$.

(vii) If $A_n \geq 0$ and $A_n \to A$ as $n \to \infty$ in the norm topology, then $\langle Ax, x \rangle = \lim_n \langle A_n x, x \rangle \geq 0$ ($x \in \mathcal{H}$), by Theorem 1.1.8(d), whence $A \geq 0$.

(viii) Since A and B commute, so do $A^{1/2}$ and $B^{1/2}$, since these operators are the limits of some sequences of polynomials in A and B, respectively. Hence

$$AB = A^{1/2}A^{1/2}B^{1/2}B^{1/2} = (A^{1/2}B^{1/2})^2.$$

It follows from Theorem 1.1.8(b) that AB is positive.

(ix) We have

$$\mathrm{sp}(AB) = \mathrm{sp}(A^{1/2}A^{1/2}B) = \mathrm{sp}(A^{1/2}BA^{1/2}) \subseteq [0, \infty),$$

since $A^{1/2}BA^{1/2} \geq 0$.

If A and B are complex $n \times n$ matrices, we can present a different proof as follows: If λ is an eigenvalue of AB, then there is a unit vector x such that $ABx = \lambda x$. Hence

$$0 \leq \langle ABx, Bx \rangle = \langle BABx, x \rangle = \langle B(\lambda x), x \rangle = \lambda \langle Bx, x \rangle$$

since $\langle Bx, x \rangle \geq 0$, we conclude that $\lambda \geq 0$. Thus, all eigenvalues of A are positive. $\square$

We use $\mathscr{A}_+$ and $\mathscr{A}_{++}$ to denote the sets of positive and strictly positive elements of $\mathscr{A}$, respectively.

There is an interesting theorem due to Rota. We state it together with a simple proof of this theorem due to Choda [215].

Theorem 1.1.10 *The spectral radius of an operator $A \in \mathbb{B}(\mathcal{H})$ is the infimum of its similarity orbit, that is*

$$r(A) = \inf\{\|BAB^{-1}\| : B \text{ is invertible}\}. \tag{1.1.6}$$

Proof Let $t > r(A)$. It suffices to show that for each $s > t$, there exists a positive invertible operator B such that $\|BAB^{-1}\| \leq s$. Since $r(A) = \lim_{n \to \infty} \|A^n\|^{\frac{1}{n}}$, we observe that $\|A^n\|^{\frac{1}{n}} < t$ for sufficiently large n. Hence, $\|s^{-2n}A^{*n}A^n\| \leq (\frac{t}{s})^{2n}$. Thus the series $\sum_{n=0}^{\infty} \|s^{-2n}A^{*n}A^n\|$ is convergent, therefore $C = \sum_{n=0}^{\infty} s^{-2n}A^{*n}A^n$ is well-defined as an operator in $\mathbb{B}(\mathcal{H})$. Furthermore, $C \geq 0$. If we put $B = C^{\frac{1}{2}}$, then B is positive invertible. In addition,

$$\|BAB^{-1}\|^2 = \|B^{-1}A^*B^2AB^{-1}\| = \|B^{-1}A^*CAB^{-1}\| \leq s^2\|B^{-1}CB^{-1}\| = s^2$$

because $0 \leq A^*CA = s^2(C - 1) \leq s^2C$. Therefore, we have $\|BAB^{-1}\| \leq s$, as required. $\square$

There are simple observations concerning 2×2 operator matrices.

Let $\mathscr{H}$ be a Hilbert space of arbitrary dimension and let $\mathscr{H} = \mathscr{H}_1 \oplus \mathscr{H}_2$ with $\mathscr{H}_2 = \mathscr{H}_1^\perp$. For $1 \leq j \leq 2$, let $\iota_j : \mathscr{H}_j \to \mathscr{H}_1 \otimes \mathscr{H}_2$ be the map that takes $x \in \mathscr{H}_j$ to the vector in $\mathscr{H}_1 \otimes \mathscr{H}_2$ that has x as its jth entry and 0 elsewhere. Then ι_j^* is the map projecting a vector in $\mathscr{H}_1 \otimes \mathscr{H}_2$ onto its jth component. As we will see in Theorem 2.1.1, $\mathbb{B}(\mathscr{H} \oplus \mathscr{H})$ is $*$-isomorphic to $\mathbb{M}_2(\mathbb{B}(\mathscr{H}))$.

The space $\mathscr{H}_j$ is invariant under $A = [A_{ij}]_{1 \leq i,j \leq 2}$, meaning $A \mathscr{H}_j \subseteq \mathscr{H}_j$, if and only if $P_j A P_j = A P_j$, where $P_j = \iota_j \iota_j^*$ is a projection in $\mathbb{B}(\mathscr{H}_1 \oplus \mathscr{H}_2)$. This occurs if and only if $A_{ij} = 0$ for all $i \neq j$. The space $\mathscr{H}_j$ reduces A, meaning $\mathscr{H}_j$ is invariant under both A and A^*, if and only if $A_{ji} = A_{ij} = 0$ for all $1 \leq i \neq j \leq 2$. In this case, we say that A is the direct sum of A_{11} and A_{22} and we write $A = A_{11} \oplus A_{22}$.

For $M = \begin{bmatrix} A & X \\ Y & B \end{bmatrix}$. The matrices $\begin{bmatrix} A & 0 \\ 0 & B \end{bmatrix}$ and $\begin{bmatrix} 0 & X \\ Y & 0 \end{bmatrix}$ are called the *pinching* and *off-diagonal part* of M, respectively. Clearly, M is self-adjoint if and only if $Y = X^*$.

Employing functional calculus, if $A, B \in \mathbb{B}_{\mathrm{sa}}(\mathscr{H})$ with spectra contained in an interval $J \subseteq \mathbb{R}$ and f is a continuous real-valued function on J, then

$$f\left(\begin{bmatrix} A & 0 \\ 0 & B \end{bmatrix}\right) = \begin{bmatrix} f(A) & 0 \\ 0 & f(B) \end{bmatrix}. \tag{1.1.7}$$

Since $\mathrm{sp}(A \oplus B) \subseteq \mathrm{sp}(A) \cup \mathrm{sp}(B)$. It follows from

$$\left\langle \begin{bmatrix} A & 0 \\ 0 & B \end{bmatrix} \begin{bmatrix} x \\ y \end{bmatrix}, \begin{bmatrix} x \\ y \end{bmatrix} \right\rangle = \langle Ax, x \rangle + \langle By, y \rangle$$

that $\begin{bmatrix} A & 0 \\ 0 & B \end{bmatrix}$ is positive if and only if so are A and B. One can see that $\|A \oplus B\| = \max(\|A\|, \|B\|)$ for all operators $A, B \in \mathbb{B}(\mathscr{H})$.

Halmos' two projections theorem [106] is one of the fundamental results in the theory of Hilbert space operators. It states that if P and Q are two projections on a Hilbert space $\mathscr{H}$ such that $\mathrm{ran}\, P$ and $\mathrm{ran}\, Q$ are in generic position, in the sense that P and Q satisfy the following conditions:

$$\mathrm{ran}\, P \cap \mathrm{ran}\, Q = \mathrm{ran}\, P \cap \ker Q = \ker P \cap \mathrm{ran}\, Q = \ker P \cap \ker Q = \{0\},$$

then there exist a unitary map W from $\mathscr{M}_1 = \mathrm{ran}\, P$ onto $\mathscr{M}_2 = \ker P$, and two positive operators C and D on $\mathscr{M}_1$ with $C^2 + D^2 = I$ and $CD = DC$ such that under the orthogonal decomposition $\mathscr{H} = \mathscr{M}_1 \oplus \mathscr{M}_2$, we have

$$P = \begin{pmatrix} I & 0 \\ 0 & 0 \end{pmatrix} \text{ and } Q = \begin{pmatrix} C^2 & CDW^{-1} \\ WCD & WD^2 W^{-1} \end{pmatrix}.$$

This result has many applications in various areas such as characterizations of the closedness of the sum of two subspaces as well as computations of various angles between two subspaces of a Hilbert space; see [33, 157].

Let us conclude this section with some simple remarks.

Inequalities play a central role in mathematics and have various applications in other disciplines. This area is active, as researchers seek to find operator or norm versions of classical real or complex inequalities. These inequalities may hold for operators that act on Hilbert spaces of either finite or infinite dimensions. This is because self-adjoint operators (Hermitian matrices) can be regarded as a generalization of real numbers. The behavior of operators in finite-dimensional Hilbert spaces is drastically different from that in infinite-dimensional Hilbert spaces [16]. There are many analogies between numbers and Hilbert space operators, some of which are illustrated in the following example.

Example 1.1.11

- Analogous to the complex conjugation $z = a + bi \mapsto \bar{z} = a - bi$, the set of real numbers

$$\mathbb{R} = \{z \in \mathbb{C} : z = \bar{z}\},$$

and the Cartesian decomposition $z = a + bi$ with $a = \frac{z+\bar{z}}{2}$ and $b = \frac{z-\bar{z}}{2i}$ for $z \in \mathbb{C}$, we can consider the adjoint operation $A \mapsto A^*$, the set of self-adjoint operators

$$\mathbb{B}_{\mathrm{sa}}(\mathscr{H}) = \{A \in \mathbb{B}(\mathscr{H}) : A = A^*\},$$

and the *Cartesian decomposition* $T = A + iB$ with $A = \frac{T+T^*}{2}$ and $B = \frac{T-T^*}{2i}$ for $T \in \mathbb{B}(\mathscr{H})$.

- Analogous to the nonnegative numbers

$$\mathbb{R}_+ = \{z \in \mathbb{C} : z = \bar{w}w \text{ for some } w \in \mathbb{C}\}$$

and the usual order $r \leq s$ on $\mathbb{R}$ (meaning $s - r \in \mathbb{R}_+$), we can consider,

$$\mathbb{B}_+(\mathscr{H}) = \{A \in \mathbb{B}(\mathscr{H}) : A = B^*B \text{ for some } B \in \mathbb{B}(\mathscr{H})\}$$

and the order $A \leq B$ on $\mathbb{B}_{\mathrm{sa}}(\mathscr{H})$ (meaning $B - A \in \mathbb{B}_+(\mathscr{H})$), respectively.

However, there are several differences between numbers and operators. For example, certain inequalities about real numbers cannot be translated to inequalities about self-adjoint operators. This is because of the noncommutative nature of operators (matrices). Several results, such as the following example, can be found in [120].

Example 1.1.12

- If $a \in \mathbb{R}$, then either $a \geq 0$ or $-a \geq 0$. This is not true in general, for operators. To see this, consider $A = \begin{bmatrix} 1 & 0 \\ 0 & -1 \end{bmatrix}$. Then neither $A \geq 0$ nor $-A \geq 0$.

- If $a, b \in \mathbb{R}$ and $a, b \geq 0$, then $a, b \geq 0$. A similar implication does not hold in general. Let
$$A = \begin{bmatrix} 1 & -1 \\ -1 & 2 \end{bmatrix} = \begin{bmatrix} 1 & 0 \\ -1 & 1 \end{bmatrix} \begin{bmatrix} 1 & -1 \\ 0 & 1 \end{bmatrix} \geq 0$$
and
$$B = \begin{bmatrix} 0 & 0 \\ 0 & 1 \end{bmatrix} = \begin{bmatrix} 0 & 0 \\ 0 & 1 \end{bmatrix} \begin{bmatrix} 0 & 0 \\ 0 & 1 \end{bmatrix} \geq 0.$$
Then $AB = \begin{bmatrix} 0 & -1 \\ 0 & 2 \end{bmatrix}$ is not positive (or even self-adjoint).

- If $a, b \in \mathbb{R}, b \geq 0$, and $-b \leq a \leq b$, then $|a| \leq b$. However, a similar assertion does not hold for operators. For instance, take $A = \begin{bmatrix} 1 & 0 \\ 0 & -1 \end{bmatrix}$ and $B = \begin{bmatrix} 3 & 2 \\ 2 & 2 \end{bmatrix}$.

- If $a, b \in \mathbb{R}$ and $0 \leq a \leq b$, then $a^2 \leq b^2$. A similar statement does not hold for the operators $A = \begin{bmatrix} 0 & 0 \\ 0 & 1 \end{bmatrix}$ and $B = \begin{bmatrix} 1 & -1 \\ -1 & 2 \end{bmatrix}$.

- If $a, b \in \mathbb{C}$, then the triangle inequality $|a + b| \leq |a| + |b|$ holds, while $|A + B| \leq |A| + |B|$ may not be valid for operators. To see this, let $A = \begin{bmatrix} 4 & -2 \\ -2 & 1 \end{bmatrix}$ and $B = \begin{bmatrix} -1 & 2 \\ 2 & -4 \end{bmatrix}$.

1.2 Some Decompositions of Operators

The theory of 2×2 operator matrices is not trivial. If $A, B, C \in \mathbb{B}(\mathcal{H})$ and both A and C are invertible, then the matrix $S = \begin{bmatrix} A & B \\ 0 & C \end{bmatrix}$ is invertible and its inverse is given by

$$S^{-1} = \begin{bmatrix} A^{-1} & -A^{-1}BC^{-1} \\ 0 & C^{-1} \end{bmatrix}.$$

However, the converse does not hold in general. For example, consider $\mathcal{H} = \ell^2$,

$$C(x_1, x_2, \ldots) = (x_2, x_3, \ldots),$$

$$A(x_1, x_2, \ldots)(0, x_1, \ldots),$$

and

$$B(x_1, x_2, \ldots) = (x_1, 0, 0, \ldots).$$

Then $\begin{bmatrix} A & B \\ 0 & C \end{bmatrix}$ is invertible but neither A nor C is invertible.

1.2 Some Decompositions of Operators

It is also observed [45] that if A is invertible, then

$$\det \begin{bmatrix} A & B \\ C & D \end{bmatrix} = \det(DA - CA^{-1}BA).$$

To see this, note that

$$\begin{bmatrix} A & B \\ C & D \end{bmatrix} = \begin{bmatrix} I & 0 \\ CA^{-1} & I \end{bmatrix} \begin{bmatrix} I & 0 \\ 0 & D - CA^{-1}B \end{bmatrix} \begin{bmatrix} I & B \\ 0 & I \end{bmatrix} \begin{bmatrix} A & 0 \\ 0 & I \end{bmatrix},$$

and therefore

$$\det \begin{bmatrix} A & B \\ C & D \end{bmatrix} = \det I \cdot \det(D - CA^{-1}B) \cdot \det I \cdot \det A = \det(DA - CA^{-1}BA).$$

Expressing an operator on a Hilbert space as a product or sum of "nice" operators, such as self-adjoint, positive, unitary, commutator, and idempotent operators, has been considered by many mathematicians. An excellent survey of factorizations into "good" operators is provided in [230]; see also [159]. In establishing the following theorems, some 2×2 matrix techniques are used.

Proposition 1.2.1 *Let $A \in \mathbb{B}(\mathcal{H})$ and $A^2 = 0$. Then A is a linear combination of 9 projections. Moreover, A can be expressed as a product of 2 self-adjoint operators and a product of 3 positive operators.*

Proof It is evident that $\mathcal{H} = \ker A \oplus \overline{\mathrm{ran}(A)}$. Since $A^2 = 0$, we can represent A as $A = \begin{bmatrix} 0 & R \\ 0 & 0 \end{bmatrix}$. In addition, R is a linear combination of four unitary operators. For a unitary U, it holds that

$$\begin{bmatrix} 0 & U \\ 0 & 0 \end{bmatrix} = \frac{1}{2} \begin{bmatrix} I & U \\ U^* & I \end{bmatrix} + \frac{i}{2} \begin{bmatrix} I & -iU \\ iU^* & I \end{bmatrix} - \frac{1}{2}(1+i)I,$$

where $\frac{1}{2}\begin{bmatrix} I & U \\ U^* & I \end{bmatrix}$ and $\frac{1}{2}\begin{bmatrix} I & -iU \\ iU^* & I \end{bmatrix}$ are projections. Hence, A is a linear combination of 9 projections [76].

Furthermore, A can be expressed as a product of 2 self-adjoint operators as:

$$A = \begin{bmatrix} 0 & R \\ 0 & 0 \end{bmatrix} = \begin{bmatrix} 1 & 0 \\ 0 & 0 \end{bmatrix} \begin{bmatrix} 0 & R \\ R^* & 0 \end{bmatrix},$$

and a product of 3 positive operators as

$$A = \begin{bmatrix} 0 & R \\ 0 & 0 \end{bmatrix} = \begin{bmatrix} \lambda & 0 \\ 0 & 0 \end{bmatrix} \begin{bmatrix} I & \frac{R}{\lambda} \\ \frac{R^*}{\lambda} & I \end{bmatrix} \begin{bmatrix} 0 & 0 \\ 0 & 1 \end{bmatrix},$$

where $\lambda > \|A\|$; see [229]. □

It is well-known that if T is an operator on a finite-dimensional space with a nonzero kernel, then it can be expressed as the product of a finite number of projections [114]. We state a related result from [168].

Theorem 1.2.2 *An operator T on a separable infinite-dimensional Hilbert space with an infinite-dimensional kernel is positive if and only if $T = \lambda PQP$ for some positive scalar λ and projections P and Q.*

Proof Let T be an operator that acts on a separable infinite-dimensional Hilbert space $\mathscr{H}$, which has an infinite-dimensional null space $\mathscr{H}_0$. If $\{e_1, e_2, \cdots\}$ is an orthonormal basis for $\mathscr{H}_0$, it can be extended to form an orthonormal basis for $\mathscr{H}$. If $\mathscr{K}$ is the closed linear span of $\{e_2, e_4, e_6, \cdots\}$, then it is clear that $\dim \mathscr{K} = \dim \mathscr{K}^\perp$. Now, suppose that V is a unitary operator from $\mathscr{K}^\perp$ onto $\mathscr{K}$ and W is the unitary from $\mathscr{H} = \mathscr{K} \oplus \mathscr{K}^\perp$ onto $\mathscr{K} \oplus \mathscr{K}$ defined by $W(y \oplus z) = y \oplus Vz$. Then, the operator WTW^* can be represented as $\begin{bmatrix} 0 & R \\ 0 & S \end{bmatrix}$ on $\mathscr{K} \oplus \mathscr{K}$. Let us identify T with WTW^*.

If T is positive, then $R = 0$ and hence, $T = \begin{bmatrix} 0 & 0 \\ 0 & S \end{bmatrix}$. Moreover $0 \leq S_1 = \frac{S}{\|T\|} \leq I$. Employing functional calculus, we can deduce that $\begin{bmatrix} I - S_1 & (S_1 - S_1^2)^{\frac{1}{2}} \\ (S_1 - S_1^2)^{\frac{1}{2}} & S_1 \end{bmatrix} = Q$ is a projection. It follows from

$$\begin{bmatrix} 0 & 0 \\ 0 & S \end{bmatrix} = \|T\| \begin{bmatrix} 0 & 0 \\ 0 & I \end{bmatrix} \begin{bmatrix} I - S_1 & (S_1 - S_1^2)^{\frac{1}{2}} \\ (S_1 - S_1^2)^{\frac{1}{2}} & S_1 \end{bmatrix} \begin{bmatrix} 0 & 0 \\ 0 & I \end{bmatrix}$$

that $T = \|T\| PQP$ is a positive scalar times a product of projections Q and $P = \begin{bmatrix} 0 & 0 \\ 0 & I \end{bmatrix}$.

The reverse statement is clear. □

In the finite-dimensional case $\mathscr{H} = \mathbb{C}^2$, however, the result does not hold. Indeed $T = \|T\| PQP$ implies that $\det(T) = \|T\|^2 \det(P)^2 \det(Q)$ which is not true for $T = \begin{bmatrix} 1 & 0 \\ 0 & 2 \end{bmatrix}$.

The following lemma allows us to extend a linear operator on a subspace to another linear operator defined on the closure of the subspace.

Lemma 1.2.3 *Let $\mathcal{X}$ be a normed space and $\mathcal{Y}$ be a Banach space. If $A : \mathcal{D} \subseteq \mathcal{X} \to \mathcal{Y}$ is a bounded linear operator on a subspace $\mathcal{D}$ of $\mathcal{X}$, then there exists a unique linear operator $\overline{A} : \overline{\mathcal{D}} \to \mathcal{Y}$ such that $\overline{A}|_{\mathcal{D}} = A$ and $\|\overline{A}\| = \|A\|$. Furthermore, if A is an isometry, so is $\overline{A}$.*

Proof Let $x \in \overline{\mathcal{D}}$. There exists a sequence (x_n) in $\mathcal{D}$ such that $x_n \to x$. Hence, (x_n) is Cauchy. Due to

$$\|Ax_n - Ax_m\| = \|A(x_n - x_m)\| \leq \|A\|\|x_n - x_m\|$$

the sequence (Ax_n) is Cauchy in $\mathcal{Y}$. It follows from the completeness of $\mathcal{Y}$ that there is an element $\overline{A}x \in \mathcal{Y}$ such that $Ax_n \to \overline{A}x$.

We therefore get a well-defined operator $\overline{A} : \overline{\mathcal{D}} \to \mathcal{Y}$. To see this, let $x'_n \to x$ with $x'_n \in \mathcal{D}$. We shall show that $\lim_n Ax'_n = \lim_n Ax_n$:

Consider the sequence $x_1, x'_1, x_2, x'_2, x_3, x'_3, \ldots$ This sequence converges to x. By the first part of the proof, $Ax_1, Ax'_1, Ax_2, Ax'_2, \ldots$ is convergent. Hence, all of its subsequences are also convergent to the same limit. Therefore, $\lim_n Ax_n = \lim_n Ax'_n$.

The operator $\overline{A}$ is linear since for $x, y \in \overline{\mathcal{D}}$ and $\lambda \in \mathbb{C}$, there exist sequences (x_n) and (y_n) in $\mathcal{D}$ such that $x_n \to x$ and $y_n \to y$. Therefore, $\lambda x_n + y_n \to \lambda x + y$, and we have

$$\overline{A}(\lambda x + y) = \lim_n A(\lambda x_n + y_n) = \lambda \lim_n Ax_n + \lim_n Ay_n = \lambda \overline{A}x + \overline{A}y.$$

We shall show that $\|\overline{A}\| = \|A\|$. Let $x \in \mathcal{D}$. Then the constant sequence (x_n) with $x_n = x$ converges to x. Therefore, $\overline{A}x = \lim_n Ax = Ax$. Thus, $\overline{A}|_{\mathcal{D}} = A$, and we have $\|A\| \leq \|\overline{A}\|$. To show that $\|A\| \geq \|\overline{A}\|$, let $y \in \overline{\mathcal{D}}$. There exists a sequence (x_n) in $\mathcal{D}$ such that $x_n \to y$ and $Ax_n \to \overline{A}y$. We have $\|Ax_n\| \leq \|A\|\|x_n\|$. Taking limits as $n \to \infty$, we get $\|\overline{A}y\| \leq \|A\|\|y\|$, whence $\|\overline{A}\| \leq \|A\|$. The uniqueness follows from the density of $\mathcal{D}$ in $\overline{\mathcal{D}}$.

If A is an isometry, then under the notation above, we have

$$\|\overline{A}x\| = \|\lim_n Ax_n\| = \lim_n \|Ax_n\| = \lim_n \|x_n\| = \|x\| \quad (x \in \overline{\mathcal{D}}).$$

□

The next theorem plays a crucial role in the operator theory; see also [52, 158].

Theorem 1.2.4 (Douglas majorization theorem [64]) *Let $A, B \in \mathbb{B}(\mathcal{H})$. Then the following statements are equivalent:*

(i) $\operatorname{ran} A \subseteq \operatorname{ran} B$;
(ii) there is an operator $C \in \mathbb{B}(\mathcal{H})$ such that $A = BC$;
(iii) $AA^* \leq r^2 BB^*$ for some $r \geq 0$.

Proof (i) $\implies$ (ii). For $x \in \mathcal{H}$ there is $z \in \mathcal{H}$ such that $Ax = Bz$. Since $\mathcal{H} = \ker B \oplus \ker B^\perp$, so there exist unique elements $y' \in \ker B$ and $y \in \ker B^\perp$ such that $z = y' + y$. Thus, $Ax = Bz = By' + By = By$. So we can define $C : \mathcal{H} \to \mathcal{H}$ by $Cx = y$. Clearly, C is linear, and $A = BC$. We use the closed graph theorem to show that C is continuous: Let $x_n \in \mathcal{H}$, $x_n \to x$, and $Cx_n \to y$. Hence, $Ax_n = BCx_n \to By$ and $Ax_n \to Ax$, since A and B are continuous. Thus, $Ax = By$. Since the elements Cx_n are in the closed subspace $\ker B^\perp$, we infer that $y \in \ker B^\perp$. We therefore conclude from the definition of C that $Cx = y$. Hence, the graph of C is closed.

(ii) $\implies$ (iii). It follows from $A = BC$ that

$$AA^* = BCC^*B^* \leq B\|CC^*\|B^* = \|C\|^2 BB^*.$$

(iii) $\implies$ (i). Let $D : \operatorname{ran} B^* \to \operatorname{ran} A^*$ be defined by $D(B^*x) = A^*x$. This gives us a well-defined linear operator since

$$\|D(B^*x)\|^2 = \|A^*x\|^2 = \langle AA^*x, x\rangle \leq r^2 \langle BB^*x, x\rangle = r^2 \|B^*x\|^2.$$

By Lemma 1.2.3, D has a unique extension to $\overline{\operatorname{ran} B^*}$. Setting $Dx = 0$ for all $x \in \overline{\operatorname{ran} B^*}^{\perp} = \ker B$, we reach a bounded linear operator D such that $DB^* = A^*$. Taking adjoints of both sides, we get $A = BD^*$ from which we immediately get $\operatorname{ran} A \subseteq \operatorname{ran} B$. $\square$

The following theorem is a helpful tool in operator theory; see [176, Theorem 2.3.4].

Theorem 1.2.5 (Polar decomposition) *Let $A \in \mathbb{B}(\mathcal{H})$. Then there is a unique partial isometry U with $\ker U = \ker A$ and $A = U|A|$. In addition, $U^*U|A| = |A|, |A| = U^*A$, and $UU^*A = A$. Furthermore, $A^* = U^*|A^*|$ is the polar decomposition of A^*.*

Proof For $|A|x \in |A|(\mathcal{H})$, put $U_0|A|x = Ax$. It follows from

$$\||A|x\|^2 = \langle |A|x, |A|x\rangle = \langle |A|^2 x, x\rangle = \langle A^*Ax, x\rangle = \langle Ax, Ax\rangle = \|Ax\|^2$$

that U_0 is a well-defined isometry from $|A|(\mathcal{H})$ onto $A(\mathcal{H})$. Using Lemma 1.2.3, we extend U_0 to an isometry U of the closure $\overline{|A|(\mathcal{H})}$ of $|A|(\mathcal{H})$. By putting $U = 0$ on $\ker A = \ker |A| = \overline{|A|(\mathcal{H})}^{\perp}$, we get a partial isometry U satisfying $U|A| = A$. The rest is easy to verify. $\square$

Corollary 1.2.6 *Let $A \in \mathbb{B}(\mathcal{H})$ be invertible. Then there is a unique unitary U such that $A = U|A|$.*

Proof By the proof of Theorem 1.2.5, there is a unique partial isometry U such that $A = U|A|$, $\ker U = \ker A$, and $\operatorname{ran} U$ is the closure of $\operatorname{ran} A$. Since A is one-to-one, we have $\ker U = \ker A = \{0\}$. Hence, U is an isometry on $(\ker U)^{\perp} = \mathcal{H}$. Therefore, $U^*U = I$. Since A is surjective, we have $U(\mathcal{H}) = \overline{A(\mathcal{H})} = \mathcal{H}$. Therefore, U is invertible. Thus, $U^{-1} = IU^{-1} = U^*UU^{-1} = U^*$. Thus, U is a unitary. $\square$

The following Lemma, together with the Cartesian decomposition, yields that each operator in $\mathbb{B}(\mathcal{H})$ is a linear combination of four unitaries.

Lemma 1.2.7 *If $A \in \mathbb{B}(\mathcal{H})$ is an invertible contraction, then there exist unitaries U_1 and U_2 such that $A = (U_1 + U_2)/2$.*

Proof First, assume that A is a self-adjoint contraction but not necessarily invertible. Since $\|A\| \leq 1$, we have $\|A^2\| = \|A\|^2 \leq 1$, hence $A^2 \leq I$. This implies that $I - A^2 \geq 0$, so

1.2 Some Decompositions of Operators

its positive square root exists and commutes with A. If we set $U := A + i(I - A^2)^{1/2}$, a straightforward computation shows that $UU^* = U^*U = I$ and $A = (U + U^*)/2$. Second, let A be an invertible contraction. Using the polar decomposition $A = V|A|$ of A with V unitary (Corollary 1.2.6), we have $|A| = V^*A$. From the first part of the proof, we know that there exist unitaries W_1 and W_2 such that $|A| = \frac{W_1}{2} + \frac{W_2}{2}$. Hence, with $U_1 = VW_1$ and $U_2 = VW_2$, we have

$$A = V|A| = \frac{VW_1 + VW_2}{2} = \frac{U_1 + U_2}{2}.$$

□

For a positive definite contraction $A \in \mathbb{M}_n$ with eigenvalues $0 \le \lambda_j \le 1$, let $A = V\,\mathrm{diag}(\lambda_1, \ldots, \lambda_n)V^*$ be its spectral decomposition. One can utilize a simple geometric argument to find $\theta_j \in \mathbb{R}$ such that $\lambda_j = \frac{\exp(i\theta_j) + \exp(-i\theta_j)}{2}$ for all $1 \le j \le n$. Then we can choose

$$U_1 = V\,\mathrm{diag}(\exp(i\theta_1), \ldots, \exp(i\theta_j))V^* \text{ and } U_2 = V\,\mathrm{diag}(\exp(-i\theta_1), \ldots, \exp(-i\theta_j))V^*.$$

Thus, $A = \frac{1}{2}(U_1 + U_2)$.

Next, we show that if the norm of an operator is less than $1 - 2/n$ for some $n > 2$, then it can be considered as the "arithmetic mean" of n unitaries.

Theorem 1.2.8 ([192, Proposition 3.2.23]) *Let $A \in \mathbb{B}(\mathcal{H})$ such that $\|A\| < 1 - 2/n$ for some $n > 2$. Then there exist unitaries $U_1, \ldots, U_n$ such that*

$$A = \frac{U_1 + \cdots + U_n}{n}.$$

Proof First, we prove that if $B \in \mathbb{B}(\mathcal{H})$ with $\|B\| < 1$ and U is a unitary, then there exist unitaries U_1 and V_1 such that

$$B + U = U_1 + V_1.$$

By replacing B with U^*B, we can assume that $U = I$. According to the spectral theorem for B, we know that $I + B$ is invertible. Furthermore, since $\|\frac{I+B}{2}\| \le 1$, we can apply Lemma 1.2.7 to conclude that there exist unitaries U_1 and V_1 such that $B + I = U_1 + V_1$.

Secondly, let us set $B = (n-1)^{-1}(nA - I)$. Then, it follows that $\|B\| < 1$ and $nA = (n-1)B + I$. By applying the aforementioned process $(n-1)$ times, we can find unitaries $V_1, \ldots, V_{n-1}, U_1, \ldots, U_n = V_{n-1}$ such that

$$\begin{aligned}
nA &= (n-1)B + I = (n-2)B + (B+I) = (n-2)B + (V_1 + U_1) \\
&= (n-3)B + (B+V_1) + U_1 = (n-3)B + (V_2 + U_2) + U_1 \\
&= (n-4)B + (B+V_2) + U_2 + U_1 \\
&= (n-4)B + (V_3 + U_3) + U_2 + U_1 = \cdots \\
&= (B + V_{n-2}) + U_{n-2} + \cdots + U_1 = U_n + U_{n-l} + U_{n-2} + \cdots + U_1.
\end{aligned}$$

□

From the latter theorem, we can conclude the *Russo–Dye theorem* [205].

Corollary 1.2.9 *The closed unit ball of $\mathbb{B}(\mathcal{H})$ is the closed convex hull of its unitaries.*

Proof If $\|A\| \leq 1$, then for each n, the operator $((1 - 3n^{-1})A)$ is a mean of n unitaries. The conclusion is obtained from Theorem 1.2.8, since A is the norm-limit of the sequence $((1 - 3n^{-1})A)$. □

When $n = 2$, Theorem 1.2.8 says that if $\|A\| < 1/2$, then A is a convex combination of four unitaries. It can be extended for every operator A with $\|A\| < 1$ as follows.

Theorem 1.2.10 *Let $A \in \mathbb{B}(\mathcal{H})$ be an operator such that $\|A\| < 1$. Then there exist four unitaries $U_1, U_2, U_3, U_4 \in \mathbb{B}(\mathcal{H})$ such that*

$$A = \frac{U_1 + U_2 + U_3 + U_n}{4}.$$

Proof Let $\|A\| < 1$. There exists a positive integer n such that $\|A\| < 1 - 1/n$. Then $\pm A + (1 - \frac{1}{n}I)$ is invertible since

$$\left\| I - \left(\pm A + \left(1 - \frac{1}{n}\right)I \right) \right\| \leq \frac{1}{n} + \|A\| < 1.$$

From Lemma 1.2.7, we conclude that there are unitaries four unitaries $U_\pm$ and $V_\pm$ such that

$$\pm A + \left(1 - \frac{1}{n}\right)I = \frac{U_\pm + V_\pm}{2}.$$

From

$$A = \frac{1}{2}\left[\left(A + \left(1 - \frac{1}{n}\right)I\right) + \left(A - \left(1 - \frac{1}{n}\right)I\right) \right],$$

we get $A = (U_+ + V_+ + U_- + V_-)/4$. □

Inspired by the above results, we can pose an interesting question as follows: "Is it true that every operator $A \in \mathbb{B}(\mathcal{H})$ with $\|A\| < 1$ can be expressed as a convex combination of n unitaries for each $n > 1$?" The answer is negative, as demonstrated by Kadison and Pedersen [133, Proposition 3]:

By a non-unitary isometry, we refer to an operator $V \in \mathbb{B}(\mathcal{H})$ such that $V^*V = I$ and $VV^* < I$. If $U_1, \ldots, U_n$ are n unitaries, then

$$\left\| V - \frac{U_1 + \cdots + U_n}{n} \right\| \geq \frac{2}{n}.$$

To prove this, suppose the opposite. Then, we have

$$\|U_1^*V - n^{-1}I\| = \|V - n^{-1}U_1\| < 2n^{-1} + \|n^{-1}(U_2 + \cdots + U_n)\|$$
$$\leq 2n^{-1} + (n-1)n^{-1} = 1 + n^{-1}.$$

Since U_1^*V is a non-unitary isometry and $-1 - n^{-1} \in \mathrm{sp}(U_1^*V - n^{-1}I)$, we have $1 + n^{-1} \leq \|U_1^*V - n^{-1}I\|$, which is a contradiction.

In addition, Choi and Li [47] showed the existence of an operator A for which $A + A^* \leq I$ but A cannot be expressed as a convex combination of unitaries U_j with $U_j + U_j^* \leq I$. Let us consider the matrix $A = \begin{bmatrix} 0 & 1 \\ 0 & 0 \end{bmatrix}$. If A is a convex combination of such matrices $U_1, \ldots, U_n$, then the $(1, 2)$-entry of U_j must be equal to one. Thus, $U_j = \begin{bmatrix} 0 & 1 \\ \mu_j & 0 \end{bmatrix}$, where $|\mu_j| = 1$. The condition $U_j + U_j^* \leq I$ entails that $\mu_j + \overline{\mu_j} \leq -1$, which is not possible since the $(2, 1)$-entry of A is a convex combination of the μ_j's.

1.3 Unitarily Invariant Norms

For vectors $x, y \in \mathcal{H}$, the *rank-one operator* $x \otimes \overline{y}$ is defined as $(x \otimes \overline{y})(z) = \langle z, y \rangle x$. By a *finite-rank operator*, we mean a finite linear combination of rank-one operators. The closure of the algebra of finite-rank operators is called the algebra of *compact operators* and is denoted by $\mathbb{K}(\mathcal{H})$. It is known that $\mathbb{K}(\mathcal{H}) = \mathbb{B}(\mathcal{H})$ if and only if $\mathcal{H}$ is finite dimensional.

The spectrum of an operator $A \in \mathbb{K}(\mathcal{H})$ is countable, and each nonzero number in $\mathrm{sp}(A)$ is an eigenvalue of A and an isolated point of $\mathrm{sp}(A)$. For any operator $A \in \mathbb{K}(\mathcal{H})$ whose all eigenvalues are real, we consider $\lambda_1(A), \lambda_2(A), \ldots$ to be the eigenvalues of A in decreasing order and repeated according to multiplicity. Similarly, let $s_1(A), s_2(A), \ldots$ be the eigenvalues of $|A| = (A^*A)^{\frac{1}{2}}$ in decreasing order and repeated according to multiplicity. If $A \in \mathbb{M}_n$, we take $s_k(A) = 0$ for $k > n$. The *singular value decomposition* states that if $A \in \mathbb{K}(\mathcal{H})$, then $A = \sum_{n=1}^{\infty} s_n(A) e_n \otimes \overline{f_n}$ for some orthonormal sets (e_n) and (f_n).

Denote the set of complex sequences that converge to zero by c_0. Consider the set $c_F \subseteq c_0$, which consists of sequences with a finite number of nonzero elements. For any $a \in c_0$, we define $\lfloor a \rfloor := (|a_n|)_{n \in \mathbb{N}} \in c_0$. According to [97, Sect. III.3], a symmetric norming function (or, in the context of matrices, a symmetric gauge function [19, p. 86]) is a function denoted as g that maps from c_F to $\mathbb{R}$ satisfying the following properties:

(i) g is a norm on c_F;
(ii) $g(a) = g(\lfloor a \rfloor)$ for every $a \in c_F$;
(iii) g is invariant under permutations.

For $a = (a_i) \in c_0$, let us define $g(a) = \sup_{n \in \mathbb{N}} g(a_1, \ldots, a_n, 0, \ldots) \in \mathbb{R} \cup \{+\infty\}$. A *unitarily invariant norm* in $\mathbb{K}(\mathcal{H})$ is a map $\|\!|\cdot|\!\| : \mathbb{K}(\mathcal{H}) \to [0, \infty]$ given by

$$\|\!|A|\!\| = g(s_1(A), s_2(A), \ldots),$$

where $A \in \mathbb{K}(\mathcal{H})$ and g is a symmetric norming function; see [97, Chap. III]. The set $\mathcal{C}_{\|\!|\cdot|\!\|} = \{A \in \mathbb{K}(\mathcal{H}) : \|\!|A|\!\| < \infty\}$ is a self-adjoint (two-sided) ideal of $\mathbb{B}(\mathcal{H})$. The *Ky Fan norms* are examples of unitarily invariant norms and are defined by $\|A\|_{(k)} = \sum_{j=1}^{k} s_j(A)$ for $k = 1, 2, \ldots$. The Ky Fan dominance theorem [19, Theorme IV.2.2] states that $\|A\|_{(k)} \leq \|B\|_{(k)}$ (for $k = 1, 2, \ldots$) if and only if $\|\!|A|\!\| \leq \|\!|B|\!\|$ for all unitarily invariant norms $\|\!|\cdot|\!\|$. An immediate consequence of this theorem is that the following assertions are equivalent:

$$\|\!|A|\!\| \leq \|\!|B|\!\| \text{ for all unitarily invariant norms;} \tag{1.3.1}$$

$$\|\!|A \oplus 0|\!\| \leq \|\!|B \oplus 0|\!\| \text{ for all unitarily invariant norms;} \tag{1.3.2}$$

$$\|\!|A \oplus A|\!\| \leq \|\!|B \oplus B|\!\| \text{ for all unitarily invariant norms.} \tag{1.3.3}$$

For a matrix $X \in \mathbb{M}_n$, it follows from the polar decomposition $X = U|X|$ of X, where U is unitary, we have

$$\|\!|X \oplus X^*|\!\| = \|\!|U|X| \oplus |X|U^*|\!\| = \|\!| |X| \oplus |X| |\!\| = \|\!|U|X| \oplus U|X| |\!\| = \|\!|X \oplus X|\!\|. \tag{1.3.4}$$

Let $\|\!|\cdot|\!\|$ be an arbitrary unitarily invariant norm. If $T = \begin{bmatrix} A & B \\ C & D \end{bmatrix}$ is self-adjoint, then $C = B^*$. Therefore, $\|\!|C|\!\| = \|\!|B|\!\|$. If T is unitary, then $AA^* + BB^* = A^*A + C^*C = I$, and so

$$\lambda_j(BB^*) = \lambda_j(I - AA^*) = 1 - \lambda_j(AA^*) = 1 - \lambda_j(A^*A) = \lambda_j(I - A^*A) = \lambda_j(C^*C).$$

Hence, $\|\!|B|\!\| = \|\!|C|\!\|$.

It is known that the *Schatten p-norms* $\|A\|_p = \left(\sum_{j=1}^{\infty} s_j^p(A) \right)^{1/p}$ defined on the ideal $\mathcal{C}_p = \{A \in \mathbb{K}(\mathcal{H}) : \|A\|_p < \infty\}$ are also unitarily invariant norms for all $p \geq 1$; see [19, Sect. IV.2]. Another significant example of a unitarily invariant norm is the usual operator norm $\|\cdot\|$.

Throughout the book, we assume that a unitarily invariant norm $\|\!|\cdot|\!\|$ is given on a two-sided ideal of bounded linear operators acting on a separable Hilbert space. The norms $\|\!|\cdot|\!\|$

1.3 Unitarily Invariant Norms

on matrix algebras $\mathbb{M}_n$ for all finite values of n are then induced by it via the equation

$$\|A\| = \|A \oplus 0\|. \tag{1.3.5}$$

Thus, we deal with a system of unitarily invariant norms $\{\|\cdot\|_s\}$ on algebras $\mathbb{M}_s$, where $s \leq N$ or on all algebras $\mathbb{M}_s$, where $s \geq 1$. These norms satisfy the relation $\|A\|_s = \|A \oplus 0_{(t-s)(t-s)}\|_t$, where $A \in \mathbb{M}_s$ and $t > s$, between norms of matrices of different sizes.

The following lemma is useful for establishing inequalities involving unitarily invariant norms.

Lemma 1.3.1 ([19, p. 75]) *Let $A, X, B \in \mathbb{M}_n$. Then*
(i) $s_j(AXB) \leq \|A\| s_j(X) \|B\|$ $(j = 1, 2, \ldots, n)$.
(ii) $\|AXB\| \leq \|A\| \|X\| \|B\|$.

The next lemma gives an estimate of $\|K\|$ when K is a contraction.

Lemma 1.3.2 *Let $\|\cdot\|$ be a unitarily invariant norm on $\mathbb{M}_n$. If K is a contraction, then*

$$\|K\| \leq \|I_n\|.$$

Proof It follows from Lemma 1.3.1(ii) that

$$\|K\| = \| |K| \| = \left\| |K|^{\frac{1}{2}} I_n |K|^{\frac{1}{2}} \right\|$$
$$\leq \| |K|^{\frac{1}{2}} \| \|I_n\| \| |K|^{\frac{1}{2}} \| = \|K\| \|I_n\| \leq \|I_n\|. \qquad \square$$

The two next lemmas deal with the positivity of block matrices.
The first lemma is known as Horn's theorem.

Lemma 1.3.3 ([239, Corollary 10.3]) *Let $A, B \in \mathbb{M}_n$. Then*

$$\prod_{i=1}^{k} s_j(AB) \leq \prod_{i=1}^{k} \left(s_j(A) s_j(B)\right) \quad (k = 1, 2, \ldots, n)$$

Lemma 1.3.4 *For any matrix $X \in \mathbb{M}_n$,*

$$\left\| \begin{bmatrix} 0 & X \\ X^* & 0 \end{bmatrix} \right\| = \|X\|.$$

Proof

$$\left\| \begin{bmatrix} 0 & X \\ X^* & 0 \end{bmatrix} \right\| = \left\| \begin{bmatrix} I_n & 0 \\ 0 & I_n \end{bmatrix} \begin{bmatrix} 0 & X \\ X^* & 0 \end{bmatrix} \begin{bmatrix} 0 & I_n \\ I_n & 0 \end{bmatrix} \right\|$$

$$= \left\| \begin{bmatrix} X & 0 \\ 0 & X^* \end{bmatrix} \right\| = \max\{\|X\|, \|X^*\|\} = \|X\|.$$

$\square$

The *Weyl monotonicity principle* in $\mathbb{M}_n$ states that

$$A \leq B \quad \text{implies} \quad \lambda_j(A) \leq \lambda_j(B) \tag{1.3.6}$$

for all $1 \leq j \leq n$; see [19, p. 63].

1.4 Moore–Penrose Inverse

The *Moore–Penrose inverse* of $A \in \mathbb{B}(\mathscr{H})$ is the unique operator $A^\dagger$, if it exists, with the following properties:
(i) $AA^\dagger A = A$;
(ii) $A^\dagger AA^\dagger = A^\dagger$;
(iii) $AA^\dagger$ is self-adjoint;
(iv) $A^\dagger A$ is self-adjoint.

Proposition 1.4.1 *An operator $A \in \mathbb{B}(\mathscr{H})$ has closed range if and only if so is A^*.*

Proof The restriction map A to $(\ker A)^\perp$, denoted by $\tilde{A}$, is a one-to-one linear map onto ran A. It follows from the open mapping theorem that $\tilde{A}$ is a bounded invertible operator. Let $y \in \overline{\operatorname{ran} A^*}$. Utilizing the Riesz representation Theorm 1.1.3 to the bounded linear functional $f : \operatorname{ran} A \to \mathbb{C}$ defined as $f(Az) = \langle \tilde{A}^{-1}Az, y \rangle$ $z \in \mathscr{H}$, we conclude that there exists $x \in$ ran A such that $f(Az) = \langle Az, x \rangle$ for all $z \in \mathscr{H}$. Therefore,

$$\langle z, A^*x \rangle = \langle Az, x \rangle = f(Az) = \langle \tilde{A}^{-1}Az, y \rangle = \langle z, y \rangle \quad (z \in \mathscr{H}).$$

Hence, $y = A^*x \in \operatorname{ran} A^*$.

Conversely, if A^* has closed range, then by the first part of the proof, $(A^*)^* = A$ also has closed range. $\square$

We can provide another proof for the closedness of ran A^* when ran A is closed. Firstly, it follows from the equivalence of (i) and (iii) in the Douglas majorization Theorem 1.2.4 and the trivial equality $AA^* = (AA^*)^{1/2}(AA^*)^{1/2}$ that $\operatorname{ran}(A) = \operatorname{ran}(AA^*)^{1/2}$.

Secondly, $\operatorname{ran}(AA^*)^{1/2} = \operatorname{ran}(AA^*)$. This follows from the fact that if $A \in \mathbb{B}(\mathscr{H})$ is positive and has closed range, then ran $A^{1/2} = \operatorname{ran} A$. To establish this, first note that ker $A =$

1.4 Moore–Penrose Inverse

$\ker A^{1/2}$ since $\langle A^{1/2}x, A^{1/2}x \rangle = \langle Ax, x \rangle$ for any $x \in \mathcal{H}$. Therefore,

$$\operatorname{ran} A = \overline{\operatorname{ran} A} = (\ker A)^\perp = \left(\ker A^{1/2}\right)^\perp = \overline{\operatorname{ran} A^{1/2}} \supseteq \operatorname{ran} A^{1/2} \supseteq \operatorname{ran} A.$$

Thirdly, let $x \in \mathcal{H}$. Then $Ax = AA^*y$ for some $y \in \mathcal{H}$. Hence, $x - A^*y \in \ker A$, and so $x = (x - A^*y) + A^*y \in \ker A + \operatorname{ran} A^*$. Since $\operatorname{ran} A^*$ is orthogonal to $\ker A$, we conclude that $\operatorname{ran} A^* = (\ker A)^\perp$ is closed. The converse is also true, that is, if $\operatorname{ran} A = \operatorname{ran}(AA^*)$, then $\operatorname{ran} A$ is closed; see [219].

An operator in $\mathbb{B}(\mathcal{H})$ has a Moore–Penrose inverse if and only if it has closed range. To establish this, let's first assume that A has the Moore–Penrose inverse $A^\dagger$. Then $AA^\dagger$ is a projection since it is self-adjoint and $(AA^\dagger)^2 = (AA^\dagger A)A^\dagger = AA^\dagger$. In addition, $\operatorname{ran}(AA^\dagger) = \operatorname{ran}(A)$ since

$$\operatorname{ran} A = \operatorname{ran}(AA^\dagger A) \subseteq \operatorname{ran}(AA^\dagger) \subseteq \operatorname{ran} A.$$

Therefore, the range of A is closed. Conversely, if A has a closed range and P is the projection onto $\operatorname{ran}(A)$, then the Douglas solution to $AX = P$ (see Theorem 1.2.4 and Exercise 1.6.5) is the Moore–Penrose inverse of A. To see this, note that $(AX)^* = P^* = P = AX$, and so AX is self-adjoint. In addition, $AXA = PA = A$ since P identically acts on the range of A. If $z \in \operatorname{ran} P$, then $XPz = Xz$, and if $z \in \ker P$, then $XPz = 0 = Xz$ since $\ker X = \ker P$. Considering $\mathcal{H} = \ker P + \operatorname{ran} P$, we arrive at $XAX = XP = X$.

To prove that XA is also self-adjoint, let us suppose that Q is the projection onto $\operatorname{ran} A^*$, which is closed since $\operatorname{ran} A$ is closed. Hence, $\ker Q = \ker A$. If $x \in \ker A$, then $(XA)(x) = 0 = Qx$. If $x \in (\ker A)^\perp = \operatorname{ran} A^*$, then $Qx = x$. In addition, $P(Ax) = Ax$. Taking the proof of (i) $\Longrightarrow$ (ii) of Theorem 1.2.4 into account and considering $x = 0 + x \in \ker A \oplus (\ker A)^\perp$, we get $X(Ax) = x = Qx$. Since $\mathcal{H} = \ker A \oplus (\ker A)^\perp$, we conclude that $XA = Q$. In particular, XA is self-adjoint.

Since all subspaces of the finite-dimensional Hilbert space $\mathbb{C}^n$ are closed, every matrix in $\mathbb{M}_n$ has a Moore–Penrose inverse. Moore–Penrose inverses are important because many matrices that are not invertible still have Moore–Penrose inverses. They are a type of generalized inverses. There are numerous results in the literature regarding spectral theory, Fredholmness, Weylness, and generalized inverses (such as the Drazin inverse and the Moore–Penrose inverse) of block matrices of operators. We refer interested readers to [52, 63, 128, 222].

One can easily observe that if A is invertible, then $A^\dagger = A^{-1}$. In addition, $(A^*)^\dagger = (A^\dagger)^*$. The *reverse order law* $(AB)^\dagger = B^\dagger A^\dagger$ does not hold, in general. A counterexample is as follows.

Example 1.4.2 Let $A = \begin{bmatrix} 1 & 0 \\ -1 & 0 \end{bmatrix}$ and $B = \begin{bmatrix} 1 & 0 \\ 1 & 0 \end{bmatrix}$. Then $A^\dagger = \begin{bmatrix} \frac{1}{2} & \frac{-1}{2} \\ 0 & 0 \end{bmatrix}$ and $B^\dagger = \begin{bmatrix} \frac{1}{2} & \frac{1}{2} \\ 0 & 0 \end{bmatrix}$. Therefore,

$$(AB)^\dagger = A^\dagger = \begin{bmatrix} \frac{1}{2} & \frac{-1}{2} \\ 0 & 0 \end{bmatrix} \neq \begin{bmatrix} \frac{1}{4} & \frac{-1}{4} \\ 0 & 0 \end{bmatrix} = B^\dagger A^\dagger.$$

However, $(A^*A)^\dagger = A^\dagger (A^*)^\dagger$; see [98, 175].

We can state a key result in the investigation of the Moore–Penrose inverse as follows.

Let $A \in \mathbb{B}(\mathscr{H})$ have closed range. Hence A admits a representation as an operator from ran $A^* \oplus \ker A$ into ran $A \oplus \ker A^*$, given by $\begin{bmatrix} A_1 & 0 \\ 0 & 0 \end{bmatrix}$ since an operator maps its kernel onto 0. Here, A_1 is invertible as it is the restriction map A from ran A^* onto ran A and ran $A = \text{ran}(AA^*)$. Then, it is straightforward to verify that the Moore-inverse of A is

$$A^\dagger = \begin{bmatrix} A_1^{-1} & 0 \\ 0 & 0 \end{bmatrix} : \text{ran } A \oplus \ker A^* \to \text{ran } A^* \oplus \ker A.$$

1.5 Differences Between Real and Complex Hilbert Spaces

It is an interesting question to ask which results on complex Hilbert spaces and their operators still hold for the real Hilbert spaces and their operators possibly under different or additional conditions. It is also worth considering which facts in the complex case do not hold when we restrict ourselves to spaces with real scalars. For a comprehensive account of the differences between operators acting on real Hilbert spaces and complex Hilbert spaces, see [173].

(i) Let $\mathscr{H}$ be a real Hilbert space. Then the spectrum of an operator $A \in \mathbb{B}(\mathscr{H})$, defined as $\{\lambda \in \mathbb{R} : A - \lambda I \text{ is not invertibe in } \mathbb{B}(\mathscr{H})\}$, may be empty. For example, take $\mathscr{H} = \mathbb{R}^2$ and $A = \begin{bmatrix} 1 & 2 \\ -2 & 0 \end{bmatrix}$. In addition, it is not true that every matrix with real entries has real eigenvalues.

(ii) The spectral theorem asserts that $\text{sp}(p(A)) = p(\text{sp}(A))$ for any bounded linear operator A acting on a complex Hilbert space $\mathscr{H}$ and any polynomial p with complex coefficients. However, this result does not hold in the suggested real spectrum in (i). For example, let us consider $\mathscr{H} = \mathbb{R}^2$ and $A = \begin{bmatrix} 0 & 1 \\ -1 & 0 \end{bmatrix}$ with $p(t) = t^2$.

(iii) Proposition 1.1.2 may not be true for real Hilbert spaces. For example, let us consider $\mathscr{H} = \mathbb{R}^2$ and $A = \begin{pmatrix} 0 & 1 \\ -1 & 0 \end{pmatrix}$ as the matrix of 90 degrees clockwise rotation.

(iv) For every pair of commuting $n \times n$ complex matrices A and B, there exists a unitary matrix $U \in \mathbb{M}_n$ such that both U^*AU and U^*BU are upper-triangular. This result does not hold for real matrices. For example, if $A = \begin{bmatrix} 0 & 1 \\ -1 & 0 \end{bmatrix}$ and $B = \begin{bmatrix} 1 & 1 \\ -1 & 1 \end{bmatrix}$, then there is no real matrix U with the desired properties [239, p. 76].

(v) A positive operator A in the setting of complex Hilbert spaces is self-adjoint. However, if $\mathcal{H}$ is a real Hilbert space, then the positivity of A in the above terms does not entail that A is self-adjoint. To see this, let $\mathcal{H} = \mathbb{R}^2$ and $A = \begin{bmatrix} 1 & 1 \\ -1 & 1 \end{bmatrix}$.

(vi) If $\mathcal{H}$ is a real Hilbert space, then the left-side inequality of (1.1.4) may not be true. For example, let $\mathcal{H} = \mathbb{R}^2$ and $A = \begin{bmatrix} 0 & -1 \\ 1 & 0 \end{bmatrix}$ to get $w(A) = 0$ and $\|A\| = 1$.

1.6 Exercises and Problems

Exercise 1.6.1 Let $\mathcal{H}$ be an infinite-dimensional Hilbert space and let $A \in \mathbb{B}(\mathcal{H})$. Prove that A is compact if and only if the range of A contains no closed infinite-dimensional subspaces. Hint: See [65].

Exercise 1.6.2 Let $[a_{ij}] \in \mathbb{M}_n$ be positive semidefinite. If $n = 2$ or $n = 3$, show that $[|a_{ij}|]$ is also positive definite. Give an example to show that this statement may not be valid for $n = 4$.

Hint: Just use the condition in Theorem 1.1.8(d).

Exercise 1.6.3 (i) Let $X = \begin{bmatrix} a & b \\ c & d \end{bmatrix}$ with $a, b, c, d \in \mathbb{R}$. Then

$$\omega(X) = \frac{|a+d| + \sqrt{(a-d)^2 + (b+c)^2}}{2}$$

holds.

(ii) Is there a formula for the numerical radius of a 3×3 matrix?
Hint: see [130, Theorem 1].

Exercise 1.6.4 Suppose that $A \in \mathbb{B}(\mathcal{H})$. Show that $r(A) = \|A\|$ if and only if $w(A) = \|A\|$.

Hint: The right implication comes from the fact that $r(A) \leq w(A)$. For the left implication set $w(A) = \|A\| = 1$. Take a sequence (x_n) in $\mathcal{H}$ with $\|x_n\| = 1$ such that $\langle Ax_n, x_n \rangle \to 1$. Then, show that $1 \in \text{sp}(A)$.

Exercise 1.6.5 Show that if (i), (ii), or (iii) in the Douglas majorization Theorem 1.2.4 holds, then there exists a unique operator C, known as the *Douglas Solution*, of $A = BX$ so that

(a) $\|C\|^2 = \inf\{\mu | AA^* \leq \mu BB^*\}$;

(b) ker A = ker C;
(c) ran $C \subseteq \overline{\text{ran } B^*}$.

Exercise 1.6.6 Let $M = \begin{bmatrix} A & C \\ 0 & B \end{bmatrix}$, let A have a dense range, and let B be one-to-one. Prove or disprove this assertion: "M has closed range if and only if B and C both do".

Exercise 1.6.7 The *Drazin inverse* of $A \in \mathbb{M}_n$ is the unique matrix A^D satisfying $A^k A^D A = A^k$, $A^D A A^D = A^D$, and $A A^D = A^D A$, where $k = \text{Ind}(A)$ denotes the index of A, that is, the smallest nonnegative integer k such that $\text{rank}(A^{k+1}) = \text{rank}(A^k)$. Prove that for every $A \in \mathbb{M}_n$, the matrix A^D exists.
Hint: See [223].

Exercise 1.6.8 Suppose that $M = \begin{bmatrix} A & B \\ B^* & C \end{bmatrix}$ is a positive partitioned matrix with $A, B, C \in \mathbb{M}_n$ and $\mathcal{S}$ is a subspace of a finite-dimensional Hilbert space $\mathcal{H}$. Let $M_{\mathcal{S}} = \begin{bmatrix} A - BC^\dagger B^* & 0 \\ 0 & 0 \end{bmatrix}$.
Show that
$$M_{\mathcal{S}} = \sup\{D : 0 \leq D \leq A \text{ and } \text{ran}(D) \subseteq \mathcal{S}\}.$$

When $\mathcal{H}$ is of arbitrary dimension, the above relation defines the so-called *shorted operator* $M_{\mathcal{S}}$.
Hint: See [4].

Exercise 1.6.9 Let $A \in \mathbb{M}_2$ with trace zero. Prove that A is unitarily equivalent to a matrix with zero-diagonal. In other words, there exists a matrix $B \in \mathbb{M}_2$ with $b_{11} = b_{22} = 0$ and a unitary $U \in \mathbb{M}_2$ such that $A = U^* B U$.
Hint: Construct a unit vector $x \in \mathbb{C}^2$ such that $\langle x, Ax \rangle = 0$. Then construct a unitary W such that the $(1, 1)$-entry of $W^* A W$ is zero.

Problem 1.6.10 (i) Consider the 2×2 matrices
$$B = \begin{bmatrix} 1 & \varepsilon \\ 0 & 0 \end{bmatrix} \quad \text{and} \quad C = \begin{bmatrix} 1 & 0 \\ 0 & \varepsilon \end{bmatrix}.$$

Show that $\|B\|_2 = \|C\|_2$. but there do not exist any 2×2 matrices A and D such that $\begin{bmatrix} A & B \\ C & D \end{bmatrix}$ is normal.
(ii) What matrix pairs B and C can be the off-diagonal entries of a normal matrix $M = \begin{bmatrix} A & B \\ C & D \end{bmatrix}$? In other words, when does $\begin{bmatrix} ? & B \\ C & ? \end{bmatrix}$ have a normal completion?

(iii) For every B, show that the matrix $\begin{bmatrix} ? & B \\ B & ? \end{bmatrix}$ has a normal completion of the form $M = \begin{bmatrix} A & B \\ B & A \end{bmatrix}$. Can one find A such that M has the least possible operator or Hilbert–Schmidt norm?

(iv) Prove that if B and C are $n \times n$ matrices with $\|B\| = \|C\|$ for every unitarily invariant norm, then the matrix $\begin{bmatrix} ? & B \\ C & ? \end{bmatrix}$ has a completion that is a scalar multiple of a unitary matrix.

Hint: See [21].

Block Matrices of Operators

In this chapter, we introduce the concept of $n \times n$ block matrices of operators, or simply operator matrices, which are of great importance. They allow us to represent an operator on a space using simpler operators on its specific subspaces. We provide an in-depth study of dilation theory and give several characterizations of the positivity of 2×2 operator matrices of the form $\begin{bmatrix} A & X \\ X^* & B \end{bmatrix}$. We also investigate the properties of 2×2 matrices with entries in a C^*-algebra. Finally, we utilize the power of operator matrices to derive a variety of inequalities related to eigenvalues and norms of matrices.

2.1 Operator Matrices of Size $n \times n$

Let
$$\oplus_{n=1}^{\infty} \mathscr{H} := \ell^2(\mathscr{H}) := \{(x_n) : x_n \in \mathscr{H}, \sum_{n=1}^{\infty} \|x_n\|^2 < \infty\}$$
be equipped with the usual addition and multiplication by scalars of sequences and the inner product
$$\langle (x_n), (y_n) \rangle := \sum_{n=1}^{\infty} \langle x_n, y_n \rangle.$$

An application of the Cauchy–Schwarz inequality in the Hilbert space $\mathscr{H}^{\oplus n}$, which is the direct sum of n copies of $\mathscr{H}$ equipped with the inner product
$$\langle (x_1, x_2, \ldots, x_n), (y_1, y_2, \ldots, y_n) \rangle = \sum_{i=1}^{n} \langle x_i, y_i \rangle$$

shows that the above series is absolutely convergent, and therefore convergent. We remark that the elements of $\mathcal{H}^{\oplus n}$ are sometimes denoted by $1 \times n$ column matrices instead of n-tuples.

For each $k \in \mathbb{N}$, we can embed $\mathcal{H}$ in $\oplus_{n=1}^{\infty} \mathcal{H}$ (and similarly in $\mathcal{H}^{\oplus n}$) via ι_k which takes x to the sequence $(0, \ldots, 0, x, 0, \ldots)$, where x appears in the kth entry.

Every bounded operator on $\oplus_{n=1}^{\infty} \mathcal{H}$ can be represented by an infinite matrix, where the (i, j)-entry of the matrix is $\langle Te_j, e_i \rangle$. However, the converse is not true in general. For example, the infinite diagonal matrix with natural numbers on its diagonal represents an unbounded operator.

Clearly the set $\mathbb{M}_n(\mathbb{B}(\mathcal{H}))$ of all $n \times n$ matrices with entries in $\mathbb{B}(\mathcal{H})$ is a $*$-algebra under the usual matrix operations and the involution $[A_{ij}]^* = [A_{ji}^*]$. A key result is as follows.

Proposition 2.1.1 $\mathbb{M}_n(\mathbb{B}(\mathcal{H}))$ *is $*$-isomorphic to* $\mathbb{B}(\mathcal{H}^{\oplus n})$.

Proof For $1 \leq j \leq n$, let $\mathcal{H}_j = \mathcal{H}$ and $\iota_j : \mathcal{H}_j \to \mathcal{H}^{\oplus n}$ be the map that takes $x \in \mathcal{H}_j$ and maps it to the vector in $\mathcal{H}^{\oplus n}$ that has x as its jth entry and 0 elsewhere. Then ι_j^* is the map that projects a vector in $\mathcal{H}^{\oplus n}$ to its jth component. This can be seen from the equation

$$\langle \iota_j x, (y_1, \ldots, y_n) \rangle = \langle x, y_j \rangle = \langle x, \iota_j^*(y_1, \ldots, y_n) \rangle.$$

These maps fulfill $\iota_i^* \iota_j = \delta_{ij} I$, where δ denotes the Kronecker delta, and $\sum_{j=1}^{n} \iota_j \iota_j^* = I_{\mathcal{H}^{\oplus n}}$. It is clear that the map $\pi : \mathbb{M}_n(\mathbb{B}(\mathcal{H})) \to \mathbb{B}(\mathcal{H}^{\oplus n})$ defined by $\pi([A_{ij}]) = \sum_{i,j=1}^{n} \iota_i A_{ij} \iota_j^*$ is a $*$-homomorphism. It is one-to-one because if $\pi([A_{ij}]) = 0$, then

$$A_{pq} = \iota_p^* \pi([A_{ij}]) \iota_q = 0$$

for all $1 \leq p, q \leq n$. It is surjective because if $A \in \mathbb{B}(\mathcal{H}^{\oplus n})$, then by considering $A_{ij} = \iota_i^* A \iota_j \in \mathbb{B}(\mathcal{H})$ we have

$$\pi([A_{ij}]) = \sum_{i,j=1}^{n} \iota_i A_{ij} \iota_j^* = \sum_{i,j=1}^{n} \iota_i (\iota_i^* A \iota_j) \iota_j^* = A.$$

$\square$

Thus, each element $A \in \mathbb{B}(\mathcal{H}^{\oplus n})$ can be identified with $[A_{ij}] \in \mathbb{M}_n(\mathbb{B}(\mathcal{H}))$, where $A_{ij} = \iota_i^* A \iota_j \in \mathbb{B}(\mathcal{H})$. In addition, we can define the norm of the matrix $[A_{ij}] \in \mathbb{M}_n(\mathbb{B}(\mathcal{H}))$ to be the operator norm of $\pi([A_{ij}]) \in \mathbb{B}(\mathcal{H}^{\oplus n})$, thereby making $\mathbb{M}_n(\mathbb{B}(\mathcal{H}))$ a C^*-algebra. In fact, it holds that $\|A^*A\| = \|\pi(A^*A)\| = \|\pi(A)^*\pi(A)\| = \|\pi(A)\|^2 = \|A\|^2$.

If $A = [A_{ij}] \in \mathbb{M}_n(\mathbb{M}_m)$, then A can be regarded as an nm-by-nm partitioned matrix. If f is a map from $\mathbb{M}_m$ to $\mathbb{M}_k$, then we can consider an nk-by-nk partitioned matrix $F(A) = [f(A_{ij})]$. It is an interesting question to ask what properties of $F(A)$ are inherited from A. In

2.1 Operator Matrices of Size $n \times n$

particular, for what functions f does the positivity of $[A_{ij}]$ ensure the positivity of $f([A_{ij}])$; see [240]. A known result of Hua [126] says that the determinant function $f(A) = \det(A)$ is one such function; see the exercises of this chapter and Corollary 2.1.7 for more information.

The following lemma provides a relationship between the norm of an operator matrix and the norms of its entries.

Lemma 2.1.2 *If $A = [A_{ij}] \in \mathbb{M}_n(\mathbb{B}(\mathcal{H}))$, then*

$$\max_{1 \leq i,j \leq n} \|A_{ij}\| \leq \|A\| \leq \left(\sum_{i,j=1}^{n} \|A_{ij}\|^2\right)^{1/2} \leq n \max_{1 \leq i,j \leq n} \|A_{ij}\|.$$

Proof Following the notation and the proof of Proposition 2.1.1, we have $\|A_{pq}\| = \|\iota_p^* A \iota_q\| \leq \|A\|$ since $\|\iota_j\| = 1$ for all $1 \leq j \leq n$. Thus, we reach the first inequality. To get the second one, let $x = (x_1, \ldots, x_n) \in \mathcal{H}^{\oplus n}$. We have

$$\|Ax\|^2 = \left\|\left(\sum_{k=1}^{n} A_{1,k} x_k, \ldots, \sum_{k=1}^{n} A_{n,k} x_k\right)\right\|^2$$

$$= \sum_{k=1}^{n} \left\langle \sum_{i=1}^{n} A_{k,i} x_i, \sum_{j=1}^{n} A_{k,j} x_j \right\rangle$$

$$= \left| \sum_{k=1}^{n} \sum_{i,j=1}^{n} \langle A_{k,i} x_i, A_{k,j} x_j \rangle \right|$$

$$\leq \sum_{k=1}^{n} \sum_{i,j=1}^{n} |\langle A_{k,i} x_i, A_{k,j} x_j \rangle|$$

$$\leq \sum_{k=1}^{n} \sum_{i,j=1}^{n} \|A_{k,i}\| \|x_i\| \|A_{k,j}\| \|x_j\|$$

$$= \sum_{k=1}^{n} \left(\sum_{j=1}^{n} \|A_{k,j}\| \|x_j\|\right)^2$$

$$\leq \sum_{k=1}^{n} \left(\sum_{j=1}^{n} \|A_{k,j}\|^2\right) \left(\sum_{j=1}^{n} \|x_j\|^2\right)$$

(by the Cauchy–Schwarz inequality for scalars)

$$= \|x\|^2 \sum_{k,j=1}^{n} \|A_{k,j}\|^2,$$

whence

$$\|A\| \le \left(\sum_{i,j=1}^{n} \|A_{ij}\|^2\right)^{1/2} \le n \max_{1 \le i,j \le n} \|A_{ij}\|.$$

□

Now, let $\mathscr{A}$ be a C^*-algebra and $\mathbb{M}_n(\mathscr{A})$ denote the $*$-algebra of all $n \times n$ matrices with entries in $\mathscr{A}$ under the ordinary operations. Using a one-to-one $*$-representation π of $\mathscr{A}$ into $\mathbb{B}(\mathscr{H})$ for some Hilbert space $\mathscr{H}$, we obtain an injective $*$-homomorphism $\pi_n : \mathbb{M}_n(\mathscr{A}) \to \mathbb{M}_n(\mathbb{B}(\mathscr{H}))$ defined by $\pi_n([A_{ij}]) = [\pi(A_{ij})]$. Thus, we can assign the norm $[\pi(A_{ij})] \in \mathbb{M}_n(\mathbb{B}(\mathscr{H}))$ to each $[A_{ij}] \in \mathbb{M}_n(\mathscr{A})$ and make $\mathbb{M}_n(\mathscr{A})$ a C^*-algebra, since $\mathbb{M}_n(\mathbb{B}(\mathscr{H}))$ is already a C^*-algebra.

The following is a well-known result. Its proof is based on the definition of the algebraic tensor product of two algebras.

Lemma 2.1.3 *Let $\mathscr{A}$ be a C^*-algebra. Then, $\mathscr{A} \otimes \mathbb{M}_n$ and $\mathbb{M}_n(\mathscr{A})$ are $*$-isomorphic.*

Proof We define $\Phi: \mathscr{A} \otimes \mathbb{M}_n \to \mathbb{M}_n(\mathscr{A})$ by $\Phi(\sum_{i,j=1}^{n} A_{ij} \otimes E_{ij}) = [A_{ij}]$, where $(E_{ij})_{1 \le i,j \le n}$ is the standard matrix units for $\mathbb{M}_n$. Then, Φ is a $*$-isomorphism. □

The following is a typical example of n-homogeneous C^*-algebra.

Lemma 2.1.4 *Let Ω be a compact Hausdorff space. Define $C(\Omega, \mathbb{M}_n)$ be a set of all matrix valued continuous functions. Then, $C(\Omega, \mathbb{M}_n)$ becomes a C^*-algebra with the C^*-norm $\|f\|_\infty = \sup\{\|f(x)\| : x \in \Omega\}$.*

Proof For any f and g in $C(\Omega, \mathbb{M}_n)$ and $\alpha \in \mathbb{C}$ we define for any $x \in \Omega$

$$(f+g)(x) = f(x) + g(x)$$
$$(\alpha \cdot f)(x) = \alpha f(x)$$
$$(f \cdot g)(x) = f(x)g(x)$$
$$f^*(x) = f(x)^*$$

Then, $C(\Omega, \mathbb{M}_n)$ becomes a unital $*$-algebra over $\mathbb{C}$.

It is evident that $\|\cdot\|_\infty$ is a norm that makes $C(\Omega, \mathbb{M}_n)$ a Banach space.

We show that $C(\Omega, \mathbb{M}_n)$ is a C^*-algebra. To do this, it is enough to prove the submultiplicativity of the norm $\|\cdot\|_\infty$ as well as the C^*-condition. For any $f, g \in C(\Omega, \mathbb{M}_n)$, we have

2.1 Operator Matrices of Size $n \times n$

$$\begin{aligned}
\|fg\|_\infty &= \sup\{\|(fg)(x)\| : x \in \Omega\} \\
&= \sup\{\|f(x)g(x)\| : x \in \Omega\} \\
&\leq \sup\{\|f(x)\|\|g(x)\| : x \in \Omega\} \\
&\leq \sup\{\|f(x)\| : x \in \Omega\}\sup\{\|g(x)\| : x \in \Omega\} \\
&= \|f\|_\infty \|g\|_\infty
\end{aligned}$$

and

$$\begin{aligned}
\|f^*f\|_\infty &= \sup\{\|(f^*f)(x)\| : x \in \Omega\} \\
&= \sup\{\|f(x)^*f(x)\| : x \in \Omega\} \\
&= \sup\{\|f(x)\|^2 : x \in \Omega\} \\
&= (\sup\{\|f(x)\| : x \in \Omega\})^2 = \|f\|_\infty^2.
\end{aligned}$$

$\square$

The following implies how to define a C^*-norm on $\mathbb{M}_n(C(\Omega))$.

Lemma 2.1.5 ([226, Proposition T.5.21]) *Let $n \in \mathbb{N}$. Then there is an isomorphism from $\mathbb{M}_n(C(\Omega))$ to $C(\Omega, \mathbb{M}_n)$.*

Proof We define $\Phi: \mathbb{M}_n(C(\Omega)) \to C(\Omega, \mathbb{M}_n)$ by

$$\Phi([f_{ij}])(x) = [f_{ij}(x)] \quad (x \in \Omega).$$

For any $A, B \in \mathbb{M}_n(C(\Omega))$ it is straightforward to verify that $\Phi(AB) = \Phi(A)\Phi(B)$, $\Phi(A)^* = \Phi(A^*)$ and that Φ is injective and surjective. $\square$

From the previous Lemma, we can define a C^*-norm on $\mathbb{M}_n(C(\Omega))$ by $\|A\| = \|\Phi(A)\|_\infty$ ($A \in \mathbb{M}_n(C(\Omega))$). Then, $\mathbb{M}_n(C(\Omega))$ becomes a C^*-algebra. Moreover, [176, Corollary 2.1.2] ensures that C^*-norm $\|\cdot\|_\infty$ is the unique norm on $C(\Omega) \otimes \mathbb{M}_n$ making it a C^*-algebra.

The following fact is interesting in its own right.

Theorem 2.1.6 *Let $\mathscr{A}$ be a C^*-algebra. Each positive element of $\mathbb{M}_n(\mathscr{A})$ is the sum of n positive elements of the form $[A_i^* A_j]$ for some $A_1, \ldots, A_n \in \mathscr{A}$.*

Proof Let $A_1, \ldots, A_n \in \mathscr{A}$. We have

$$\begin{bmatrix} A_1^* A_1 & \cdots & A_1^* A_n \\ \vdots & & \vdots \\ A_n^* A_1 & \cdots & A_n^* A_n \end{bmatrix} = \begin{bmatrix} A_1 & \cdots & A_n \\ 0 & \cdots & 0 \\ \vdots & & \vdots \\ 0 & \cdots & 0 \end{bmatrix}^* \begin{bmatrix} A_1 & \cdots & A_n \\ 0 & \cdots & 0 \\ \vdots & & \vdots \\ 0 & \cdots & 0 \end{bmatrix} \geq 0.$$

Next, let $A \in \mathbb{M}_n(\mathscr{A})$ be positive. Therefore, there exists $B = [B_{ij}] \in \mathbb{M}_n(\mathscr{A})$ such that $A = B^* B$. Hence, $A_{ij} = \sum_{k=1}^n B_{ki}^* B_{kj}$. Set $C_k = [B_{ki}^* B_{kj}]$ $(1 \leq k \leq n)$. Then, we have $A = \sum_{k=1}^n C_k$. □

In a Hilbert space $\mathscr{H}$, the square matrix $[\langle x_i, x_j \rangle]$ is called the *Gram matrix* associated with $x_1, \ldots, x_n \in \mathscr{H}$.

Corollary 2.1.7 ([240, Theorem 2.1]) *Let $A = [A_{ij}]_{n \times n}$ be a positive operator matrix with $A_{ij} \in \mathbb{M}_n$. Then $[\mathrm{tr}(A_{ij})]$ is a positive semidefinite matrix in $\mathbb{M}_n$.*

Proof By Theorem 2.1.6, without loss of generality, we can assume that each A_{ij} is of the form $A_i^* A_j$ with A_i's in $\mathbb{M}_n$. Hence,

$$[\mathrm{tr}(A_{ij})] = [\mathrm{tr}(A_i^* A_j)] = [\langle A_i, A_j \rangle],$$

which represents a Gram matrix in the Hilbert space $\mathbb{M}_n$ equipped with the inner product $\langle A, B \rangle = \mathrm{tr}(A^* B)$. We deduce the result from the fact that the Gram matrix is positive semidefinite; see [11]. □

Recall that $T = [t_{ij}]_{n \times n}$ is said to be a *nonnegative matrix* if each entry t_{ij} is a nonnegative number. A nonnegative matrix is not necessarily a positive semidefinite matrix.

Our next result concerns the relationships between the numerical radii, operator norm, and spectral radii of A and $\tilde{A}$, respectively. The remaining theorems in this section are adapted from [125].

Theorem 2.1.8 *Let $A = [A_{ij}]_{n \times n}$ be an operator matrix and $\tilde{A} = [\|A_{ij}\|]_{n \times n}$ be its so-called block-norm matrix. Then*
 (i) $\omega(A) \leq \omega(\tilde{A})$.
 (ii) $\|A\| \leq \|\tilde{A}\|$.
 (iii) $r(A) \leq r(\tilde{A})$.

Proof For any vector $x = (x_1, \ldots, x_n) \in \mathscr{H}^{\oplus n}$ we write $|x| := (\|x_1\|, \ldots, \|x_n\|)$. Then $|x|$ is a unit vector in Hilbert space $\mathbb{C}^n$ if x is a unit vector in $\mathscr{H}^{\oplus n}$.

(i) Let $x = (x_1, \ldots, x_n) \in \mathcal{H}^{\oplus n}$ be a unit vector. Since

$$|\langle Ax, x\rangle| = \left|\sum_{i,j}^{n}(A_{ij}x_j, x_i)\right|$$

$$\leq \sum_{i,j}^{n} \|A_{ij}\| \|x_i\| \|x_j\| = \langle \tilde{A}|x|, |x|\rangle,$$

we have

$$\omega(A) = \sup_{\|x\|=1} |\langle Ax, x\rangle| \leq \sup_{\|x\|=1} \langle \tilde{A}|x|, |x|\rangle \leq \omega(\tilde{A}).$$

(ii) We have

$$\|A\|^2 = \sup_{\|x\|=1} \|Ax\|^2 = \sup_{\|x\|=1} \langle A^*Ax, x\rangle$$

$$= \sup_{\|x\|=1} \left|\sum_{i=1}^{n}\sum_{j=1}^{n}\sum_{k=1}^{n}\langle A_{kj}x_j, A_{ki}x_i\rangle\right|$$

$$\leq \sup_{\|x\|=1} \sum_{i=1}^{n}\sum_{j=1}^{n}\sum_{k=1}^{n} \|A_{kj}\| \|A_{ki}\| \|x_j\| \|x_i\|$$

$$= \sup_{\|x\|=1} \langle \tilde{A}^*\tilde{A}|x|, |x|\rangle \leq \|\tilde{A}\|^2.$$

Hence, $\|A\| \leq \|\tilde{A}\|$.

(iii) Consider operators $A = [A_{ij}]_{n\times n}$ and $B = [B_{ij}]_{n\times n}$. We observe that $\tilde{A}\tilde{B} - \widetilde{AB}$ is a nonnegative matrix, since

$$\widetilde{AB} = [\|\sum_{k=1}^{n} A_{ik}B_{kj}\|]_{n\times n} \quad \text{and} \quad \tilde{A}\tilde{B} = [\sum_{k=1}^{n} \|A_{ik}\| \|B_{kj}\|]_{n\times n}.$$

Therefore, by the norm monotonicity of nonnegative matrices (Exercise 2.9.1), we get $\|\widetilde{AB}\| \leq \|\tilde{A}\tilde{B}\|$. Using induction, we have

$$\|A^m\| \leq \|\widetilde{A^m}\| \leq \|\tilde{A}^m\|$$

for every positive integer m. This leads to

$$r(A) = \lim_{m\to\infty} \|A^m\|^{\frac{1}{m}} \leq \lim_{m\to\infty} \|\tilde{A}^m\|^{\frac{1}{m}} = r(\tilde{A}).$$

□

The following theorem gives a relationship between the norm of a block-norm matrix and its entries.

Theorem 2.1.9 *Let $A = [A_{ij}]_{n \times n}$ be a positive operator matrix and let $\tilde{A}$ be its block-norm matrix. Then*

$$\max_{1 \leq i \leq n} \|A_{ii}\| \leq \|A\| \leq \|\tilde{A}\| \leq \sum_{i=1}^{n} \|A_{ii}\|.$$

Proof The first two left-hand side inequalities are consequences of Lemma 2.1.2 and Theorem 2.1.8(ii).

To prove the right-hand side, let $\tilde{A}_0$ be the matrix $[\|A_{ii}\|^{\frac{1}{2}} \|A_{jj}\|^{\frac{1}{2}}]_{n \times n}$. Since

$$\tilde{A}_0 = \left(\|A_{11}\|^{\frac{1}{2}}, \|A_{22}\|^{\frac{1}{2}}, \ldots, \|A_{nn}\|^{\frac{1}{2}} \right)^* \left(\|A_{11}\|^{\frac{1}{2}}, \|A_{22}\|^{\frac{1}{2}}, \ldots, \|A_{nn}\|^{\frac{1}{2}} \right),$$

we have

$$\|\tilde{A}_0\| = \left\| \left(\|A_{11}\|^{\frac{1}{2}}, \|A_{22}\|^{\frac{1}{2}}, \ldots, \|A_{nn}\|^{\frac{1}{2}} \right) \left(\|A_{11}\|^{\frac{1}{2}}, \|A_{22}\|^{\frac{1}{2}}, \ldots, \|A_{nn}\|^{\frac{1}{2}} \right)^* \right\|$$

$$= \left\| \begin{bmatrix} \sum_{i=1}^{n} \|A_{ii}\| & 0 & \ldots & 0 \\ 0 & 0 & \ldots & 0 \\ \vdots & \vdots & & \vdots \\ 0 & 0 & \ldots & 0 \end{bmatrix} \right\|$$

$$= \sum_{i=1}^{n} \|A_{ii}\|.$$

On the other hand, since A is positive, we have $\|A_{ij}\| \leq \|A_{ii}\|^{\frac{1}{2}} \|A_{jj}\|^{\frac{1}{2}}$. Now, the nonnegativity of matrices $\tilde{A}_0, \tilde{A}$ and $\tilde{A}_0 - \tilde{A}$ ensures that $\|\tilde{A}\| \leq \|\tilde{A}_0\|$. Thus $\|\tilde{A}\| \leq \sum_{i=1}^{n} \|A_{ii}\|$. □

Applying Theorem 2.1.9 to the positive matrix $A^*A = \left[\sum_{k=1}^{n} A_{ki}^* A_{kj} \right]$ we arrive at the following result.

Corollary 2.1.10 *Let $A = [A_{ij}]_{n \times n}$ be an operator matrix. Then*

$$\max_{1 \leq j \leq n} \left\| \sum_{i=1}^{n} A_{ij}^* A_{ij} \right\|^{1/2} \leq \|A\| \leq \left(\sum_{j=1}^{n} \left\| \sum_{i=1}^{n} A_{ij}^* A_{ij} \right\| \right)^{1/2}.$$

We now present some norm estimates for finite sums of positive operators.

2.1 Operator Matrices of Size $n \times n$

Theorem 2.1.11 *Let $A_k \in \mathbb{B}(\mathscr{H}_k, \mathscr{H})$, $k = 1, 2, \ldots, n$. Then*

$$\left\| \sum_{k=1}^{n} A_k A_k^* \right\| \leq \| [\|A_i^* A_j\|] \|. \tag{2.1.1}$$

In particular,

$$\|A_1 A_1^* + A_2 A_2^*\| \leq \frac{1}{2} \left(\|A_1\|^2 + \|A_2\|^2 + \sqrt{(\|A_1\|^2 - \|A_2\|^2)^2 + 4\|A_1^* A_2\|^2} \right). \tag{2.1.2}$$

Proof Let $A = (A_1, A_2, \ldots, A_n) \in \mathbb{B}(\bigoplus_{k=1}^n \mathscr{H}_k, \mathscr{H})$. Then

$$\left\| \sum_{k=1}^{n} A_k A_k^* \right\| = \|AA^*\| = \|A^* A\| = \|[A_i^* A_j]\|.$$

In addition, by $\|A\| \leq \|\tilde{A}\| = \|[\|A_{ij}\|]\|$, we have

$$\|[A_i^* A_j]\| \leq \| [\|A_i^* A_j\|] \|.$$

The matrix on the right-hand side of (2.1.1) is Hermitian. Its norm equals its spectral radius, which for $n = 2$ turns into

$$\frac{1}{2} \left(\|A_1\|^2 + \|A_2\|^2 + \sqrt{(\|A_1\|^2 - \|A_2\|^2)^2 + 4\|A_1^* A_2\|^2} \right).$$

$\square$

For an $n \times n$ block matrix $A = [A_{ij}]$, let $\mathcal{O}A$ be the block matrix obtained from A by replacing all of its diagonals with the zero matrix. Similarly, let $\mathcal{D}A = A - \mathcal{O}A$, which is the block matrix obtained from A by replacing all of its off-diagonals with the zero matrix. Bhatia et al. [22] established the following theorem. An invariant norm $\|\cdot\|$ is called a *weakly invariant norm* if $\|U^* AU\| = \|A\|$ for all operators $A \in \mathcal{C}_{\|\cdot\|}$ and all unitaries U.

Theorem 2.1.12 *For every weakly invariant norm $\|\cdot\|$,*

$$\|\mathcal{O}A\| \leq \left(2 - \frac{2}{n} \right) \|A\| \quad (A \in \mathbb{M}_n).$$

Proof Let $U = \operatorname{diag}(1, w_0, w_0^2, \ldots, w_0^{n-1})$, where w_0 is a primitive nth root of unity. Then U is a unitary and

$$\mathcal{D}A = \frac{1}{n} \sum_{k=0}^{n-1} U^{*k} A U^k,$$

since the right-hand sum is $1/n$ times the Schur product of A with the matrix whose (i, j)-entry is $\sum_{k=0}^{n-1} w_0^{k(i-j)} = n\delta_{ij}$, where δ_{ij} denotes the Kronecker delta. Therefore,

$$\mathcal{O}A = A - \mathcal{D}A = \frac{n-1}{n}A - \frac{1}{n}\sum_{k=1}^{n-1} U^{*k}AU^*. \tag{2.1.3}$$

Hence,

$$\|\mathcal{O}A\| \leq \frac{n-1}{n}\|A\| + \frac{1}{n}\sum_{k=1}^{n-1}\left\|U^{*k}AU^*\right\| = 2\frac{n-1}{n}\|A\|.$$

□

Corollary 2.1.13 *Let $A \in \mathbb{M}_n$ be positive semidefinite. Then,*
(i) $\|\mathcal{O}A\| \leq \left(1 - \frac{1}{n}\right)\|A\|$;
(ii) $\|\mathcal{O}A\|_2 \leq \left(1 - \frac{1}{n}\right)^{1/2}\|A\|_2$,

for every weakly invariant norm $\|\cdot\|$.

Proof (i) Since $A \geq 0$, we observe from (2.1.3) that $\mathcal{O}A$ is the difference of two positive semidefinite matrices $\frac{n-1}{n}A$ and $\frac{1}{n}\sum_{k=1}^{n-1} U^{*k}AU^*$. Hence,

$$\|\mathcal{O}A\| \leq \max\left\{\left\|\frac{n-1}{n}A\right\|, \left\|\frac{1}{n}\sum_{k=1}^{n-1} U^{*k}AU^*\right\|\right\} = \left(1 - \frac{1}{n}\right)\|A\|.$$

(ii) Since $A = [a_{ij}] \geq 0$, the submatrix $\begin{bmatrix} a_{ii} & a_{ij} \\ a_{ij} & a_{jj} \end{bmatrix}$ is positive semidefinite. Therefore, we have

$$|a_{ij}|^2 \leq a_{ii}a_{jj} \leq \frac{a_{ii}^2 + a_{jj}^2}{2}.$$

Hence, by Lemma 2.1.2

$$\|\mathcal{O}A\|_2^2 \leq \sum_{i \neq j} |a_{ij}|^2 \leq (n-1)\sum_{i=1}^{n} |a_{ii}|^2 = (n-1)\|\mathcal{D}A\|_2^2.$$

Therefore,

$$\|A\|_2^2 = \|\mathcal{O}A\|_2^2 + \|\mathcal{D}A\|_2^2 \geq \left(1 + \frac{1}{n-1}\right)\|\mathcal{O}A\|_2^2 = \frac{n}{n-1}\|\mathcal{O}A\|_2^2.$$

□

2.2 Dilation

Let $A \in \mathbb{B}(\mathcal{H})$. An operator $B \in \mathbb{B}(\mathcal{K})$ is called a *dilation* of A if $\mathcal{H} \subseteq \mathcal{K}$ and $A = PB|_{\mathcal{H}}$ in which $P \in \mathbb{B}(\mathcal{K})$ is the projection onto $\mathcal{H}$. This means that A can be represented as the $(1, 1)$-entry of a matrix on $\mathcal{K} = \mathcal{H} \oplus \mathcal{H}^\perp$ of the form $B = \begin{bmatrix} A & * \\ * & * \end{bmatrix}$. Moreover, if $A^n = PB^n|_{\mathcal{H}}$ for all positive integers n, then B is called a *power dilation* of A.

The next two results appeared in [107].

Proposition 2.2.1 *Suppose that $A \in \mathbb{B}(\mathcal{H})$. Then $0 \leq A \leq I$ if and only if A has a projection dilation.*

Proof If $0 \leq A \leq I$, then for all $t \in \text{sp}(A)$ we have $t(1-t) \geq 0$. By functional calculus, $A(I - A)$ is positive, and so it has a positive square root. One can verify that
$$P = \begin{bmatrix} A & (A(I-A))^{\frac{1}{2}} \\ (A(I-A))^{\frac{1}{2}} & I - A \end{bmatrix} \in \mathbb{B}(\mathcal{H} \oplus \mathcal{H}) \text{ is the desired dilation. The converse}$$
follows from $\|A\| \leq \|P\|$. $\square$

Proposition 2.2.2 *Suppose that $A \in \mathbb{B}(\mathcal{H})$. Then $\|A\| \leq 1$ if and only if A has a unitary dilation.*

Proof If $\|A\| \leq 1$, then $U = \begin{bmatrix} A & (I - AA^*)^{\frac{1}{2}} \\ (I - A^*A)^{\frac{1}{2}} & -A^* \end{bmatrix} \in \mathbb{B}(\mathcal{H} \oplus \mathcal{H})$ is a unitary dilation of A. The converse is clear. $\square$

Proposition 2.2.3 *Every contraction has an isometric power dilation.*

Proof Let $\|A\| \leq 1$. Then $\|A^*A\| = \|A\|^2 \leq 1$. Hence, $A^*A \leq I$ and so $B := (I - A^*A)^{1/2}$ is a well-defined positive operator and $B^*B + A^*A = I$. The infinite matrix
$$V = \begin{bmatrix} A & 0 & 0 & 0 & \cdots \\ B & 0 & 0 & 0 & \cdots \\ 0 & I & 0 & 0 & \cdots \\ 0 & 0 & I & 0 & \cdots \\ 0 & 0 & 0 & \ddots \\ \vdots & \vdots & \vdots & & \cdots \end{bmatrix}$$
gives an operator via

$$V = \begin{bmatrix} A & 0 & 0 & 0 & \cdots \\ B & 0 & 0 & 0 & \cdots \\ 0 & I & 0 & 0 & \cdots \\ 0 & 0 & I & 0 & \cdots \\ 0 & 0 & 0 & \ddots & \\ \vdots & \vdots & \vdots & & \cdots \end{bmatrix} \begin{bmatrix} x_1 \\ x_2 \\ x_3 \\ x_4 \\ \vdots \end{bmatrix} = \begin{bmatrix} Ax_1 \\ Bx_1 \\ x_2 \\ x_3 \\ \vdots \end{bmatrix}.$$

It is a contraction in $\mathbb{B}(\oplus_{n=1}^{\infty} \mathcal{H})$, since $V^*V = I_{\oplus_{n=1}^{\infty} \mathcal{H}}$. Thus, V is an isometric dilation of A. In addition,

$$V^n = \begin{bmatrix} A^n & * & * & \cdots \\ * & * & * & \cdots \\ * & * & * & \cdots \\ \vdots & \vdots & \vdots & \cdots \end{bmatrix} \tag{2.2.1}$$

for all $n = 1, 2, \ldots$, in which $*$ denotes certain operators. $\square$

In the next proposition, the introduced unitary has a useful property related to the powers of operators.

Proposition 2.2.4 *Every isometry has a unitary power dilation.*

Proof Let V be an isometry. Then

$$U = \begin{bmatrix} V & I - VV^* \\ 0 & V^* \end{bmatrix} \in \mathbb{B}(\mathcal{H} \oplus \mathcal{H}).$$

is a unitary dilation of V, since

$$U^*U = \begin{bmatrix} V^* & 0 \\ I - VV^* & V \end{bmatrix} \begin{bmatrix} V & I - VV^* \\ 0 & V^* \end{bmatrix} = I_{\mathcal{H} \oplus \mathcal{H}} = UU^*.$$

Moreover,

$$U^n = \begin{bmatrix} V^n & * \\ * & * \end{bmatrix} \tag{2.2.2}$$

for all $n \in \mathbb{N}$. $\square$

2.2 Dilation

Theorem 2.2.5 (Sz.-Nagy's dilation theorem) *Let $A \in \mathbb{B}(\mathcal{H})$ be a contraction. Then there exists a Hilbert space $\mathcal{K}$ containing $\mathcal{H}$ as a subspace and a unitary operator U on $\mathcal{K}$ such that*

$$A^n = P_{\mathcal{H}} U^n|_{\mathcal{H}}, \qquad (2.2.3)$$

where $P_{\mathcal{H}} \in \mathbb{B}(\mathcal{K})$ denotes the projection onto $\mathcal{H}$. In other words, every contraction has a unitary power dilation.

Proof It follows from Proposition 2.2.3 that there is an isometry $V \in \mathbb{B}(\oplus_{n=1}^{\infty} \mathcal{H})$ satisfying (2.2.1). For this isometry V, Proposition 2.2.4 implies that there is a unitary $U \in \mathbb{B}((\oplus_{n=1}^{\infty} \mathcal{H}) \oplus (\oplus_{n=1}^{\infty} \mathcal{H}))$ satisfying (2.2.2). Thus, $\mathcal{K} = (\oplus_{n=1}^{\infty} \mathcal{H}) \oplus (\oplus_{n=1}^{\infty} \mathcal{H})$ is the required dilation of $\mathcal{H}$ when we identify $\mathcal{H}$ with $(\mathcal{H} \oplus 0 \oplus 0 \oplus \cdots) \oplus 0$. □

Ando established that every two commuting contractions have a common (same) unitary power dilation [5]. However, Parrott [188] showed that this result is not true for more than two contractions.

The following theorem is a consequence of the Nagy theorem; see [95].

Theorem 2.2.6 (von Neumann inequality) *Let $A \in \mathbb{B}(\mathcal{H})$ be a contraction. Then for any polynomial p,*

$$\|p(A)\| \leq \sup\{|p(z)| : |z| \leq 1\}.$$

Proof It follows from the spectral theorem for a unitary operator U, whose spectrum is contained in the unit circle, that

$$\|p(U)\| = \sup\{|p(z)| : z \in \text{sp}(U)\} \leq \sup\{|p(z)| : |z| \leq 1\}.$$

By Nagy's dilation Theorem 2.2.5, we have

$$\|p(A)\| = \|P_{\mathcal{H}} p(U)|_{\mathcal{H}}\| \leq \|p(U)\| \leq \sup\{|p(z)| : |z| \leq 1\}.$$

□

An operator $T \in \mathbb{B}(\mathcal{H})$, where $\mathcal{H}$ is a separable Hilbert space is said to be *diagonalizable* if there exist an orthonormal basis (e_n) of $\mathcal{H}$ and scalars (λ_n) such that $Te_n = \lambda_n e_n$ for all n.

Proposition 2.2.7 ([224, Theorem 2]) *An operator $T \in \mathbb{B}(\mathcal{H})$ can be expressed as the direct sum of operators each of which is of the form*

$$\begin{bmatrix} A_0 & B_0 & & & \\ C_0 & A_1 & B_1 & & \\ & C_1 & A_2 & B_2 & \\ & & C_2 & A_3 & \ddots \\ & & & \ddots & \ddots \end{bmatrix} \qquad (2.2.4)$$

on $\mathcal{K}_0 \oplus \mathcal{K}_1 \oplus \cdots$, where the unspecified entries are zeros and the $\mathcal{K}_j$'s are all finite dimensional (with possibly different dimensions).

Proof Suppose $e \in \mathcal{H}$ is any nonzero vector. Let $\mathcal{H}_1$ be the reducing closed subspace generated by

$$e, Te, T^*e, T^2e, TT^*e, T^*Te, T^{*2}e, \ldots.$$

By removing the sequence vectors which are linear combinations of the ones preceding them and then applying the well-known Gram–Schmidt process to the resulting sequence, we obtain an orthonormal basis (x_n) for $\mathcal{H}$. For each $j \geq 1$, assume that n_j is the index such that $x_1, \ldots, x_{n_j}$ and vectors of the form $T_1 \cdots T_k e$ with $T_i = T$ or T^*, and $k \leq j$, generate the same subspace, and let $\mathcal{K}_j$ be the (closed) subspace generated by $x_{n_{j-1}}, \ldots, x_{n_j}$. One can verify that the matrix representation of $T|\mathcal{H}_1$ with respect to this basis is of the form (2.2.4). If $\mathcal{H}_1 \neq \mathcal{H}$, we can repeat the above process with the orthogonal complement of $\mathcal{H}_1$. The rest of the proof is straightforward. □

The next result is due to Fong and Wu [80].

Theorem 2.2.8 *Every operator has a diagonal dilation. If the given operator is self-adjoint, then the diagonalizable dilation can be chosen to be self-adjoint.*

Proof Let T be of the form (2.2.4) on $\mathcal{K}_0 \oplus \mathcal{K}_1 \oplus \cdots$. Suppose that

$$D := \begin{bmatrix} A_0 & \sqrt{2}B_0 & & & & & \\ \sqrt{2}C_0 & A_1 & & & & & \\ & & A_1 & 2B_1 & & & \\ & & 2C_1 & A_2 & & & \\ & & & & A_2 & 2B_2 & \\ & & & & 2C_2 & A_3 & \\ & & & & & & \ddots \end{bmatrix}$$

on $\mathcal{K}_0 \oplus \mathcal{K}_1 \oplus \mathcal{K}_1 \oplus \mathcal{K}_2 \oplus \mathcal{K}_2 \ldots$. Define $V : \mathcal{H} \to \mathcal{K}$ by

$$V(x_0, x_1, x_2, \ldots) = (x_0, x_1/\sqrt{2}, x_1/\sqrt{2}, x_2/\sqrt{2}, x_2/\sqrt{2}, \ldots).$$

2.2 Dilation

Then V is an isometry and $V^* : \mathcal{K} \to \mathcal{H}$ is given by

$$V^*(u_0, u_1, v_1, u_2, v_2, \ldots) = (u_0, (u_1 + v_1)/\sqrt{2}, (u_2 + v_2)/\sqrt{2}, \ldots).$$

A straightforward verification shows that $V^*DV = T$, and D is a dilation of T. We can assume that $\|T\| \leq 1/2$. Hence, each of the finite blocks

$$\begin{bmatrix} A_0 & \sqrt{2}B_0 \\ \sqrt{2}C_0 & A_1 \end{bmatrix} \text{ and } \begin{bmatrix} A_j & \sqrt{2}B_j \\ \sqrt{2}C_j & A_{j+1} \end{bmatrix}, \quad j = 1, 2, \ldots$$

appearing in D is a contraction. Hence, each of them has a unitary dilation and, since it acts on a finite-dimensional space, is diagonal. Thus, D has a diagonal dilation and so does T. □

Corollary 2.2.9 *An operator of the form $T \oplus 0$, where 0 acts on an infinite-dimensional space, is the product of three diagonal operators.*

Proof It follows from Theorem 2.2.8 that T appears in the $(1, 1)$-entry of some diagonal operator $\begin{bmatrix} T & * \\ * & * \end{bmatrix}$. The identity $T \oplus 0 = (1 \oplus 0)D(1 \oplus 0)$ provides a representation of $T \oplus 0$ as the product of three diagonal operators. □

An operator $T : \mathcal{H} \to \mathcal{K}$ is said to be *Fredholm*, if $\ker T$ is finite dimensional and $T(\mathcal{H})$ has finite codimension.

Theorem 2.2.10 *Suppose that $T : \mathcal{H} \to \mathcal{K}$ is a Fredholm operator. Then, there exists $\varepsilon > 0$ such that if $S \in \mathbb{B}(\mathcal{H}, \mathcal{K})$ and $\|S\| < \varepsilon$, then $T + S$ is Fredholm.*

Proof Since $\mathcal{H} = (\ker T)^\perp \oplus \ker T$, $\mathcal{K} = \overline{\operatorname{ran} T} \oplus \ker T^*$, and T is Fredholm, we can write T as $\begin{bmatrix} T_1 & 0 \\ 0 & 0 \end{bmatrix}$ where $T_1 : (\ker T)^\perp \to \overline{\operatorname{ran} T}$ is invertible. The group of invertible operators is open, so there exists $\varepsilon > 0$ such that $\|S_1\| < \varepsilon$ implies that $T_1 + S_1$ is invertible. For $S \in \mathbb{B}(\mathcal{H}, \mathcal{K})$ with $\|S\| < \varepsilon$, represented as $S = \begin{bmatrix} S_1 & S_2 \\ S_3 & S_4 \end{bmatrix}$, we have

$$T + S = \begin{bmatrix} T_1 + S_1 & 0 \\ 0 & 0 \end{bmatrix} + \begin{bmatrix} 0 & S_2 \\ S_3 & S_4 \end{bmatrix}.$$

Since the first matrix is Fredholm and the second is a finite rank operator, $T + S$ is Fredholm. □

The next result is due to Bunce [38].

Theorem 2.2.11 *Let Λ be either $\{1, 2, \ldots, n\}$ or $\mathbb{N}$, and let $\{A_i : i \in \Lambda\}$ be a family of operators in $\mathbb{B}(\mathcal{H})$. Then the following two conditions are equivalent.*

(i) $\sum_{i \in \Lambda} A_i^* A_i \leq I$.

(ii) There exist a Hilbert space $\mathcal{K}$ containing $\mathcal{H}$ and coisometries $\{S_i : i \in \Lambda\}$ acting on $\mathcal{K}$ such that $S_i S_j^* = 0$ for all $i \neq j$, $S_i(\mathcal{H}) \subseteq \mathcal{H}$, and the restriction $S_i|_{\mathcal{H}}$ is A_i for each i.

Proof (ii) $\Longrightarrow$ (i) The family $\{S_i^* S_i : i \in \Lambda\}$ is an orthogonal family of projections on $\mathcal{K}$, so
$$Q = \sum_{i \in \Lambda} S_i^* S_i \leq I.$$

Let P be the projection of $\mathcal{K}$ onto $\mathcal{H}$. Then
$$\sum_{i \in \Lambda} A_i^* A_i = \sum P S_i^* S_i|_{\mathcal{H}} = PQP|_{\mathcal{H}} \leq I.$$

(i) $\Longrightarrow$ (ii) Suppose that
$$\sum_{i \in \Lambda} A_i^* A_i \leq I.$$

Let $\mathcal{H}_\Lambda = \oplus_{i \in \Lambda} \mathcal{H}$ be the Hilbert space direct sum of the cardinality of Λ many copies of $\mathcal{H}$. Let $T \in \mathbb{B}(\mathcal{H}_\Lambda)$ be defined by
$$T(x_1, x_2, x_3, \ldots) = (A_1 x_1, A_2 x_1, A_3 x_1, \ldots).$$

Hence, T is an operator-valued matrix with the A_i down the first column and zeros elsewhere. Then T^*T is the operator matrix with $\sum_{i \in \Lambda} A_i^* A_i$ in the upper left corner and zeroes elsewhere, so $T^*T \leq I$, and T is a contraction. By Propositions 2.2.3 and 2.2.4, there is a Hilbert space $\mathcal{K}_0$ containing $\mathcal{H}_\Lambda$, and a unitary $S \in \mathbb{B}(\mathcal{K}_0)$ with $S(\mathcal{H}_\Lambda) \subseteq \mathcal{H}_\Lambda$ and $S|_{\mathcal{H}_\Lambda} = T$. Let L be the orthogonal complement of $\mathcal{H}_\Lambda$ in $\mathcal{K}_0$, so $\mathcal{K}_0 = \mathcal{H}_\Lambda \oplus L$. With respect to this decomposition of $\mathcal{K}_0$, we can write
$$S = \begin{bmatrix} A_1 & 0 & 0 & \cdots & X_1 \\ A_2 & 0 & 0 & \cdots & X_2 \\ \vdots & & \ddots & & \vdots \\ 0 & 0 & 0 & \cdots & Y \end{bmatrix},$$

where for $i \in \Lambda$, $X_i : L \to \mathcal{H}$, and $Y : L \to L$ are suitable maps. Since $SS^* = I$ it is easily seen that $A_i A_i^* + X_i X_i^* = I$ for $i \in \Lambda$, $A_i A_j^* + X_i X_j^* = 0$ for $i \neq j$, $X_i Y^* = 0$ for $i \in \Lambda$, and $YY^* = I$.

If $Y^*Y = I$, then $X_i = 0$ for all i and the $S_i = A_i$'s are the desired operators. Otherwise, $Y^*Y \neq I$ and L must be of infinite dimension. Then there exists a family $\{Z_i : i \in \Lambda\}$ of

2.2 Dilation

coisometries acting on L such that the Z_i's have orthogonal initial spaces, $Z_i Z_i^* = I$ for $i \in \Lambda$, and $Z_i Z_j^* = 0$ for $i \neq j$. Let $\mathcal{K} = \mathcal{H} \oplus L$ and define $S_i \in \mathbb{B}(\mathcal{K})$ by

$$S_i = \begin{bmatrix} A_i & X_i \\ 0 & Z_i Y \end{bmatrix}.$$

An easy computation shows that the S_i's are the desired coisometries. □

The following result provides a condition on the powers of an operator T in order for T to be a direct sum of a nilpotent operator and a contraction operator.

Proposition 2.2.12 *Let $T \in \mathbb{B}(\mathcal{H})$, and let n be a positive integer. Then the following assertions are equivalent:*

(i) *T^n and T^{n+1} are positive contractions.*
(ii) *T is a direct sum of operators N and S such that $N^n = 0$ and $0 \le S \le I$.*

Proof (ii) $\Longrightarrow$ (i) is obvious, so it suffices to show (i) $\Longrightarrow$ (ii). First, we remark that $\ker T^n = \ker T^{n+1}$, because

$$\ker T^n \subseteq \ker T^{n+1} \subseteq \cdots \subseteq \ker T^{2n} = \ker(T^n)^* T^n = \ker T^n.$$

We decompose the Hilbert space $\mathcal{H}$ into the direct sum of $\ker T^n$ and $(\ker T^n)^\perp = \overline{\operatorname{ran}(T^n)}$. Then T, T^n and T^{n+1} have the form

$$T = \begin{bmatrix} A & B \\ C & D \end{bmatrix}, \quad T^n = \begin{bmatrix} 0 & 0 \\ 0 & E \end{bmatrix}, \quad T^{n+1} = \begin{bmatrix} 0 & 0 \\ 0 & F \end{bmatrix}$$

with respect to the decomposition $\mathcal{H} = (\ker T^n) \oplus (\ker T^n)^\perp$. It follows from

$$T^{n+1} = TT^n = \begin{bmatrix} 0 & BE \\ 0 & DE \end{bmatrix} = T^n T = \begin{bmatrix} 0 & 0 \\ EC & ED \end{bmatrix}$$

that $BE = 0$, $EC = 0$ and $DE = ED = F = F^* = (ED)^* = D^*E$. Since $\overline{\operatorname{ran}(E)} = (\ker E)^\perp = (\ker T^n)^\perp$, we have $B = 0$, $C = 0$, $D = D^*$, and $A^n = 0$. By the positivity and the contractivity of T^n and T^{n+1}, we conclude that D is a positive contraction, because $\|D^n\| \le 1$. □

To simplify certain problems, such as unitary equivalence, we can utilize dilations.

If T is a contraction, then $M_T = \begin{bmatrix} T & (I - TT^*)^{\frac{1}{2}} \\ 0 & 0 \end{bmatrix}$ is a partial isometry. If two contractions T and S are unitarily equivalent, that is there exists a unitary U such that $U^*TU = S$, then

$$\begin{bmatrix} U & 0 \\ 0 & U \end{bmatrix}^* M_T \begin{bmatrix} U & 0 \\ 0 & U \end{bmatrix} = M_S.$$

If T and S are invertible contraction and M_T and M_S are unitarily equivalent, then T and S are also unitarily equivalent. Hence, the unitary equivalence problem for partial isometries is equivalent to the problem for invertible contractions. This is also equivalent to the unitary equivalence problem for arbitrary elements, since suitable maps $T \mapsto \lambda T$ and $T \mapsto T + \mu I$ transform T into invertible contractions and also preserve unitary equivalence; see [107, Problem 132].

Theorem 2.2.13 *There is a partial isometry whose spectrum is a given compact subset Ω of the closed unit disc containing the origin; see [107, Problem 133].*

Proof Let $\mathcal{H}$ be a separable Hilbert space with an orthonormal basis (e_n) and let (λ_n) be a sequence that is dense in Ω. Then $Te_n = \lambda_n e_n$ is a contraction, and its spectrum is Ω. If we set $S = \begin{bmatrix} T & (I - TT^*)^{\frac{1}{2}} \\ 0 & 0 \end{bmatrix}$, then $SS^*S = S$. By applying the above discussions to $S - \lambda I$, $\lambda \neq 0$, we conclude that $\mathrm{sp}(S) = \mathrm{sp}(T) = \Omega$. □

2.3 2 × 2 Matrices with Entries in a C^*-Algebra

The algebra of all 2×2 matrices with entries in $\mathbb{B}(\mathcal{H})$ is a typical example of the C^*-algebra $\mathbb{M}_2(\mathscr{A})$. Clearly, if C^*-algebras $\mathscr{A}$ and $\mathscr{B}$ are $*$-isomorphic, then $\mathbb{M}_2(\mathscr{A})$ and $\mathbb{M}_2(\mathscr{B})$ are also $*$-isomorphic. There exist two non-isomorphic unital C^*-algebras $\mathscr{A}$ and $\mathscr{B}$ such that $\mathscr{A} \simeq \mathbb{M}_2(\mathscr{A}) \simeq \mathbb{M}_2(\mathscr{B})$. For example, consider

$$\mathscr{A} = \{T \oplus T : T \in \mathbb{B}(\mathcal{H})\} + \mathbb{K}(\mathcal{H} \oplus \mathcal{H})$$

and

$$\mathscr{B} = \{T \oplus T \oplus 0 : T \in \mathbb{B}(\mathcal{H}), 0 \in \mathbb{B}(\mathcal{H}_0)\} + \mathbb{K}(\mathcal{H} \oplus \mathcal{H} \oplus \mathcal{H}_0),$$

where $\mathcal{H}$ is a separable infinite-dimensional Hilbert space, and $\mathcal{H}_0$ is one dimensional; see [195].

If both $\mathscr{A}$ and $\mathscr{B}$ belong to one of the following classes of C^*-algebras, then $\mathbb{M}_2(\mathscr{A}) \simeq \mathbb{M}_2(\mathscr{B})$ ensures that $\mathscr{A} \simeq \mathscr{B}$:

(i) commutative C^*-algebras, since the center of $\mathbb{M}_2(\mathscr{A})$ is $\left\{ \begin{bmatrix} A & 0 \\ 0 & A \end{bmatrix} : A \in \mathscr{A} \right\}$;

(ii) UHF algebras [96, Theorem 1];

(iii) perturbed block diagonal algebras [196].

We should mention that there are two non-isomorphic C^*-algebras, $\mathscr{A}_1$ and $\mathscr{A}_2$, such that $\mathbb{K}(\mathcal{H}) \subseteq \mathscr{A}_i$ and $\frac{\mathscr{A}_i}{\mathbb{K}(\mathcal{H})} \simeq \mathbb{M}_2$, where $i = 1, 2$; see [197].

2.3 2 × 2 Matrices with Entries in a C*-Algebra

The following theorem gives a condition under which a C^*-algebra $\mathscr{A}$ can be represented in the form $\mathbb{M}_2(\mathscr{B})$ for some C^*-algebra $\mathscr{B}$.

Theorem 2.3.1 *Let $\mathscr{A}$ be a C^*-algebra containing $\mathbb{M}_2$ as a unital C^*-subalgebra. Then $\mathscr{A} \simeq \mathbb{M}_2(\mathscr{B})$ for some C^*-algebra $\mathscr{B}$.*

Proof Suppose that $(E_{ij})_{1 \leq i,j \leq 2}$ is the standard system of matrix units of $\mathbb{M}_2$ and let $\mathscr{B} = E_{11} \mathscr{A} E_{11}$. Then $\Phi : \mathscr{A} \to \mathbb{M}_2(E_{11} \mathscr{A} E_{11})$ is defined by

$$\Phi(A) = \begin{bmatrix} E_{11}AE_{11} & E_{11}AE_{21} \\ E_{12}AE_{11} & E_{12}AE_{21} \end{bmatrix}$$

and $\psi : \mathbb{M}_2(E_{11} \mathscr{A} E_{11}) \to \mathscr{A}$ is defined by

$$\psi\left(\begin{bmatrix} E_{11}AE_{11} & E_{11}BE_{11} \\ E_{11}CE_{11} & E_{11}DE_{11} \end{bmatrix} \right) = E_{11}AE_{11} + E_{11}BE_{12} + E_{21}CE_{11} + E_{21}DE_{12}.$$

These are $*$-homomorphisms, and are inverses of each other. □

Moreover, for unital C^*-algebras we have the following characterization of C^*-algebras that are $*$-isomorphic to $\mathbb{M}_2$; see [166].

Theorem 2.3.2 *A unital C^*-algebra $\mathscr{A}$ is $*$-isomorphic to $\mathbb{M}_2$ if and only if there exists a projection $P \in \mathscr{A}$ such that*

$$P \mathscr{A} P = \mathbb{C} P, \ (I - P) \mathscr{A} (I - P) = \mathbb{C}(I - P), \ (I - P) \mathscr{A} P \neq 0, \ \text{and} \ P \mathscr{A} (I - P) \neq 0. \tag{2.3.1}$$

Proof If $\mathscr{A} = \mathbb{M}_2$, then $P = \begin{bmatrix} 1 & 0 \\ 0 & 0 \end{bmatrix}$ is an appropriate projection. Conversely, suppose that P is a projection satisfying (2.3.1). For $0 \neq U \in P\mathscr{A}(I - P)$ we evidently have $U = PU(I - P)$ and so $UU^* \in P\mathscr{A} P$ and $U^*U \in (I - P)\mathscr{A}(I - P)$. Hence, there exists $r > 0$ such that $UU^* = rP$ and $U^*U = r(I - P)$. Replacing U with $\frac{U}{\sqrt{r}}$, we can assume that $UU^* = P$ and $U^*U = I - P$. If $A \in P\mathscr{A}(I - P)$, then there is $\lambda \in \mathbb{C}$ such that $A = A(I - P) = A(U^*U) = (AU^*)U = (\lambda P)U = \lambda U$.
Similarly, if $B \in (I - P)\mathscr{A} P$, we have $B = \mu U^*$ for some $\mu \in \mathbb{C}$. It follows from

$$\mathscr{A} = P\mathscr{A} P \oplus (I - P)\mathscr{A} P \oplus P\mathscr{A}(I - P) \oplus (I - P)\mathscr{A}(I - P)$$

that each $X \in \mathscr{A}$ is of the form $\mu_1 P + \mu_2 U + \mu_3 U^* + \mu_4(I - P)$ with $\mu_i \in \mathbb{C}, 1 \leq i \leq 4$.

It is straightforward to show that $\Phi : \mathscr{A} \to \mathbb{M}_2$ defined by $\Phi(X) = \begin{bmatrix} \mu_1 & \mu_2 \\ \mu_3 & \mu_4 \end{bmatrix}$ is a $*$-isomorphism. □

Evidently, $P\mathscr{A}(I-P) = 0$ if and only if $(I-P)\mathscr{A}P = 0$. If this occurs and $P\mathscr{A}P = \mathbb{C}p$ and $(I-P)\mathscr{A}(I-P) = \mathbb{C}(I-P)$, we have

$$\mathscr{A} = \mathbb{C}P \oplus \mathbb{C}(I-P) \simeq \mathbb{C}^2 \simeq \left\{ \begin{bmatrix} \lambda & 0 \\ 0 & \mu \end{bmatrix} : \lambda, \mu \in \mathbb{C} \right\}.$$

It is known that every C^*-algebra $\mathscr{A}$ with a projection P can be embedded in $\mathbb{M}_2(\mathscr{A})$. Indeed the map $\Phi : \mathscr{A} \to \mathbb{M}_2(\mathscr{A})$ defined by

$$\Phi(A) = \begin{bmatrix} PAP & PA(I-P) \\ (I-P)AP & (I-P)A(I-P) \end{bmatrix}$$

is an injective $*$-homomorphism. However, we are interested in C^*-algebras $\mathscr{A}$ for which $\mathscr{A} \simeq \mathbb{M}_2(\mathscr{A})$:

Definition 2.3.3 A projection P in a unital C^*-algebra $\mathscr{A}$ is called *halving* if $P \sim I$ and $I - P \sim I$, meaning that there exist partial isometries $U, V \in \mathscr{A}$ such that $P = UU^*$, $I - P = VV^*$, and $U^*U = I = V^*V$.

Every properly infinite von Neumann algebra contains a halving projection; see [216, Proposition V.1.36].

Theorem 2.3.4 ([226, Corollary 5.3.6]) *If $\mathscr{A}$ is a unital C^*-algebra containing a halving projection P, then $\mathscr{A} \simeq \mathbb{M}_2(\mathscr{A})$.*

Proof In the notation above, it is straightforward to show that the map

$$A \mapsto \begin{bmatrix} U^*PAPU & U^*PA(I-P)V \\ V^*(I-P)APU & V^*(I-P)A(I-P)V \end{bmatrix}$$

is an isomorphism between $\mathscr{A}$ and $\mathbb{M}_2(\mathscr{A})$. □

There is also a factorization theorem in C^*-algebra theory:

Theorem 2.3.5 *For each element A in a unital C^*-algebra $\mathscr{A}$, there exists a unique $B \in \mathscr{A}$ such that $A = BB^*B$.*

Proof Let $A \in \mathscr{A}$. Then $\begin{bmatrix} 0 & A^* \\ A & 0 \end{bmatrix}$ is a self-adjoint operator in $\mathbb{M}_2(\mathscr{A})$ that anti-commutes with $\begin{bmatrix} I & 0 \\ 0 & -I \end{bmatrix}$. Using functional calculus, we conclude that $\begin{bmatrix} 0 & A^* \\ A & 0 \end{bmatrix}^{\frac{1}{3}} = \begin{bmatrix} U & B^* \\ B & V \end{bmatrix}$

anti-commutes with $\begin{bmatrix} I & 0 \\ 0 & -I \end{bmatrix}$. It follows that $U = V = 0$. Next, $\begin{bmatrix} 0 & A^* \\ A & 0 \end{bmatrix} = \begin{bmatrix} 0 & B^* \\ B & 0 \end{bmatrix}^3$ implies that $A = BB^*B$; see [201, Proposition 2.31]. □

The *Fuglede theorem* [83] asserts that if A is a normal element of a C^*-algebra $\mathscr{A}$ and C is an element of $\mathscr{A}$ such that $AC = CA$, then $A^*C = CA^*$. This theorem can be extended using the so-called Berberian's technique [15] as follows.

Theorem 2.3.6 (Putnam–Fuglede theorem) *If A and B are normal elements of a C^*-algebra $\mathscr{A}$ and C is an element of $\mathscr{A}$ such that $AC = CB$, then $A^*C = CB^*$.*

Proof The normal element $\begin{bmatrix} A & 0 \\ 0 & B \end{bmatrix}$ commutes with $\begin{bmatrix} 0 & C \\ 0 & 0 \end{bmatrix}$ in the C^*-algebra $\mathbb{M}_2(\mathscr{A})$. It follows from Fuglede's theorem that

$$\begin{bmatrix} A & 0 \\ 0 & B \end{bmatrix}^* \begin{bmatrix} 0 & C \\ 0 & 0 \end{bmatrix} = \begin{bmatrix} 0 & C \\ 0 & 0 \end{bmatrix} \begin{bmatrix} A & 0 \\ 0 & B \end{bmatrix}^*.$$

Therefore, we have $A^*C = CB^*$. □

If $\mathcal{M}$ is a subspace of $\mathbb{B}(\mathscr{H})$, then

$$\mathscr{M} = \left\{ \begin{bmatrix} \alpha & T \\ 0 & \beta \end{bmatrix} : T \in \mathcal{M}, \alpha, \beta \in \mathbb{C} \right\}$$

is a subalgebra of $\mathbb{B}(\mathscr{H} \oplus \mathscr{H})$. Then

(i) $\mathcal{M}$ is weak* closed (weak operator closed, respectively) if and only if so is $\mathscr{M}$.
(ii) $\mathcal{M}$ is reflexive if and only if $\mathscr{M}$ is reflexive [50, p. 320]. Recall that a linear subspace $\mathcal{M}$ of $\mathbb{B}(\mathscr{H})$ is called *reflexive* if $\mathcal{M} = \{T \in \mathbb{B}(\mathscr{H}) : Th \in [\mathcal{M}h] \text{ for all } h \in H\}$ where $[\mathcal{M}h]$ is the closed linear span of $\{Th : T \in \mathcal{M}\}$.

This observation helps us provide some interesting examples:

Example 2.3.7 Suppose S is a trace class operator with infinite rank. Then $\mathcal{M} = \{T \in \mathbb{B}(\mathscr{H}) : \text{tr}(TS) = 0\}$ is a weak*-closed subspace of $\mathbb{B}(\mathscr{H})$, which is not weak operator closed.

$$\mathscr{M} = \left\{ \begin{bmatrix} \alpha & T \\ 0 & \beta \end{bmatrix} : T \in \mathcal{M}, \alpha, \beta \in \mathbb{C} \right\}$$

is therefore a weak**-closed subalgebra of $\mathbb{B}(\mathscr{H} \oplus \mathscr{H})$, which is not operator closed; see [50, p. 108].

In fact, many counterexamples in the theory of reflexive algebras are produced by first finding an appropriate space and afterward placing it in a corner of a matrix [81].

Kadison [132] asked whether every unital bounded representation of a unital C^*-algebra $\mathscr{A}$ is similar to a $*$-representation of $\mathscr{A}$. More precisely, if $\Phi : \mathscr{A} \to \mathbb{B}(\mathscr{H})$ is a bounded unital homomorphism, is there an invertible bounded operator S such that $A \mapsto S^{-1}\Phi(A)S$ defines a $*$-homomorphism? This is called the *Kadison's similarity problem*. Paulsen [189] proved that this is equivalent to showing that Φ is completely bounded. In particular, the Kadison similarity problem holds when $\mathscr{A}$ is commutative or $\mathscr{A} = \mathbb{B}(\mathscr{H})$.

The derivation problem asserts that if $\pi : \mathscr{A} \to \mathbb{B}(\mathscr{H})$ is a $*$-representation and $\delta : \mathscr{A} \to \mathbb{B}(\mathscr{H})$ is a bounded π-*derivation*, that is $\delta(AB) = \pi(A)\delta(B) + \delta(A)\pi(B)$, is δ an π-*inner derivation*? In other words, is there a $T \in \mathbb{B}(\mathscr{H})$ such that $\delta(A) = \pi(A)T - T\pi(A) = \delta_T(A)$?

Clearly, the map $\Phi : \mathscr{A} \to \mathbb{M}_2(\mathbb{B}(\mathscr{H}))$ defined by $\Phi(A) = \begin{bmatrix} \pi(A) & \delta(A) \\ 0 & \pi(A) \end{bmatrix}$ is a unital homomorphism.

One known result of the connection between these problems is as follows. By a *derivation*, we mean an id-derivation.

Theorem 2.3.8 ([199, Proposition 8.9]) *Let $\mathscr{A} \subseteq \mathbb{B}(\mathscr{H})$ be a C^*-algebra. A derivation $\delta : \mathscr{A} \to \mathbb{B}(\mathscr{H})$ is inner if and only if the associated homomorphism $\Phi : \mathscr{A} \to \mathbb{B}(\mathscr{H} \oplus \mathscr{H})$ defined by $\Phi(A) = \begin{bmatrix} A & \delta(A) \\ 0 & A \end{bmatrix}$ is similar to a $*$-homomorphism.*

It was proved by Kirchberg [136] that the C^*-algebras satisfying the derivation problem are exactly those satisfying Kadison's similarity problem.

Example 2.3.9 Suppose that $\mathscr{A}$ is the algebra of upper triangular 2×2 matrices with entries in $\mathbb{C}$ and $\mathscr{B} = \mathbb{M}_2$. The map $T : \mathscr{A} \to \mathscr{B}$ defined by

$$T\left(\begin{bmatrix} \alpha & \beta \\ 0 & \gamma \end{bmatrix}\right) = \begin{bmatrix} \alpha & \alpha + \beta \\ 0 & \gamma \end{bmatrix}$$

preserves the determinant and invertibility. Nevertheless, T is not multiplicative because

$$T\left(\begin{bmatrix} 1 & 1 \\ 0 & 1 \end{bmatrix}^2\right) \neq T\left(\begin{bmatrix} 1 & 1 \\ 0 & 1 \end{bmatrix}\right)^2; \text{ see [127]}.$$

2.4 Positivity of 2 × 2 Matrices of Operators

In this section, we present several characterizations of the positivity of 2×2 matrices of operators of the form

$$M = \begin{bmatrix} A & X \\ X^* & B \end{bmatrix}, \qquad (2.4.1)$$

where $A, B, X \in \mathbb{B}(\mathcal{H})$. In addition, we provide some related implications.

The first characterization of the positivity of a 2×2 operator matrix M is as follows:

Theorem 2.4.1 *Let* $M = \begin{bmatrix} A & X \\ X^* & B \end{bmatrix}$ *such that* $A \in \mathbb{B}_{++}(\mathcal{H})$. *Then* M *is positive if and only if* $B \geq X^* A^{-1} X$.

Proof We provide three proofs with different methods.

The first proof ([20, Theorem 1.3.3]): Put $C = \begin{bmatrix} I & 0 \\ -X^* A^{-1} & I \end{bmatrix}$. Then C is invertible and we have

$$A \oplus (B - X^* A^{-1} X) = C \cdot M \cdot C^*.$$

Therefore, we can use the fact that "if $0 \leq D_1 \leq D_2$, then $Z^* D_1 Z \leq Z^* D_2 Z$ for all $Z \in \mathbb{B}(\mathcal{H})$" to conclude that $M \geq 0$ if and only if $B \geq X^* A^{-1} X$.

The second proof: Suppose that $M \geq 0$. Let

$$M^{\frac{1}{2}} = \begin{bmatrix} Y_{11} & Y_{12} \\ Y_{21} & Y_{22} \end{bmatrix} = [Y_1 \ Y_2], \text{ where } Y_1 = \begin{bmatrix} Y_{11} \\ Y_{21} \end{bmatrix} \text{ and } Y_2 = \begin{bmatrix} Y_{12} \\ Y_{22} \end{bmatrix}.$$

Then

$$M = \begin{bmatrix} Y_1^* \\ Y_2^* \end{bmatrix} \cdot [Y_1, Y_2] = \begin{bmatrix} Y_1^* Y_1 & Y_1^* Y_2 \\ Y_2^* Y_1 & Y_2^* Y_2 \end{bmatrix}.$$

Hence, $A = Y_1^* Y_1$, $X = Y_1^* Y_2$ and $B = Y_2^* Y_2$. Set

$$Z = Y_1 \cdot A^{-1} \cdot Y_1^* = Y_1 \cdot (Y_1^* Y_1)^{-1} \cdot Y_1^*.$$

Then Z is positive since A^{-1} is positive, and $Z^2 = Z$. Therefore, Z is a projection, which gives $I - Z \geq 0$. It follows that $B - X^* A^{-1} X = Y_2^* (I - Z) Y_2 \geq 0$.

Conversely, assume that $B \geq X^* A^{-1} X$. Let us set

$$Y = (Y_1, Y_2) \text{ with } Y_1 = A^{\frac{1}{2}} \text{ and } Y_2 = A^{-\frac{1}{2}} X. \qquad (2.4.2)$$

Then $M = Y^* Y + 0 \oplus (B - X^* A^{-1} X) \geq 0$.

The third proof: Assume that $M \geq 0$. Then for any $x, y \in \mathcal{H}$, we have

$$\left\langle M(x, -y)^T, (x, -y)^T \right\rangle \geq 0.$$

Hence,
$$\langle Ax, x\rangle - 2\mathrm{Re}\langle Xy, x\rangle + \langle By, y\rangle \geq 0.$$

Replacing x with $A^{-1}Xy$ and noting that A^{-1} is positive, we get

$$\begin{aligned}
\langle By, y\rangle &\geq 2\mathrm{Re}\langle Xy, A^{-1}Xy\rangle - \langle AA^{-1}Xy, A^{-1}Xy\rangle \\
&= 2\mathrm{Re}\langle X^*A^{-1}Xy, y\rangle - \langle X^*A^{-1}Xy, y\rangle \\
&= \langle X^*A^{-1}Xy, y\rangle,
\end{aligned}$$

since $\langle X^*A^{-1}Xy, y\rangle \geq 0$. Thus, $B \geq X^*A^{-1}X$.

Conversely, assume that $B \geq X^*A^{-1}X$. Let Y be defined by (2.4.2). Then for any $x, y \in \mathcal{H}$, we have

$$\langle M(x, y)^T, (x, y)^T\rangle = \|Y_1 x + Y_2 y\|^2 + \langle (B - X^*A^{-1}X)y, y\rangle \geq 0,$$

whence $M \geq 0$. $\square$

Remark 2.4.2 Let $M = \begin{bmatrix} A & X \\ X^* & B \end{bmatrix}$ such that $B \in \mathbb{B}_{++}(\mathcal{H})$. Then Theorem 2.4.1 turns to

$$M \geq 0 \iff A \geq XB^{-1}X^*. \qquad (2.4.3)$$

To see this, note that the matrix $V = \begin{bmatrix} 0 & I \\ I & 0 \end{bmatrix}$ is unitary. Therefore, the operator matrix M is positive if and only if $VMV^* = \begin{bmatrix} B & X^* \\ X & A \end{bmatrix}$ is positive. Theorem 2.4.1 gives the equivalent condition $A \geq XB^{-1}X^*$.

We can add a positive operator of the form εI to a positive operator A to get a strictly positive operator $A_\varepsilon = A + \varepsilon I$. Then we can use Theorem 2.4.1, and finally let $\varepsilon \to 0$. The following result is an implication of Theorem 2.4.1.

Theorem 2.4.3 Let $A, B \geq 0$. Then matrix $\begin{bmatrix} A & X \\ X^* & B \end{bmatrix} \geq 0$ if and only if $B \geq X^*A_\varepsilon^{-1}X$ for every $\varepsilon > 0$.

The following result has been established in the proof of Theorem 2.6.1. However, we prove it in a slightly different way:

Corollary 2.4.4 If $\begin{bmatrix} A & X^* \\ X & B \end{bmatrix} \geq 0$, then so is $\begin{bmatrix} \|A\| & \|X\| \\ \|X\| & \|B\| \end{bmatrix} \geq 0$

2.4 Positivity of 2 × 2 Matrices of Operators

Proof By Theorem 2.4.3, we have

$$A \geq X^* B_\varepsilon^{-1} X,$$

for arbitrary $\varepsilon > 0$. The inequality $B_\varepsilon^{-1} \geq \|B_\varepsilon\|^{-1} I$ shows that

$$A \geq X^* \|B_\varepsilon\|^{-1} X.$$

Hence, $\|A\| \|B_\varepsilon\| \geq \|X^* X\| = \|X\|^2$, which entails that

$$\begin{bmatrix} \|A\| & \|X\| \\ \|X\| & \|B_\varepsilon\| \end{bmatrix} \geq 0.$$

Letting $\varepsilon \to 0$, we arrive at $\begin{bmatrix} \|A\| & \|X\| \\ \|X\| & \|B\| \end{bmatrix} \geq 0.$ □

As a consequence, we can derive a Cauchy–Schwarz inequality for operators $A_1, \ldots, A_n$ and $B_1, \ldots, B_n$ in $\mathbb{B}(\mathcal{H})$ as follows. Since

$$\begin{bmatrix} \sum_{i=1}^n A_i^* A_i & \sum_{i=1}^n A_i^* B_i \\ \sum_{i=1}^n B_i^* A_i & \sum_{i=1}^n B_i^* B_i \end{bmatrix} = \sum_{i=1}^n [A_i \ B_i]^* [A_i \ B_i] \geq 0,$$

we get

$$\begin{bmatrix} \left\|\sum_{i=1}^n A_i^* A_i\right\| & \left\|\sum_{i=1}^n A_i^* B_i\right\| \\ \left\|\sum_{i=1}^n B_i^* A_i\right\| & \left\|\sum_{i=1}^n B_i^* B_i\right\| \end{bmatrix} \geq 0,$$

and therefore,

$$\left\|\sum_{i=1}^n A_i^* B_i\right\| \leq \left\|\sum_{i=1}^n A_i^* A_i\right\|^{1/2} \left\|\sum_{i=1}^n B_i^* B_i\right\|^{1/2} \qquad \text{(an operator Cauchy–Schwarz inequality)}.$$

Next, we show that for X and Y in $\mathbb{B}_{\mathrm{sa}}(\mathcal{H})$

$$X \geq Y \text{ if and only if } \begin{bmatrix} X & Y \\ Y & Y \end{bmatrix} \geq 0. \tag{2.4.4}$$

For arbitrary $\varepsilon > 0$ we have

$$\begin{bmatrix} X & Y \\ Y & Y \end{bmatrix} + \varepsilon \begin{bmatrix} I & I \\ I & I \end{bmatrix} \geq 0 \iff \begin{bmatrix} X_\varepsilon & Y_\varepsilon \\ Y_\varepsilon & Y_\varepsilon \end{bmatrix} \geq 0,$$

$$\iff X_\varepsilon \geq Y_\varepsilon Y_\varepsilon^{-1} Y_\varepsilon,$$

(by Theorem (2.4.1))

$$\iff X_\varepsilon \geq Y_\varepsilon,$$
$$\iff X \geq Y.$$

Hence, by letting $\varepsilon \longrightarrow 0$, we get the desired implication.

An application of (2.4.3) can be found in a work by Najafi [178].

We can deduce the classical Cauchy–Schwarz inequality for scalars from Theorem 2.4.1. To do this, let $a_1, \ldots, a_n$ and $b_1, \ldots, b_n$ be finite sequences of complex numbers. From $A_i = \begin{bmatrix} |a_i|^2 & \bar{a}_i b_i \\ \bar{b}_i a_i & |b_i|^2 \end{bmatrix} \geq 0$ we conclude that

$$A = \sum_{i=1}^n A_i = \begin{bmatrix} \sum_{i=1}^n |a_i|^2 & \sum_{i=1}^n \bar{a}_i b_i \\ \sum_{i=1}^n \bar{b}_i a_i & \sum_{i=1}^n |b_i|^2 \end{bmatrix} \geq 0.$$

It follows from the equivalence of (i) and (iii) in Theorem 2.4.1 that

$$\left| \sum_{i=1}^n a_i \bar{b}_i \right|^2 \leq \left(\sum_{i=1}^n |a_i|^2 \right) \left(\sum_{i=1}^n |b_i|^2 \right).$$

For an arbitrary operator matrix $M = \begin{bmatrix} A & B \\ C & D \end{bmatrix}$, the operator $D - CA^{-1}B$ is called the *Schur complement* of A in M. The following assertion is a readily consequence of Theorem 2.4.1.

Corollary 2.4.5 *Let $A \in \mathbb{B}_{++}(\mathscr{H})$ and $X \in \mathbb{B}(\mathscr{H})$. Then*

$$X^* A^{-1} X = \min \left\{ B \in \mathbb{B}_+(\mathscr{H}) : \begin{bmatrix} A & X \\ X^* & B \end{bmatrix} \geq 0 \right\},$$

or equivalently

$$B - X^* A^{-1} X = \sup \left\{ Y \geq 0 : \begin{bmatrix} 0 & 0 \\ 0 & Y \end{bmatrix} \leq \begin{bmatrix} A & X \\ X^* & B \end{bmatrix} \right\}.$$

We observe that if $\begin{bmatrix} A & X \\ X^* & B \end{bmatrix} \geq 0$, then

2.4 Positivity of 2 × 2 Matrices of Operators

$$\begin{bmatrix} I & -I \\ 0 & 0 \end{bmatrix} \begin{bmatrix} A & X \\ X^* & B \end{bmatrix} \begin{bmatrix} I & 0 \\ -I & 0 \end{bmatrix} = \begin{bmatrix} A + B - X - X^* & 0 \\ 0 & 0 \end{bmatrix} \geq 0.$$

Therefore, $X + X^* \leq A + B$.

Corollary 2.4.6 *Let $M = \begin{bmatrix} A & X \\ X^* & B \end{bmatrix} \geq 0$ and $B \in \mathbb{B}_{++}(\mathcal{H})$. Then there exists a partial isometry U such that $|X| \leq |A^{1/2} U B^{1/2}|$. Furthermore, if $\overline{\mathrm{ran}(X)} = \overline{\mathrm{ran}(X^*)} = \mathcal{H}$, then U can be chosen to be a unitary.*

Proof Let $T = XB^{-\frac{1}{2}}$ and $T = U|T|$ be the polar decomposition of T. Then $|T| = U^*T$. Hence, we have $|T|^2 = U^*T \cdot T^*U$. By Remark 2.4.2, we can conclude that

$$\begin{aligned} |X|^2 &= X^*X = B^{\frac{1}{2}} \cdot |T|^2 \cdot B^{\frac{1}{2}} = B^{\frac{1}{2}} \cdot U^*TT^*U \cdot B^{\frac{1}{2}} \\ &= B^{\frac{1}{2}} U^* \cdot XB^{-1}X^* \cdot (B^{\frac{1}{2}} U^*)^* \\ &\leq B^{\frac{1}{2}} U^* \cdot A \cdot (B^{\frac{1}{2}} U^*)^* = |A^{1/2} U B^{1/2}|^2, \end{aligned}$$

which gives $|X| \leq |A^{1/2} U B^{1/2}|$ by the operator monotonicity of $t \mapsto t^{1/2}$.
Since $\overline{\mathrm{ran}(U)} = \overline{\mathrm{ran}(T)} = \overline{\mathrm{ran}(X)}$ and

$$\overline{\mathrm{ran}(U^*)} = \overline{\mathrm{ran}(T^*)} = \mathcal{H} \iff \overline{\mathrm{ran}(X^*)} = \mathcal{H},$$

we conclude that U is a unitary if and only if $\overline{\mathrm{ran}(X)} = \overline{\mathrm{ran}(X^*)} = \mathcal{H}$. □

It should be noted that the converse of Corollary 2.4.6 does not hold: $|X| \leq |A^{1/2} U B^{1/2}|$ does not imply the positivity of the operator matrix M. For example, take $X = I$, $A = \begin{bmatrix} 2 & 0 \\ 0 & 1 \end{bmatrix}$, $B^{-1} = \begin{bmatrix} 1 & 0 \\ 0 & 2 \end{bmatrix}$, and $U = \begin{bmatrix} 0 & 1 \\ 1 & 0 \end{bmatrix}$.

A necessary condition for the positivity of M can be obtained by using the Douglas majorization theorem 1.2.4. Here is the procedure:

Lemma 2.4.7 ([88, Lemma 6]) *Let $M = \begin{bmatrix} A & X \\ X^* & B \end{bmatrix} \geq 0$. Then*

$$\mathrm{ran}(X) \subseteq \mathrm{ran}(A^{\frac{1}{2}}) \text{ and } \mathrm{ran}(X^*) \subseteq \mathrm{ran}(B^{\frac{1}{2}}).$$

Proof Since $M \geq 0$, we have $A, B \in \mathbb{B}_+(\mathcal{H})$. Let $M^{\frac{1}{2}} = \begin{bmatrix} Y & W \\ W^* & Z \end{bmatrix}$. Then direct computation yields

$$M = \begin{bmatrix} Y^2 + WW^* & YW + WZ \\ W^*Y + ZW^* & W^*W + Z^2 \end{bmatrix},$$

hence
$$A = Y^2 + WW^* \text{ and } X = YW + WZ,$$
which indicate by the Douglas majorization theorem 1.2.4 that $\operatorname{ran}(A^{\frac{1}{2}})$ contains both $\operatorname{ran}(Y)$ and $\operatorname{ran}(W)$. As a result, we have $\operatorname{ran}(X) \subseteq \operatorname{ran}(A^{\frac{1}{2}})$ since it is clear that $\operatorname{ran}(X) \subseteq \operatorname{ran}(Y) + \operatorname{ran}(W)$. Similarly, we have $\operatorname{ran}(X^*) \subseteq \operatorname{ran}(B^{\frac{1}{2}})$. □

Based on the preceding lemma, another characterization of the positivity of M can be derived as follows:

Theorem 2.4.8 ([88, Theorem 7]) *Let* $M = \begin{bmatrix} A & X \\ X^* & B \end{bmatrix}$ *such that* $A, B \in \mathbb{B}_+(\mathcal{H})$. *Then* $M \geq 0$ *if and only if there exists an operator* $C \in \mathbb{B}(\mathcal{H})$ *such that either* $X = C^* B^{1/2}$ *and* $C^*C \leq A$ *or* $X = B^{1/2}C$ *and* $C^*C \leq B$.

Proof We prove that (i) $M \geq 0$ if and only if (ii) there is an operator $C \in \mathbb{B}(\mathcal{H})$ such that $X = C^* B^{1/2}$ and $C^*C \leq A$. The other assertion can be proved similarly.

(i) $\Longrightarrow$ (ii). Suppose that $M \geq 0$. By Lemma 2.4.7 and the Douglas majorization theorem 1.2.4, there exists $C \in \mathbb{B}(\mathcal{H})$ such that
$$X^* = B^{\frac{1}{2}}C \text{ and } \overline{\operatorname{ran}(C)} \subseteq \overline{\operatorname{ran}(B^{\frac{1}{2}})} = \overline{\operatorname{ran}(B)}.$$

Hence, $PC = C$ and $C^*P = C^*$, where P is the projection from $\mathcal{H}$ onto $\overline{\operatorname{ran}(B)}$.

Given any $y \in \operatorname{ran}(B)$ and $z \in \mathcal{H}$, we have $y = Bu$ for some $u \in \mathcal{H}$. Put $x = B^{\frac{1}{2}}u$. Then $y = B^{\frac{1}{2}}x$ and thus

$$\left\langle \begin{bmatrix} A & C^* \\ C & P \end{bmatrix} \begin{bmatrix} z \\ y \end{bmatrix}, \begin{bmatrix} z \\ y \end{bmatrix} \right\rangle = \langle Az, z \rangle + \langle Xx, z \rangle + \langle X^*z, x \rangle + \langle Bx, x \rangle$$
$$= \left\langle M \begin{bmatrix} z \\ x \end{bmatrix}, \begin{bmatrix} z \\ x \end{bmatrix} \right\rangle \geq 0.$$

It follows that $\begin{bmatrix} A & C^* \\ C & P \end{bmatrix} \geq 0$, and so

$$\begin{bmatrix} A - C^*C & 0 \\ 0 & P \end{bmatrix} = \begin{bmatrix} I & -C^* \\ 0 & I \end{bmatrix} \begin{bmatrix} A & C^* \\ C & P \end{bmatrix} \begin{bmatrix} I & 0 \\ -C & I \end{bmatrix} \geq 0,$$

that is, $A \geq C^*C$, as desired.

2.4 Positivity of 2 × 2 Matrices of Operators

(ii) $\Longrightarrow$ (i). This follows from the identity

$$M = \begin{bmatrix} A - C^*C & 0 \\ 0 & 0 \end{bmatrix} + \begin{bmatrix} C & B^{1/2} \\ 0 & 0 \end{bmatrix}^* \begin{bmatrix} C & B^{1/2} \\ 0 & 0 \end{bmatrix}.$$

□

The following theorem, due to Fiedler [78], provides a bound for $A \circ A^{-1}$.

Theorem 2.4.9 (Fiedler's inequality) *If $A \in \mathbb{M}_n$ is positive definite, then $A \circ A^{-1} \geq I$.*

Proof Let $A > 0$. By functional calculus, we have $A^{-1} > 0$. Theorem 2.4.1 entails that $\begin{bmatrix} A & I \\ I & A^{-1} \end{bmatrix} \geq 0$ and $\begin{bmatrix} A^{-1} & I \\ I & A \end{bmatrix} \geq 0$. Using the Schur product theorem, we conclude that the Hadamard product $\begin{bmatrix} A \circ A^{-1} & I \circ I \\ I \circ I & A^{-1} \circ A \end{bmatrix}$ of the above two matrices is positive semidefinite. Since $I \circ I = I$ and the Schur product is commutative, we reach $\begin{bmatrix} A \circ A^{-1} & I \\ I & A \circ A^{-1} \end{bmatrix} \geq 0$. An application of Theorem 2.4.1 again shows that $A \circ A^{-1} \geq (A \circ A^{-1})^{-1}$. Therefore, $t \geq t^{-1}$ for any $t \in \text{sp}(A \circ A^{-1})$. Hence, $t \geq 1$ ($t \in \text{sp}(A \circ A^{-1})$). Using the functional calculus for the positive semidefinite matrix $A \circ A^{-1}$, we conclude that $A \circ A^{-1} \geq I$. □

We can also provide another characterization of the positivity of M in terms of contractions. To this end, we need the following lemma (see [20, Proposition 1.3.1] for another proof for matrices):

Lemma 2.4.10 *An operator $K \in \mathbb{B}(\mathcal{H})$ is contraction if and only if $\begin{bmatrix} I & K \\ K^* & I \end{bmatrix} \geq 0$.*

Proof The operator K is a contraction if and only if $\|K^*K\| = \|K\|^2 \leq 1$. The latter is equivalent to $0 \leq t \leq 1$ for all $t \in \text{sp}(K^*K)$. This fact, by the functional calculus for the self-adjoint operator K^*K, is equivalent to $K^*K \leq I$. Eventually, it follows from Theorem 2.4.1 that $K^*K \leq I$ holds if and only if $\begin{bmatrix} I & K \\ K^* & I \end{bmatrix} \geq 0$. □

Corollary 2.4.11 ([44, Lemma 2.2]) *Let $T, A \in \mathbb{B}(\mathcal{H})$ commute and $T \geq A^*A$. Then $T \geq AA^*$.*

Proof We can assume that T is invertible. It follows from Theorem 2.4.1 that the inequality $T \geq A^*A$ gives $\begin{bmatrix} T & A^* \\ A & I \end{bmatrix} \geq 0$. Then

$$\begin{bmatrix} I & T^{-1/2}A^*T^{-1/2} \\ T^{-1/2}AT^{-1/2} & I \end{bmatrix} = \begin{bmatrix} T^{-1/2} & 0 \\ 0 & T^{-1/2} \end{bmatrix} \begin{bmatrix} T & A^* \\ A & T \end{bmatrix} \begin{bmatrix} T^{-1/2} & 0 \\ 0 & T^{-1/2} \end{bmatrix} \geq 0.$$

It follows from Lemma 2.4.10 that $\|T^{-1/2}A^*T^{-1/2}\| \leq 1$ and so

$$\|T^{-1/2}AT^{-1/2}\| = \|(T^{-1/2}A^*T^{-1/2})^*\| \leq 1.$$

By an inverse argument, we get $\begin{bmatrix} T & A \\ A^* & I \end{bmatrix} \geq 0$, and hence $T \geq AA^*$. $\square$

The following theorem is due to Ando [6]. Based on Theorem 2.4.8, an alternative proof of this theorem can be found in [88, Theorem 8]. A proof for matrices is stated in [121, Theorem 7.7.9]; see also [20, Proposition 1.3.2] for a proof of Theorem 2.4.1.

Theorem 2.4.12 *Let $A, B \in \mathbb{B}_+(\mathcal{H})$ and $X \in \mathbb{B}(\mathcal{H})$. Then the following are equivalent:*

(i) *The operator matrix $M = \begin{bmatrix} A & X \\ X^* & B \end{bmatrix}$ is positive.*

(ii) *There exists a contraction K such that $X = A^{1/2}KB^{1/2}$.*

Proof (i) $\Longrightarrow$ (ii). First, assume that $A > 0$ and $B > 0$. By Theorem 2.4.1, we have $B \geq X^*A^{-1}X$. Hence

$$I \geq B^{-1/2}X^*A^{-1}XB^{-1/2} = (A^{-1/2}XB^{-1/2})^*(A^{-1/2}XB^{-1/2}).$$

Hence, $\|A^{-1/2}XB^{-1/2}\|^2 = \|(A^{-1/2}XB^{-1/2})^*(A^{-1/2}XB^{-1/2})\| \leq \|I\| = 1$. Therefore, the operator $K = A^{-1/2}XB^{-1/2}$ is a contraction and

$$A^{1/2}KB^{1/2} = A^{1/2}(A^{-1/2}XB^{-1/2})B^{1/2} = X.$$

Second, assume that $A \geq 0$ and $B \geq 0$. Considering $\begin{bmatrix} A + \frac{1}{n} & X \\ X^* & B + \frac{1}{n} \end{bmatrix}$ and employing the first case, we get a sequence of contractions (K_n) such that

$$\left(A + \frac{1}{n}\right)^{1/2} K_n \left(B + \frac{1}{n}\right)^{1/2} = X, \text{ for any } n \in \mathbb{N}.$$

Since the closed unit ball of $\mathbb{B}(\mathcal{H})$ is weakly compact, we can assume that (K_n) converges weakly to a contraction $K \in \mathbb{B}(\mathcal{H})$. Taking limits as n tends to infinity, we arrive at $A^{1/2}KB^{1/2} = X$.

(ii) $\Longrightarrow$ (i). Since K is a contraction, by Lemma 2.4.10, we have $\begin{bmatrix} I & K \\ K^* & I \end{bmatrix} \geq 0$. Hence

2.4 Positivity of 2 × 2 Matrices of Operators

$$\begin{bmatrix} A & X \\ X^* & B \end{bmatrix} = \begin{bmatrix} A & A^{1/2}KB^{1/2} \\ B^{1/2}K^*A^{1/2} & B \end{bmatrix}$$

$$= \begin{bmatrix} A^{1/2} & 0 \\ 0 & B^{1/2} \end{bmatrix} \begin{bmatrix} I & K \\ K^* & I \end{bmatrix} \begin{bmatrix} A^{1/2} & 0 \\ 0 & B^{1/2} \end{bmatrix} \geq 0.$$

□

Remark 2.4.13 When dealing with operators in a von Neumann algebra $\mathscr{A}$, the contraction K can be chosen to be in $\mathscr{A}$; see [77, Proposition 2.5].

An extension of Lemma 2.4.10 is mentioned in [23] inspired by [191] as follows.

Theorem 2.4.14 *Let $A \in \mathbb{M}_n$. The following conditions are equivalent:*
(i) $\|A\| \leq 1$.
(ii) *The $(n+1) \times (n+1)$ block matrix*

$$\begin{bmatrix} I & A^* & A^{*2} & \cdots & A^{*n} \\ A & I & A^* & \cdots & A^{*n-1} \\ \vdots & \vdots & \vdots & \vdots & \vdots \\ A^n & A^{n-1} & A^{n-2} & \cdots & I \end{bmatrix}$$

is positive for every $n \geq 1$.

Proof First note that for every operator X with $X^{n+1} = 0$,

$$(I - X)^{-1} + (I - X^*)^{-1} - I$$
$$= (I - X)^{-1} \left[(I - X^*) + (I - X) - (I - X)(I - X^*) \right] (I - X^*)^{-1}$$
$$= (I - X)^{-1} \left[I - XX^* \right] (I - X^*)^{-1}$$

is positive if and only if $I - XX^* \geq 0$, that is, $\|X\| \leq 1$.

Set

$$R_n(A) := \begin{bmatrix} 0 & 0 & \cdots & 0 & 0 \\ A & 0 & \cdots & 0 & 0 \\ 0 & A & \cdots & 0 & 0 \\ 0 & 0 & \cdots & A & 0 \end{bmatrix}.$$

For $1 \leq k \leq n$, $R_n(A)^k$ is the block matrix with entries A on the kth subdiagonal and 0's elsewhere, and $R_n(A)^{n+1} = 0$. Therefore,

$$(I - R_n(A))^{-1} = I + R_n(A) + \cdots + R_n(A)^n$$

$$= \begin{bmatrix} I & 0 & 0 & \cdots & 0 \\ A & I & 0 & \cdots & 0 \\ A^2 & A & I & \cdots & 0 \\ \vdots & \vdots & \vdots & \ddots & \vdots \\ A^n & A^{n-1} & A^{n-2} & \cdots & I \end{bmatrix}.$$

In addition, $\|R_n(A)\| = \|A\|$. Hence, by the first part of the proof, $\|A\| \leq 1$ if and only if

$$(I - R_n(A))^{-1} + (I - R_n(A)^*)^{-1} - I \geq 0.$$

This is equivalent to the positivity of the matrix

$$\begin{bmatrix} I & A^* & A^{*2} & \cdots & A^{*n} \\ A & I & A^* & \cdots & A^{*n-1} \\ \vdots & \vdots & \vdots & \ddots & \vdots \\ A^n & A^{n-1} & A^{n-2} & \cdots & I \end{bmatrix}.$$

□

Now, we present another condition equivalent to the positivity of a 2×2 operator matrix.

Theorem 2.4.15 *Let $M = \begin{bmatrix} A & X \\ X^* & B \end{bmatrix}$. Then the following statements are equivalent:*

(i) The operator matrix M is positive.

(ii) $A = Y_1^* Y_1$, $B = Y_2^* Y_2$, and $X = Y_1^* Y_2$ for some $Y_1, Y_2 \in \mathbb{B}(\mathscr{H}, \mathscr{H} \oplus \mathscr{H})$

Proof *The first proof:* A proof can be carried out using the following outlines in the second proof of Theorem 2.4.1.

The second proof (T. Ando): By Theorem 2.4.12, there is a contraction K such that $X = A^{1/2} K B^{1/2}$. Let V be the unitary dilation of K constructed in Proposition 2.2.2. Let $Y_1, Y_2 \in \mathbb{B}(\mathscr{H}, \mathscr{H} \oplus \mathscr{H})$ be defined as follows:

$$Y_1 = V^* \begin{bmatrix} A^{1/2} \\ 0 \end{bmatrix} \quad \text{and} \quad Y_2 = \begin{bmatrix} B^{1/2} \\ 0 \end{bmatrix}.$$

Direct computation yields $A = Y_1^* Y_1$, $B = Y_2^* Y_2$, and $X = Y_1^* Y_2$. □

Utilizing the polar decompositions $Y_k = U_k |Y_k|$, where $k = 1, 2$ and U_k's are partial isometries, we get $X = Y_1^* Y_2 = |Y_1| U_1^* U_2 |Y_2| = A^{1/2} K B^{1/2}$, where $K = U_1^* U_2$ is a contraction. This gives an alternative method for finding the contraction in Theorem 2.4.12.

The next result yields a generalized Cauchy–Schwarz inequality.

2.4 Positivity of 2 × 2 Matrices of Operators

Proposition 2.4.16 (Generalized Cauchy–Schwarz inequality) *Let* $M = \begin{bmatrix} A & X \\ X^* & B \end{bmatrix}$ *such that* $A, B \in \mathbb{B}_+(\mathcal{H})$. *Then M is positive if and only if*

$$|\langle x, Xy \rangle|^2 \leq \langle x, Ax \rangle \langle y, By \rangle \tag{2.4.5}$$

for all $x, y \in \mathcal{H}$.

Proof Suppose that M is positive. Then it follows from Theorem 2.4.12 that $X = A^{1/2} K B^{1/2}$ for some contraction K. Since $K^* K \leq I$, we know that for any $x, y \in \mathcal{H}$,

$$\begin{aligned}|\langle x, Xy \rangle|^2 &= |\langle x, A^{1/2} K B^{1/2} y \rangle|^2 = |\langle A^{1/2} x, K B^{1/2} y \rangle|^2 \\ &\leq |\langle A^{1/2} x, A^{1/2} x \rangle| |\langle K B^{1/2} y, K B^{1/2} y \rangle| \\ &\quad \text{(by the usual Cauchy–Schwarz inequality in Hilbert spaces)} \\ &= |\langle x, Ax \rangle| |\langle y, B^{1/2} K^* K B^{1/2} y \rangle| \\ &\leq \langle x, Ax \rangle \langle y, By \rangle.\end{aligned}$$

To prove the converse, suppose that (2.4.5) holds. For any $x, y \in \mathcal{H}$, we have

$$\begin{aligned}\left\langle (x, y)^T, M(x, y)^T \right\rangle &= \langle x, Ax \rangle + 2\mathrm{Re}\langle x, Xy \rangle + \langle y, By \rangle \\ &\geq \langle x, Ax \rangle - 2|\langle x, Xy \rangle| + \langle y, By \rangle \\ &\geq \langle x, Ax \rangle - 2\langle x, Ax \rangle^{\frac{1}{2}} \langle y, By \rangle^{\frac{1}{2}} + \langle y, By \rangle \\ &= \left(\langle x, Ax \rangle^{\frac{1}{2}} - \langle y, By \rangle^{\frac{1}{2}} \right)^2 \geq 0.\end{aligned}$$

Hence, $M \geq 0$. □

The following result reads as follows. Two operators A and B are called *unitarily congruent* if there is a unitary U such that $B = U^* AU$. Then $U^* AU$ is called a *unitary congruence* of A.

Proposition 2.4.17 *Let* $M = \begin{bmatrix} A & X \\ X & B \end{bmatrix}$, *where* $A, B,$ *and* X *are self-adjoint and* $A + B \in \mathbb{B}_{++}(\mathcal{H})$. *Then*

$$M \geq 0 \iff \|(B + A)^{-1/2} (B - A + 2iX)(B + A)^{-1/2}\| \leq 1.$$

Proof Using the unitary congruent by the unitary $U = \frac{1}{\sqrt{2}}\begin{bmatrix} -iI & iI \\ I & I \end{bmatrix}$, we can deduce that M is positive if and only if

$$U^*MU = \frac{1}{2}\begin{bmatrix} B+A & B-A+2iX \\ B-A-2iX & B+A \end{bmatrix}$$

is positive. In light of Theorem 2.4.1, the latter is positive if and only if

$$(B+A) \geq (B-A+2iX)(B+A)^{-1}(B-A-2iX)$$

and this is equivalent to $\|(B+A)^{-1/2}(B-A+2iX)(B+A)^{-1/2}\| \leq 1$. □

Proposition 2.4.18 *Let* $M = \begin{bmatrix} A & X \\ X & B \end{bmatrix}$, *where* $A, B, X \in \mathbb{B}(\mathscr{H})$ *with* X *self-adjoint and* $M \geq 0$. *Then*

$$\pm X \leq \frac{1}{2\sqrt{t(1-t)}}\bigl[tB + (1-t)A\bigr], \text{ for every } t \in (0,1). \tag{2.4.6}$$

Proof For every $t \in (0, 1)$, let

$$U_t = \begin{bmatrix} (1-t)^{1/2}I & -t^{1/2}I \\ t^{1/2}I & (1-t)^{1/2}I \end{bmatrix}.$$

Then U_t is a unitary and

$$U_t^*MU_t = \begin{bmatrix} M_{11}(t) & M_{12}(t) \\ M_{21}(t) & M_{22}(t) \end{bmatrix},$$

where

$$M_{11}(t) = 2\sqrt{t(1-t)}X + (1-t)A + tB,$$
$$M_{12}(t) = -tX + (1-t)X + \sqrt{t(1-t)}(B-A),$$
$$M_{21}(t) = (1-t)X - tX + \sqrt{t(1-t)}(B-A),$$
$$M_{22}(t) = -2\sqrt{t(1-t)}X + tA + (1-t)B.$$

From $M_{11}(t) \geq 0$ and $M_{22}(t) \geq 0$ we conclude that

$$X \leq \frac{1}{2\sqrt{t(1-t)}}\bigl[tA + (1-t)B\bigr] \text{ and } -X \leq \frac{1}{2\sqrt{t(1-t)}}\bigl[(1-t)A + tB\bigr].$$

2.4 Positivity of 2 × 2 Matrices of Operators

These inequalities lead to

$$X \le \frac{1}{2\sqrt{t(1-t)}}[(1-t)A + tB] \quad \text{and} \quad -X \le \frac{1}{2\sqrt{t(1-t)}}[tA + (1-t)B].$$

Replacing t with $(1-t)$ yields (2.4.6). □

Substituting $t = \frac{1}{2}$ into inequality (2.4.6) yields $\pm X \le \frac{A+B}{2}$. However, this does not imply the positivity of the operator matrix $M = \begin{bmatrix} A & X \\ X^* & B \end{bmatrix}$, which in turn implies that $A = A^*$, $B = B^*$, $X = X^*$, and $A + B \in \mathbb{B}_{++}(\mathcal{H})$. An immediate example is $X = 0$, $A = 2I$, and $B = -I$. However, in the converse direction, we have the following equivalent conditions:

$$\pm X \le \frac{A+B}{2} \iff (B+A)^{-1/2}(\pm 2X)(B+A)^{-1/2} \le I,$$

$$\iff \|(2B+2A)^{-1/2}(4iX)(2B+2A)^{-1/2}\| \le 1,$$

$$\iff \begin{bmatrix} A & X \\ X & B \end{bmatrix} + \begin{bmatrix} B & X \\ X & A \end{bmatrix} \ge 0. \tag{2.4.7}$$

The last equivalence follows from Proposition 2.4.17. However, the last operator matrix is $M + UMU^*$ with $U = \begin{bmatrix} 0 & I \\ I & 0 \end{bmatrix}$. Therefore,

$$\pm X \le \frac{A+B}{2} \iff M + UMU^* \ge 0.$$

In the special case where $A = B$, we have the following proposition:

Proposition 2.4.19 *Let* $A, X \in \mathbb{B}(\mathcal{H})$ *be such that X is self-adjoint and A is positive. Then*

$$M = \begin{bmatrix} A & X \\ X & A \end{bmatrix} \ge 0 \iff \pm X \le A.$$

Proof Clearly, $M \ge 0$ if and only if $\begin{bmatrix} A + \frac{1}{n}I & X \\ X & A + \frac{1}{n}I \end{bmatrix} \ge 0$ for any $n \in \mathbb{N}$. Similarly $\pm X \le A$ if and only if $\pm X \le A + \frac{1}{n}I$ for any $n \in \mathbb{N}$. So it suffices to prove that for any $n \in \mathbb{N}$,

$$\begin{bmatrix} A + \frac{1}{n}I & X \\ X & A + \frac{1}{n}I \end{bmatrix} \ge 0 \iff \pm X \le A + \frac{1}{n}I.$$

This follows immediately from (2.4.7) since $A + \frac{1}{n}I$ is strictly positive. □

For example, it follows from Proposition 2.4.19 that if X is a Hermitian matrix, then $\begin{bmatrix} |X| & X \\ X & |X| \end{bmatrix} \geq 0$. Therefore, if X and Y are Hermitian, then

$$\begin{bmatrix} |X| \circ |Y| & X \circ Y \\ X \circ Y & |X| \circ |Y| \end{bmatrix} \geq 0.$$

We then obtain $\pm X \circ Y \leq |X| \circ |Y|$. Note that the inequality $|X \circ Y| \leq |X| \circ |Y|$ does not hold in general. To see this consider $X = \begin{bmatrix} 2 & 2 \\ 2 & 1 \end{bmatrix}$ and $Y = \begin{bmatrix} 1 & 3 \\ 3 & 3 \end{bmatrix}$.

In a similar way, if $X, Y \in \mathbb{M}_n$, then $\begin{bmatrix} X^*X & X^*Y \\ Y^*X & Y^*Y \end{bmatrix}$ and $\begin{bmatrix} Y^*Y & Y^*X \\ X^*Y & X^*X \end{bmatrix}$ are positive. It follows from the Schur product theorem that

$$\begin{bmatrix} (X^*X) \circ (Y^*Y) & (X^*Y) \circ (Y^*X) \\ (X^*Y) \circ (Y^*X) & (X^*X) \circ (Y^*Y) \end{bmatrix} \geq 0,$$

which yields that

$$\pm (X^*Y) \circ (Y^*X) \leq (X^*X) \circ (Y^*Y).$$

Another consequence of Proposition 2.4.19 reads as follows.

Corollary 2.4.20 *Let $A \in \mathbb{B}(\mathcal{H})$ be self-adjoint. Then A is positive if and only if the operator matrix $\begin{bmatrix} A & A \\ A & A \end{bmatrix}$ is positive.*

Proposition 2.4.21 *Let $A \in \mathbb{B}(\mathcal{H})$. Then the operator matrix $\begin{bmatrix} |A| & A^* \\ A & |A^*| \end{bmatrix}$ is positive.*

Proof We use the polar decomposition $A = U|A|$ of A. Then $A^* = |A|U^*$ is the polar decomposition of A^*. It follows from Corollary 2.4.20 that $\begin{bmatrix} |A| & |A| \\ |A| & |A| \end{bmatrix}$ is positive. Hence,

$$\begin{bmatrix} |A| & A^* \\ A & |A^*| \end{bmatrix} = \begin{bmatrix} |A| & |A|U^* \\ U|A| & U|A|U^* \end{bmatrix} = \begin{bmatrix} I & 0 \\ 0 & U \end{bmatrix} \begin{bmatrix} |A| & |A| \\ |A| & |A| \end{bmatrix} \begin{bmatrix} I & 0 \\ 0 & U^* \end{bmatrix} \geq 0.$$

□

2.5 Diagonal Blocks and Unitary Orbits

For partitions of positive semidefinite matrices, the diagonal blocks play a special role. This is evident in a remarkable decomposition due to Bourin and Lee [34]:

Lemma 2.5.1 *For every positive semidefinite matrix in $\mathbb{M}_{n+m}$ partitioned into blocks, there is a decomposition*

$$\begin{bmatrix} A & X \\ X^* & B \end{bmatrix} = U \begin{bmatrix} A & 0 \\ 0 & 0 \end{bmatrix} U^* + V \begin{bmatrix} 0 & 0 \\ 0 & B \end{bmatrix} V^*$$

for some unitaries $U, V \in \mathbb{M}_{n+m}$.

Proof To obtain the decomposition, factorize the block matrix as a square of positive semidefinite matrices,

$$\begin{bmatrix} A & X \\ X^* & B \end{bmatrix} = \begin{bmatrix} C & Y \\ Y^* & D \end{bmatrix} \begin{bmatrix} C & Y \\ Y^* & D \end{bmatrix}$$

and observe that it can be written with $T = \begin{bmatrix} C & Y \\ 0 & 0 \end{bmatrix}$ and $S = \begin{bmatrix} 0 & 0 \\ Y^* & D \end{bmatrix}$ as

$$\begin{bmatrix} C & 0 \\ Y^* & 0 \end{bmatrix} \begin{bmatrix} C & Y \\ 0 & 0 \end{bmatrix} + \begin{bmatrix} 0 & Y \\ 0 & D \end{bmatrix} \begin{bmatrix} 0 & 0 \\ Y^* & D \end{bmatrix} = T^*T + S^*S.$$

The polar decompositions of T and S show that T^*T and S^*S are unitarily congruent to

$$TT^* = \begin{bmatrix} A & 0 \\ 0 & 0 \end{bmatrix} \quad \text{and} \quad SS^* = \begin{bmatrix} 0 & 0 \\ 0 & B \end{bmatrix}.$$

This gives the decomposition. □

A simple consequence of this decomposition is the following result, implicit in [37].

Theorem 2.5.2 *Given any positive semidefinite matrix in $\mathbb{M}_{2n}$ partitioned into blocks in $\mathbb{M}_n$ with Hermitian off-diagonal blocks, it holds that*

$$\begin{bmatrix} A & X \\ X & B \end{bmatrix} = \frac{1}{2} \left\{ U(A+B)U^* + V(A+B)V^* \right\}$$

for some isometries $U, V \in \mathbb{M}_{2n,n}$.

Here $V \in \mathbb{M}_{p,q}$ is an isometry if $p \geq q$ and $V^*V = I_q$.
Non-trivial generalizations of Theorem 2.5.2 for higher numbers of blocks are given in [34].

Proof Taking the unitary matrix

$$W = \frac{1}{\sqrt{2}} \begin{bmatrix} -iI & iI \\ I & I \end{bmatrix},$$

where I is the identity of $\mathbb{M}_n$, we have

$$W^* \begin{bmatrix} A & X \\ X & B \end{bmatrix} W = \frac{1}{2} \begin{bmatrix} A+B & * \\ * & A+B \end{bmatrix}$$

where $*$ stands for unspecified entries. According to Lemma 2.5.1, there exist two unitaries $U, V \in \mathbb{M}_{2n}$ partitioned into equally sized matrices,

$$U = \begin{bmatrix} U_{11} & U_{12} \\ U_{21} & U_{22} \end{bmatrix}, \quad V = \begin{bmatrix} V_{11} & V_{12} \\ V_{21} & V_{22} \end{bmatrix}$$

such that

$$\frac{1}{2} \begin{bmatrix} A+B & * \\ * & A+B \end{bmatrix} = \frac{1}{2} \left\{ U \begin{bmatrix} A+B & 0 \\ 0 & 0 \end{bmatrix} U^* + V \begin{bmatrix} 0 & 0 \\ 0 & A+B \end{bmatrix} V^* \right\}.$$

Therefore

$$\frac{1}{2} \begin{bmatrix} A+B & * \\ * & A+B \end{bmatrix} = \frac{1}{2} \{ \tilde{U}(A+B)\tilde{U}^* + \tilde{V}(A+B)\tilde{V}^* \}$$

where

$$\tilde{U} = \begin{bmatrix} U_{11} \\ U_{21} \end{bmatrix} \quad \text{and} \quad \tilde{V} = \begin{bmatrix} V_{12} \\ V_{22} \end{bmatrix}$$

are isometries. The proof is complete by assigning $W\tilde{U}$ and $W\tilde{V}$ to new isometries U and V, respectively. □

Corollary 2.5.3 *Let* $H = \begin{bmatrix} A & X \\ X & B \end{bmatrix} \in \mathbb{M}_{2n}$ *be a positive block partitioned into Hermitian blocks in* $\mathbb{M}_n$. *Then,*

$$\lambda_{1+2k}(H) \leq \lambda_{1+k}(A+B)$$

for all $k = 0, \ldots, n-1$.

Proof Together with Theorem 2.5.2, the inequalities follow immediately from a simple fact known as *Weyl's theorem*: if $Y, Z \in \mathbb{M}_m$ are Hermitian, then

$$\lambda_{r+s+1}(Y+Z) \leq \lambda_{r+1}(Y) + \lambda_{s+1}(Z)$$

for all nonnegative integers r and s such that $r + s \leq m - 1$. □

2.5 Diagonal Blocks and Unitary Orbits

Some generalizations of Corollary 2.5.3 for a larger number of blocks can be found in [35]. The next series of corollaries are from [37].

Corollary 2.5.4 *Let $S, T \in \mathbb{M}_n$ be Hermitian. Then,*

$$\|T^2 + ST^2S\| \leq \|T^2 + TS^2T\|$$

for all unitarily invariant norms. Furthermore,

$$\lambda_{1+2k}(T^2 + ST^2S) \leq \lambda_{1+k}(T^2 + TS^2T)$$

for all $k = 0, \ldots, n-1$.

Proof The nonzero eigenvalues of $T^2 + ST^2S = \begin{bmatrix} T & ST \end{bmatrix}\begin{bmatrix} T & ST \end{bmatrix}^*$ are the same as those of

$$\begin{bmatrix} T & ST \end{bmatrix}^* \begin{bmatrix} T & ST \end{bmatrix} = \begin{bmatrix} T^2 & TST \\ TST & TS^2T \end{bmatrix}.$$

This block matrix is positive and its off-diagonal blocks are Hermitian. Therefore, the norm inequality follows from Theorem 2.5.2, and the eigenvalue inequalities follow from Corollary 2.5.3. □

If we use a unitary congruence with

$$J = \frac{1}{\sqrt{2}} \begin{bmatrix} I & -I \\ I & I \end{bmatrix}$$

where I is the identity of $\mathbb{M}_n$, we observe that

$$J^* \begin{bmatrix} A & X \\ X^* & B \end{bmatrix} J = \begin{bmatrix} \frac{A+B}{2} + \mathrm{Re}\, X & * \\ * & \frac{A+B}{2} - \mathrm{Re}\, X \end{bmatrix}$$

where $*$ stands for unspecified entries. Thus, Lemma 2.5.1 yields:

Corollary 2.5.5 *For every positive matrix in $\mathbb{M}_{2n}$ written in blocks of the same size, there is a decomposition*

$$\begin{bmatrix} A & X \\ X^* & B \end{bmatrix} = U \begin{bmatrix} \frac{A+B}{2} + \mathrm{Re}\, X & 0 \\ 0 & 0 \end{bmatrix} U^* + V \begin{bmatrix} 0 & 0 \\ 0 & \frac{A+B}{2} - \mathrm{Re}\, X \end{bmatrix} V^*$$

for some unitaries $U, V \in \mathbb{M}_{2n}$.

This is equivalent to Corollary 2.5.6 below since $\begin{bmatrix} A & X \\ X^* & B \end{bmatrix}$ is unitarily congruent to $\begin{bmatrix} A & iX \\ -iX^* & B \end{bmatrix}$. One may also use the unitary matrix $W = \frac{1}{\sqrt{2}} \begin{bmatrix} I & -iI \\ -iI & I \end{bmatrix}$ to see

$$W^* \begin{bmatrix} A & X \\ X^* & B \end{bmatrix} W = \begin{bmatrix} \frac{A+B}{2} + \operatorname{Im} X & * \\ * & \frac{A+B}{2} - \operatorname{Im} X \end{bmatrix}.$$

Corollary 2.5.6 *For every positive semidefinite matrix in $\mathbb{M}_{2n}$ written in blocks of the same size, there exists a decomposition*

$$\begin{bmatrix} A & X \\ X^* & B \end{bmatrix} = U \begin{bmatrix} \frac{A+B}{2} + \operatorname{Im} X & 0 \\ 0 & 0 \end{bmatrix} U^* + V \begin{bmatrix} 0 & 0 \\ 0 & \frac{A+B}{2} - \operatorname{Im} X \end{bmatrix} V^*$$

for some unitaries $U, V \in \mathbb{M}_{2n}$.

Corollary 2.5.6 is a generalization of Theorem 2.5.2. Noticing that $\operatorname{Im} X \leq |\operatorname{Im} X| = \frac{1}{2}|X - X^*|$, we have:

Corollary 2.5.7 *For every positive semidefinite matrix in $\mathbb{M}_{2n}$ written in blocks of the same size, it holds that*

$$\begin{bmatrix} A & X \\ X^* & B \end{bmatrix} \leq \frac{1}{2} \left\{ U \begin{bmatrix} A + B + |X - X^*| & 0 \\ 0 & 0 \end{bmatrix} U^* + V \begin{bmatrix} 0 & 0 \\ 0 & A + B + |X - X^*| \end{bmatrix} V^* \right\}$$

for some unitaries $U, V \in \mathbb{M}_{2n}$.

Recall that a matrix (operator, respectively) is said to be *accretive* if its real part is positive semidefinite (positive, respectively).

Corollary 2.5.8 *If a positive semidefinite matrix in $\mathbb{M}_{2n}$ is written in blocks of the same size such that the right upper block X is accretive, then*

$$\left\| \begin{bmatrix} A & X \\ X^* & B \end{bmatrix} \right\| \leq \|A + B\| + \|\operatorname{Re} X\|$$

for all unitarily invariant norms.

Proof By Corollary 2.5.5, for all Ky Fan k-norms $\|\cdot\|_k$, $k = 1, \ldots, 2n$, we have

$$\left\| \begin{bmatrix} A & X \\ X^* & B \end{bmatrix} \right\|_k \leq \left\| \begin{bmatrix} \frac{A+B}{2} + \operatorname{Re} X & 0 \\ 0 & 0 \end{bmatrix} \right\|_k + \left\| \begin{bmatrix} 0 & 0 \\ 0 & \frac{A+B}{2} \end{bmatrix} \right\|_k.$$

Equivalently,

$$\left\| \begin{bmatrix} A & X \\ X^* & B \end{bmatrix} \right\|_k \leq \left\| \left(\frac{A+B}{2} + \mathrm{Re}X \right)^{\downarrow} \right\|_k + \left\| \left(\frac{A+B}{2} \right)^{\downarrow} \right\|_k$$

where $Z^{\downarrow}$ stands for the diagonal matrix listing the eigenvalues of a positive semidefinite matrix $Z \in \mathbb{M}_n$ in decreasing order. By using the triangle inequality for $\|\cdot\|_k$ and the fact that

$$\|Z_1^{\downarrow}\|_k + \|Z_2^{\downarrow}\|_k = \|Z_1^{\downarrow} + Z_2^{\downarrow}\|_k$$

for all positive semidefinite matrices $Z_1, Z_2 \in \mathbb{M}_n$, we infer

$$\left\| \begin{bmatrix} A & X \\ X^* & B \end{bmatrix} \right\|_k \leq \left\| (A+B)^{\downarrow} + (\mathrm{Re}X)^{\downarrow} \right\|_k.$$

Hence

$$\left\| \begin{bmatrix} A & X \\ X^* & B \end{bmatrix} \right\| \leq \left\| (A+B)^{\downarrow} + (\mathrm{Re}X)^{\downarrow} \right\|$$

for any symmetric norm. The triangle inequality completes the proof. □

2.6 Inequalities Follow from the Positivity of 2×2 Block Matrices

In this section, we present several inequalities that follow from the positivity of 2×2 block matrices. The first result and its related remark appeared in [125].

Theorem 2.6.1 *Let* $T = \begin{bmatrix} A & X \\ X^* & B \end{bmatrix} \in \mathbb{M}_2(\mathbb{M}_k)$ *be positive. Then*

(i)
$$\max\{\|A\|, \|B\|\} \leq \|T\| \leq \frac{1}{2}(\|A\| + \|B\| + \sqrt{(\|A\| - \|B\|)^2 + 4\|X\|^2})$$
$$\leq \|A\| + \|B\|;$$

(ii)
$$\max\{\|A\|, \|B\|\} \leq \|T\| \leq \max\{\|A\|, \|B\|\} + \|X\|.$$

Proof (i) It follows from Lemma 2.1.2 that

$$\max\{\|A\|, \|B\|\} \leq \|T\|.$$

It follows from Theorem 2.4.12 that there is a contraction K such that $X = A^{1/2} K B^{1/2}$. Hence, $\|X\| \leq \|A\|^{1/2} \|B\|^{1/2}$. Again, by Theorem 2.4.12,

$$\widetilde{T} = \begin{bmatrix} \|A\| & \|X\| \\ \|X^*\| & \|B\| \end{bmatrix} \geq 0. \quad (2.6.1)$$

Hence,

$$\|T\| \leq \|\widetilde{T}\| = r(\widetilde{T}) = \frac{1}{2}(\|A\| + \|B\| + \sqrt{(\|A\| - \|B\|)^2 + 4\|X\|^2})$$
$$\leq \frac{1}{2}(\|A\| + \|B\| + \sqrt{(\|A\| - \|B\|)^2 + 4\|A\|\,\|B\|})$$
$$= \|A\| + \|B\|.$$

(ii) The inequalities

$$\sqrt{(\|A\| - \|B\|)^2 + 4\|X\|^2} \leq |\,\|A\| - \|B\|\,| + 2\|X\|$$

and

$$\|A\| + \|B\| + |\,\|A\| - \|B\|\,| = 2\max\{\|A\|, \|B\|\}$$

yield $\|T\| \leq \max\{\|A\|, \|B\|\} + \|X\|$. □

The proof of Theorem 2.6.1(i) shows that the 2×2 block-norm matrix $\widetilde{T}$ is positive when T is positive; see Corollary 2.4.4. However, the result does not hold for 3×3 block-norm matrices. To illustrate this, consider the operator matrix $T = [T_{ij}]_{3\times 3}$, where $T_{11} = \begin{bmatrix} 10 & 0 \\ 0 & 10 \end{bmatrix}$, $T_{12} = \begin{bmatrix} 3 \\ -2 \end{bmatrix}$, $T_{13} = \begin{bmatrix} 9 \\ 4 \end{bmatrix}$, $T_{21} = [3\ -2]$, $T_{22} = [10]$, $T_{23} = [1]$, $T_{31} = [9\ 4]$, $T_{32} = [1]$, $T_{33} = [10]$. Then

$$\widetilde{T} = [\|T_{ij}\|]_{3\times 3} = \begin{bmatrix} 10 & \sqrt{13} & \sqrt{97} \\ \sqrt{13} & 10 & 1 \\ \sqrt{97} & 1 & 10 \end{bmatrix}.$$

There are interesting inequalities involving the trace of matrices and trace class operators. One may consult the comprehensive book [32] for such inequalities.

Zhang [239] answered a question posed by Besenyei in the IMAGE problem 50-3 as follows.

Theorem 2.6.2 *Let* $\begin{bmatrix} A & B \\ B^* & C \end{bmatrix} \in \mathbb{M}_2(\mathbb{M}_k)$ *be positive. Then*

$$\operatorname{tr}(AC) - \operatorname{tr}(B^*B) \leq \operatorname{tr} A \operatorname{tr} C - \operatorname{tr} B^* \operatorname{tr} B = \operatorname{tr} A \operatorname{tr} C - |\operatorname{tr} B|^2. \quad (2.6.2)$$

Proof We can assume that matrix A is diagonal by replacing A and B, and C with U^*AU, U^*BU, and U^*CU, respectively. Here, U is a unitary matrix that makes UAU^* diagonal.

2.6 Inequalities Follow from the Positivity of 2 × 2 Block Matrices

Therefore, the inequality is equivalent to

$$\sum_i a_{ii} c_{ii} - \sum_{i,j} |b_{ij}|^2 \leq \sum_i a_{ii} \sum_j c_{jj} - \sum_i \bar{b}_{ii} \sum_i b_{ii}.$$

This can be further simplified to

$$\sum_{i \neq j} \bar{b}_{ii} b_{jj} - \sum_{i \neq j} |b_{ij}|^2 \leq \sum_{i \neq j} a_{ii} c_{jj}.$$

To prove the latter inequality, it is enough to show that

$$\sum_{i \neq j} \bar{b}_{ii} b_{jj} \leq \sum_{i \neq j} a_{ii} c_{jj}.$$

This inequality holds since the matrix $\begin{bmatrix} a_{ii} c_{jj} & b_{ii} \bar{b}_{jj} \\ \bar{b}_{ii} b_{jj} & c_{ii} a_{jj} \end{bmatrix}$, which is a principal submatrix of $\begin{bmatrix} A & B \\ B^* & C \end{bmatrix} \otimes \begin{bmatrix} C & B^* \\ B & A \end{bmatrix} \geq 0$ is positive. It should be noted that $\begin{bmatrix} x & y \\ \bar{y} & z \end{bmatrix} \geq 0$ implies $y + \bar{y} \leq x + z$. □

Kittaneh and Lin [139] extended inequality (2.6.2) as follows.

Theorem 2.6.3 *Let* $\begin{bmatrix} A & B \\ B^* & C \end{bmatrix} \in \mathbb{M}_2(\mathbb{M}_k)$ *be positive. Then*

$$|\operatorname{tr}(AC) - \operatorname{tr}(B^*B)| \leq \operatorname{tr} A \operatorname{tr} C - |\operatorname{tr} B|^2.$$

Proof By (2.6.2), It is enough to prove that

$$\operatorname{tr}(B^*B) - \operatorname{tr}(AC) \leq \operatorname{tr} A \operatorname{tr} C - |\operatorname{tr} B|^2.$$

It follows from Theorem 2.4.15 that there exist matrices $X, Y \in \mathbb{M}_{n \times 2n}$ such that

$$\begin{bmatrix} A & B \\ B^* & C \end{bmatrix} = \begin{bmatrix} XX^* & XY^* \\ YX^* & YY^* \end{bmatrix} \geq 0.$$

From (2.6.2) we conclude that

$$\operatorname{tr}(X^*XY^*Y) - \operatorname{tr}(Y^*XX^*Y) \leq \operatorname{tr}(X^*X)\operatorname{tr}(Y^*Y) - |\operatorname{tr}(X^*Y)|^2.$$

By the cyclic (tracial) property of tr($\cdot$), we get

$$\text{tr}(YX^*XY^*) - \text{tr}(YY^*XX^*) \leq \text{tr}(X^*X)\,\text{tr}(Y^*Y) - |\text{tr}(X^*Y)|^2,$$

which yields the required inequality. □

There are other types of trace inequalities for entries of a positive block matrix. For example, it is proved in [151] that $\text{tr}(AC) + |\text{tr}\, B|^2 \leq \text{tr}\, A\, \text{tr}\, C + \text{tr}(B^*B)$. It is possible for $\begin{bmatrix} A & B \\ B^* & C \end{bmatrix}$ to be positive, but $\begin{bmatrix} A & B^* \\ B & C \end{bmatrix}$ to not be positive. If both of the above matrices are positive, then we say that $\begin{bmatrix} A & B \\ B^* & C \end{bmatrix}$ is *positive partial transpose (PPT)*. Utilizing the technique in [82], we get the following result.

Theorem 2.6.4 *Let* $M = \begin{bmatrix} A & B \\ B^* & C \end{bmatrix} \in \mathbb{M}_2(\mathbb{M}_k)$ *be PPT. Then*

(i) $\text{tr}(B^*B) + \text{tr}\, B^*\, \text{tr}\, B \leq \text{tr}(AC) + \text{tr}\, A\, \text{tr}\, C;$
(ii) $\text{tr}(B^*B) \leq \text{tr}(AC).$

Proof (i) Since $M' = \begin{bmatrix} A & B^* \\ B & C \end{bmatrix} \geq 0$ and the trace functional is completely positive, we have

$$\begin{bmatrix} \text{tr}(A) & \text{tr}(B^*) \\ \text{tr}(B) & \text{tr}(C) \end{bmatrix} \geq 0.$$

Therefore,

$$N_1 = \begin{bmatrix} \text{tr}(A) + A & \text{tr}(B^*) + B^* \\ \text{tr}(B) + B & \text{tr}(C) + C \end{bmatrix} \geq 0.$$

In addition,

$$N_2 = \begin{bmatrix} C & -B^* \\ -B & A \end{bmatrix} = \begin{bmatrix} 0 & I \\ -I & 0 \end{bmatrix} \begin{bmatrix} A & B \\ B^* & C \end{bmatrix} \begin{bmatrix} 0 & -I \\ I & 0 \end{bmatrix} \geq 0.$$

Hence

$$2\,\text{tr}(A)\,\text{tr}(C) + 2\,\text{tr}(AC) - 2\,\text{tr}(B^*B) - 2\,\text{tr}(B^*)\,\text{tr}(B) = \text{tr}(N_1 N_2)$$
$$= \text{tr}(N_1^{1/2} N_2 N_1^{1/2}) \geq 0.$$

(ii) It follows from $2\,\text{tr}(AC) - 2\,\text{tr}(B^*B) = \text{tr}(M' N_2) = \text{tr}(M'^{1/2} N_2 M'^{1/2}) \geq 0$. □

If B is normal, then we can remove the condition PPT in Theorem 2.6.4 to obtain (ii). In fact, as shown in the proof of [36, Proposition 3.4], if $M \geq 0$, by Theorem 2.4.15, there is a contraction K such that $B = A^{1/2}KC^{1/2}$. It is a known result that if the matrix XY is normal, then $\|XY\| \leq \|YX\|$; see [19, p. 253]. Hence, for the Schatten 2-norm $\|\cdot\|_2$, we have $\|B\|_2 = \|A^{1/2}KC^{1/2}\|_2 \leq \|KC^{1/2}A^{1/2}\|_2$. Therefore,

$$\operatorname{tr}(B^*B) = \|B\|_2^2 \leq \|KC^{1/2}A^{1/2}\|_2^2 = \operatorname{tr}(A^{1/2}C^{1/2}K^*KC^{1/2}A^{1/2})$$
$$\leq \operatorname{tr}(A^{1/2}C^{1/2}C^{1/2}A^{1/2}) \leq \operatorname{tr}(AC),$$

since $K^*K \leq I$.

We denote by $\otimes^r A := A \otimes \cdots \otimes A$ the *r-fold tensor power* of A. A slight extension of Theorem 2.6.3 is given by Li [146] as follows; see also [82].

Theorem 2.6.5 *Let* $\begin{bmatrix} A & B \\ B^* & C \end{bmatrix} \in \mathbb{M}_2(\mathbb{M}_k)$ *be positive. Then for* $r \geq 1$

$$(\operatorname{tr} A \operatorname{tr} C)^r - (\operatorname{tr} B^* \operatorname{tr} B)^r \geq |(\operatorname{tr} AC)^r - (\operatorname{tr} B^*B)^r|.$$

Proof We have $\begin{bmatrix} \otimes^r A & \otimes^r B \\ \otimes^r B^* & \otimes^r C \end{bmatrix} \geq 0$ since it is a principal submatrix of $\otimes^r \begin{bmatrix} A & B \\ B^* & C \end{bmatrix}$. Employing Theorem 2.6.3, we have

$$\operatorname{tr} \otimes^r A \operatorname{tr} \otimes^r C - \operatorname{tr} \otimes^r B^* \operatorname{tr} \otimes^r B \geq |\operatorname{tr}(\otimes^r A)(\otimes^r C) - \operatorname{tr}(\otimes^r B^*)(\otimes^r B)|.$$

Using the properties $(\otimes^r X)(\otimes^r Y) = \otimes^r(XY)$ and $\operatorname{tr}(\otimes^r X) = (\operatorname{tr} X)^r$ (see [239, Chap. 2]), we derive the desired inequality. □

2.7 Inequalities Related to Unitarily Invariant Norms and Numerical Radius

We start this section by discussing some inequalities involving unitarily invariant norms on $\mathbb{M}_n$. The first inequality can be stated as follows; see [7]:

Theorem 2.7.1 *Let* $A, B \in \mathbb{M}_n$ *be positive operators. Then*

$$\frac{1}{2}\|(A+B) \oplus (A+B)\| \leq \|A \oplus B\| \leq \|A + B\|$$

for any unitarily invariant norm $\|\cdot\|$.

Proof We have

$$\frac{1}{2} \|(A+B) \oplus (A+B)\| = \frac{1}{2} \|(A \oplus B) + (B \oplus A)\| \leq \|A \oplus B\|,$$

which yields the left inequality. To achieve the right inequality, we put $X = \begin{bmatrix} A^{1/2} & B^{1/2} \\ 0 & 0 \end{bmatrix}$, $Y = \begin{bmatrix} A & A^{1/2}B^{1/2} \\ B^{1/2}A^{1/2} & B \end{bmatrix}$, and $U = \begin{bmatrix} I & 0 \\ 0 & -I \end{bmatrix}$. We have

$$\left\| \begin{bmatrix} A & 0 \\ 0 & B \end{bmatrix} \right\| = \left\| \frac{1}{2}(Y + UYU^*) \right\| \leq \frac{1}{2}(\|Y\| + \|UYU^*\|) = \|Y\|$$

$$= \|X^*X\|$$

$$= \|XX^*\| \qquad \text{(since } X^*X \text{ and } XX^* \text{ have the same eigenvalues)}$$

$$= \left\| \begin{bmatrix} A+B & 0 \\ 0 & 0 \end{bmatrix} \right\| = \|A+B\|.$$

$\square$

As a consequence of the above theorem for the Schatten p-norm and the equation

$$\|A \oplus B\|_p^p = \|A\|_p^p + \|B\|_p^p,$$

we obtain *McCarthy's inequalities* [164].

Corollary 2.7.2 *Let $p \geq 1$ and let $A, B \in \mathbb{M}_n$ be positive operators. Then*

$$2^{1-p}\|A+B\|_p^p \leq \|A\|_p^p + \|B\|_p^p \leq \|A+B\|_p^p.$$

We aim to prove a unitarily invariant version of the arithmetic-geometric mean inequality. To prove it, we need the following lemma that appeared in [137, Lemma 1].

Lemma 2.7.3 *Let $A, B \in \mathbb{M}_n$ be positive operators. If AB is self-adjoint. Then,*

$$\|\text{Re}\,(BA)\| \geq \|AB\|. \tag{2.7.1}$$

To establish an operator version of the following arithmetic-geometric mean inequality due to Kittaneh [137], we employ 4×4 block matrices.

Theorem 2.7.4 *Suppose that $A, B \in \mathbb{M}_n$ are positive operators. Then*

$$\|AA^*X + XBB^*\| \geq 2\|A^*XB\| \tag{2.7.2}$$

for any unitarily invariant norm $\|\cdot\|$.

2.7 Inequalities Related to Unitarily Invariant Norms and Numerical Radius

Proof Consider

$$T := \begin{bmatrix} 0 & 0 & A & 0 \\ 0 & 0 & 0 & B \\ A^* & 0 & 0 & 0 \\ 0 & B^* & 0 & 0 \end{bmatrix} \quad \text{and} \quad S := \begin{bmatrix} 0 & X & 0 & 0 \\ X^* & 0 & 0 & 0 \\ 0 & 0 & 0 & 0 \\ 0 & 0 & 0 & 0 \end{bmatrix}$$

as operators in $\mathbb{B}(\mathscr{H} \oplus \mathscr{H} \oplus \mathscr{H} \oplus \mathscr{H})$. Since T and S are self-adjoint, the operator matrix TST is self-adjoint, and hence we can utilize Lemma 2.7.3 to get

$$\|T^2 S + ST^2\| = 2 \|\mathrm{Re}(T^2 S)\| \geq 2 \|TST\|.$$

Some computations yield that

$$T^2 S + ST^2 = \begin{bmatrix} 0 & AA^*X + XBB^* & 0 & 0 \\ X^*AA^* + BB^*X^* & 0 & 0 & 0 \\ 0 & 0 & 0 & 0 \\ 0 & 0 & 0 & 0 \end{bmatrix}$$

and

$$TST = \begin{bmatrix} 0 & 0 & 0 & 0 \\ 0 & 0 & 0 & 0 \\ 0 & 0 & 0 & A^*XB \\ 0 & 0 & B^*X^*A & 0 \end{bmatrix}.$$

It follows from the equivalence of inequalities (1.3.1), (1.3.2), and (1.3.3) that

$$\left\| \begin{bmatrix} 0 & AA^*X + XBB^* \\ X^*AA^* + BB^*X^* & 0 \end{bmatrix} \right\| \geq 2 \left\| \begin{bmatrix} 0 & A^*XB \\ B^*X^*A & 0 \end{bmatrix} \right\|.$$

We have

$$\left\| \begin{bmatrix} AA^*X + XBB^* & 0 \\ 0 & X^*AA^* + BB^*X^* \end{bmatrix} \right\|$$

$$= \left\| \begin{bmatrix} AA^*X + XBB^* & 0 \\ 0 & X^*AA^* + BB^*X^* \end{bmatrix} \begin{bmatrix} 0 & I \\ I & 0 \end{bmatrix} \right\|$$

$$= \left\| \begin{bmatrix} 0 & AA^*X + XBB^* \\ X^*AA^* + BB^*X^* & 0 \end{bmatrix} \right\|$$

$$\geq 2 \left\| \begin{bmatrix} 0 & A^*XB \\ B^*X^*A & 0 \end{bmatrix} \right\|$$

$$= 2 \left\| \begin{bmatrix} 0 & A^*XB \\ B^*X^*A & 0 \end{bmatrix} \begin{bmatrix} 0 & I \\ I & 0 \end{bmatrix} \right\|$$

$$= 2 \left\| \begin{bmatrix} A^*XB & 0 \\ 0 & B^*X^*A \end{bmatrix} \right\|.$$

Hence,
$$\left\|(AA^*X + XBB^*) \oplus (X^*AA^* + BB^*X^*)\right\| \geq 2\left\|A^*XB \oplus B^*X^*A\right\|.$$

It follows from (1.3.4) that
$$\left\|(AA^*X + XBB^*) \oplus (AA^*X + XBB^*)\right\| \geq 2\left\|A^*XB \oplus A^*XB\right\|.$$

This inequality, together with the equivalence (1.3.1) and (1.3.3), ensures that
$$\left\|AA^*X + XBB^*\right\| \geq 2\left\|A^*XB\right\|.$$

□

The following theorem gives an upper bound for the norm of a *commutator* $[X, A] = XA - AX$ due to Kittaneh [138].

Theorem 2.7.5 *If $A, B \in \mathbb{M}_n$ are positive and $X \in \mathbb{M}_n$, then*
$$\|AX - XB\| \leq \max\{\|A\|, \|B\|\} \, \|X\|. \tag{2.7.3}$$

Furthermore, the inequality is sharp.

Proof First, assume that A is a positive contraction. We show that
$$\|AX - XA\| \leq \|X\|.$$

From dilation theory, $P = \begin{bmatrix} A & (A - A^2)^{1/2} \\ (A - A^2)^{1/2} & I - A \end{bmatrix} \in \mathbb{B}(\mathcal{H} \oplus \mathcal{H})$ is a projection. Let $Y = \begin{bmatrix} X & 0 \\ 0 & 0 \end{bmatrix}, Q = \begin{bmatrix} I & 0 \\ 0 & 0 \end{bmatrix}$, and $R = 2P - I$. Then R is a unitary operator and $PY - YP = \frac{1}{2}(RY - YR)$. Then

$$\left\| \begin{bmatrix} AX - XA & 0 \\ 0 & 0 \end{bmatrix} \right\| = \left\| \begin{bmatrix} I & 0 \\ 0 & 0 \end{bmatrix} \begin{bmatrix} AX - XA & -X(A - A^2)^{1/2} \\ (A - A^2)^{1/2}X & 0 \end{bmatrix} \begin{bmatrix} I & 0 \\ 0 & 0 \end{bmatrix} \right\|$$
$$= \|Q(PY - YP)Q\|$$
$$\leq \|Q\| \, \|PY - YP\| \, \|Q\| \qquad \text{(by Lemma 1.3.1)}$$
$$= \|PY - YP\|$$
$$= \frac{1}{2}\|RY - YR\| \leq \frac{1}{2}(\|RY\| + \|YR\|)$$
$$\leq \frac{1}{2}(\|R\| \, \|Y\| + \|R\| \, \|Y\|) = \|Y\|$$
$$= \left\| \begin{bmatrix} X & 0 \\ 0 & 0 \end{bmatrix} \right\|.$$

2.7 Inequalities Related to Unitarily Invariant Norms and Numerical Radius

We conclude from the equivalence of inequalities (1.3.2) and (1.3.1) that

$$\|AX - XA\| \leq \|X\|. \qquad (2.7.4)$$

Replace A in (2.7.4) with the contraction $A/\|A\|$ for an arbitrary nonzero positive operator to get

$$\|AX - XA\| \leq \|A\| \, \|X\|. \qquad (2.7.5)$$

Second, put $C = \begin{bmatrix} A & 0 \\ 0 & B \end{bmatrix} \geq 0$ and $Z = \begin{bmatrix} 0 & X \\ 0 & 0 \end{bmatrix}$. Then $\|C\| = \max\{\|A\|, \|B\|\}$ and

$$\|Z\| = \left\| \begin{bmatrix} 0 & X \\ 0 & 0 \end{bmatrix} \right\| = \left\| \begin{bmatrix} 0 & X \\ 0 & 0 \end{bmatrix} \begin{bmatrix} 0 & I \\ I & 0 \end{bmatrix} \right\| = \left\| \begin{bmatrix} X & 0 \\ 0 & 0 \end{bmatrix} \right\|.$$

Thus,

$$\left\| \begin{bmatrix} AX - XB & 0 \\ 0 & 0 \end{bmatrix} \right\| = \|CZ - ZC\|$$

$$\leq \|C\| \, \|Z\| \qquad \text{(by inequality (2.7.5))}$$

$$= \max\{\|A\|, \|B\|\} \, \|X \oplus 0\|.$$

To show that the inequality is sharp it is enough to consider $A = B = \begin{bmatrix} 0 & 0 \\ 0 & 1 \end{bmatrix}$ and $X = \begin{bmatrix} 0 & 0 \\ 1 & 0 \end{bmatrix}$. Then equality holds in (2.7.3) for the operator norm. □

Bhatia and Kittaneh [25] showed that if A and B are positive operators and m is any positive integer, then

$$\|A^m + B^m\| \leq \|(A + B)^m\|, \qquad (2.7.6)$$

and then extended it for the operator norm as follows.

Theorem 2.7.6 *Let A and B be positive operators. Then*

$$\|A^r + B^r\| \leq \|(A + B)^r\| \qquad (2.7.7)$$

for all $1 \leq r < \infty$. For $0 \leq r \leq 1$, the reverse inequality holds.

Proof Assume that $1 \leq r < \infty$. Let m be any positive integer and let Ω_m be the set of all real numbers r with $1 \leq r < \infty$ for which inequality (2.7.7) holds. It follows from (2.7.6) that $1, m \in \Omega_m$, so it is enough to show that Ω_m is convex. Assume that $r, s \in \Omega_m$. Set $t = (r + s)/2$. We have

$$\|A^t + B^t\| = \left\|\begin{bmatrix} A^t + B^t & 0 \\ 0 & 0 \end{bmatrix}\right\| = \left\|\begin{bmatrix} A^{r/2} & B^{r/2} \\ 0 & 0 \end{bmatrix}\begin{bmatrix} A^{s/2} & 0 \\ B^{s/2} & 0 \end{bmatrix}\right\|$$

$$\leq \left\|\begin{bmatrix} A^{r/2} & B^{r/2} \\ 0 & 0 \end{bmatrix}\right\| \left\|\begin{bmatrix} A^{s/2} & 0 \\ B^{s/2} & 0 \end{bmatrix}\right\|$$

$$= \|A^r + B^r\|^{1/2}\|A^s + B^s\|^{1/2} \quad \text{(by (1.1.2))}$$

$$\leq \|A+B\|^{r/2}\|A+B\|^{s/2} \quad (\text{since } r, s \in \Omega_m)$$

$$= \|A+B\|^t = \|(A+B)^t\|,$$

from which we deduce that $t \in \Omega_m$. Thus Ω_m is convex.

Assuming that $0 \leq r \leq 1$, we can replace r with $1/r$ in addition to replacing A and B with A^r and B^r respectively in (2.7.7). This yields the desired inequality. □

Continuing our work, we discuss some inequalities related to the numerical radius. Numerous mathematicians have applied techniques with block matrices to get inequalities involving numerical radius; see [31] and the references therein.

Theorem 2.7.7 ([79]) *Let $A, B, C, D \in \mathbb{B}(\mathcal{H})$. Then the following results hold:*

(i) $w\left(\begin{bmatrix} A & 0 \\ 0 & B \end{bmatrix}\right) = \max\{w(A), w(B)\}$.

(ii) $w\left(\begin{bmatrix} A & 0 \\ 0 & B \end{bmatrix}\right) \leq w\left(\begin{bmatrix} A & C \\ D & B \end{bmatrix}\right)$ *and* $w\left(\begin{bmatrix} 0 & C \\ D & 0 \end{bmatrix}\right) \leq w\left(\begin{bmatrix} A & C \\ D & B \end{bmatrix}\right)$.

(iii) $w\left(\begin{bmatrix} 0 & D \\ C & 0 \end{bmatrix}\right) = w\left(\begin{bmatrix} 0 & C \\ D & 0 \end{bmatrix}\right)$.

(iv) $w\left(\begin{bmatrix} 0 & C \\ D & 0 \end{bmatrix}\right) = \frac{1}{2}\sup_{\theta \in \mathbb{R}} \|e^{i\theta}C + e^{-i\theta}D^*\|$.

(v) *For any $\theta \in \mathbb{R}$*,

$$w\left(\begin{bmatrix} 0 & A \\ e^{i\theta}B & 0 \end{bmatrix}\right) = w\left(\begin{bmatrix} 0 & A \\ B & 0 \end{bmatrix}\right). \tag{2.7.8}$$

(vi) $w\left(\begin{bmatrix} A & B \\ B & A \end{bmatrix}\right) = \max\{w(A+B), w(A-B)\}$

(vii) $w\left(\begin{bmatrix} 0 & B \\ B & 0 \end{bmatrix}\right) = w(B)$

2.7 Inequalities Related to Unitarily Invariant Norms and Numerical Radius

Proof (i) It follows from (1.1.5) that

$$w\left(\begin{bmatrix} A & 0 \\ 0 & B \end{bmatrix}\right) = \sup_{\theta \in \mathbb{R}} \left\| \operatorname{Re}\left(e^{i\theta}\begin{bmatrix} A & 0 \\ 0 & B \end{bmatrix}\right) \right\| = \sup_{\theta \in \mathbb{R}} \left\| \begin{bmatrix} \operatorname{Re}(e^{i\theta}A) & 0 \\ 0 & \operatorname{Re}(e^{i\theta}B) \end{bmatrix} \right\|$$

$$= \sup_{\theta \in \mathbb{R}} \max\left\{ \left\| \operatorname{Re}(e^{i\theta}A) \right\|, \left\| \operatorname{Re}(e^{i\theta}B) \right\| \right\}$$

$$= \max\left\{ \sup_{\theta \in \mathbb{R}} \left\| \operatorname{Re}(e^{i\theta}A) \right\|, \sup_{\theta \in \mathbb{R}} \left\| \operatorname{Re}(e^{i\theta}B) \right\| \right\}$$

$$= \max\{w(A), w(B)\}.$$

(ii) Let $U = \begin{bmatrix} I & 0 \\ 0 & -I \end{bmatrix}$ and $T = \begin{bmatrix} A & C \\ D & B \end{bmatrix}$. Then $\begin{bmatrix} A & 0 \\ 0 & B \end{bmatrix} = \frac{T + U^*TU}{2}$. Therefore,

$$w\left(\begin{bmatrix} A & 0 \\ 0 & B \end{bmatrix}\right) \leq \frac{w(T) + w(U^*TU)}{2} = w(T).$$

For the second inequality, it is sufficient to use the identity $\begin{bmatrix} 0 & C \\ D & 0 \end{bmatrix} = \frac{T - U^*TU}{2}$.

(iii) Utilizing the unitary operator $U = \begin{bmatrix} 0 & I \\ I & 0 \end{bmatrix}$, we arrive at

$$w\left(\begin{bmatrix} 0 & D \\ C & 0 \end{bmatrix}\right) = w\left(\begin{bmatrix} 0 & I \\ I & 0 \end{bmatrix}\begin{bmatrix} 0 & C \\ D & 0 \end{bmatrix}\begin{bmatrix} 0 & I \\ I & 0 \end{bmatrix}\right) = w\left(\begin{bmatrix} 0 & C \\ D & 0 \end{bmatrix}\right).$$

(iv) Employing the fact that

$$\left\| \begin{bmatrix} 0 & X \\ X^* & 0 \end{bmatrix} \right\| = \left\| \begin{bmatrix} 0 & X \\ X^* & 0 \end{bmatrix}\begin{bmatrix} 0 & X \\ X^* & 0 \end{bmatrix} \right\|^{\frac{1}{2}} = \left\| \begin{bmatrix} X^*X & 0 \\ 0 & XX^* \end{bmatrix} \right\|^{\frac{1}{2}}$$

$$= \max\{\|X^*X\|, \|XX^*\|\}^{\frac{1}{2}} = \|X\|$$

and (1.1.5), we infer that

$$w\left(\begin{bmatrix} 0 & C \\ D & 0 \end{bmatrix}\right) = \frac{1}{2}\sup_{\theta \in \mathbb{R}} \left\| e^{i\theta}\begin{bmatrix} 0 & C \\ D & 0 \end{bmatrix} + e^{-i\theta}\begin{bmatrix} 0 & D^* \\ C^* & 0 \end{bmatrix} \right\|$$

$$= \frac{1}{2}\sup_{\theta \in \mathbb{R}} \left\| \begin{bmatrix} 0 & e^{i\theta}C + e^{-i\theta}D^* \\ e^{-i\theta}C^* + e^{i\theta}D & 0 \end{bmatrix} \right\|$$

$$= \frac{1}{2}\sup_{\theta \in \mathbb{R}} \|e^{i\theta}C + e^{-i\theta}D^*\|.$$

(v) It is deduced from the weakly unitarily invariant property $w(X) = w(U^*XU)$ for $X = \begin{bmatrix} 0 & A \\ B & 0 \end{bmatrix}$ and the unitary $U = \begin{bmatrix} I & 0 \\ 0 & e^{i\theta/2}I \end{bmatrix}$

(vi) It is enough to use the property $w(X) = w(U^*XU)$ for $X = \begin{bmatrix} A & B \\ B & A \end{bmatrix}$ and the unitary $U = \frac{1}{\sqrt{2}} \begin{bmatrix} I & I \\ -I & I \end{bmatrix}$

(vii) It follows from (vi) with $A = 0$.

□

As a corollary, we deduce the following result.

Corollary 2.7.8 ([79]) *If $A, X \in \mathbb{B}(\mathcal{H})$, then*

$$w(AX + X^*A) \leq 2\|X\|w(A).$$

Proof Let U be a unitary. Then

$$V = \frac{1}{2}\begin{bmatrix} I + U & I - U \\ I - U & I + U \end{bmatrix}$$

is also unitary and $V^*(A \oplus (-A))V = \begin{bmatrix} W' & X' \\ Y' & Z' \end{bmatrix}$, where $W' = \frac{1}{2}(AU + U^*A)$ and $Z' = -W'$. It follows from Theorem 2.7.7(ii), (v) and (vii) that

$$w(AU + U^*A) \leq 2w(V^*(A \oplus (-A))V) = 2w(A \oplus (-A)) = 2w(A).$$

We can assume that $\|X\| < 1$. Using Theorem 1.2.8 we can write $X = \sum_{i=1}^n \alpha_i U_i$ for some n, where $\sum_{i=1}^n \alpha_i = 1$, $\alpha_i \geq 0$, and U_i's are unitary operators. Therefore,

$$w(AX + X^*A) \leq \sum_{i=1}^n \alpha_i w(AU_i + U_i^*A) \leq 2\sum_{i=1}^n \alpha_i w(A) = 2w(A).$$

□

The next theorem and its two subsequent corollaries are due to Hirzallah et al. [119].

Theorem 2.7.9 *Suppose that $A, B \in \mathbb{B}(\mathcal{H})$. Then,*

$$\frac{\max\{w(A + B), w(A - B)\}}{2} \leq w\left(\begin{bmatrix} 0 & A \\ B & 0 \end{bmatrix}\right) \leq \frac{w(A + B) + w(A - B)}{2}. \quad (2.7.9)$$

2.7 Inequalities Related to Unitarily Invariant Norms and Numerical Radius

Proof To establish the theorem, we frequently use Theorem 2.7.7. We have

$$w(A+B) = w\left(\begin{bmatrix} 0 & A+B \\ A+B & 0 \end{bmatrix}\right)$$

$$= w\left(\begin{bmatrix} 0 & A \\ B & 0 \end{bmatrix} + \begin{bmatrix} 0 & B \\ A & 0 \end{bmatrix}\right)$$

$$\leq w\left(\begin{bmatrix} 0 & A \\ B & 0 \end{bmatrix}\right) + w\left(\begin{bmatrix} 0 & B \\ A & 0 \end{bmatrix}\right)$$

$$= 2w\left(\begin{bmatrix} 0 & A \\ B & 0 \end{bmatrix}\right),$$

whence

$$\frac{w(A+B)}{2} \leq w\left(\begin{bmatrix} 0 & A \\ B & 0 \end{bmatrix}\right). \quad (2.7.10)$$

Replacing B with $-B$ in (2.7.10), we get

$$\frac{w(A-B)}{2} \leq w\left(\begin{bmatrix} 0 & A \\ -B & 0 \end{bmatrix}\right) = w\left(\begin{bmatrix} 0 & A \\ B & 0 \end{bmatrix}\right). \quad (2.7.11)$$

The left side of inequality (2.7.9) is deduced from (2.7.10) and (2.7.11). To prove the right-hand side of (2.7.9), consider the unitary operator $U = \frac{1}{\sqrt{2}}\begin{bmatrix} I & I \\ -I & I \end{bmatrix}$. We have

$$w\left(\begin{bmatrix} 0 & A \\ B & 0 \end{bmatrix}\right) = w\left(U^*\begin{bmatrix} 0 & A \\ B & 0 \end{bmatrix}U\right) = \frac{1}{2}w\left(\begin{bmatrix} A+B & A-B \\ -(A-B) & -(A+B) \end{bmatrix}\right)$$

$$= \frac{1}{2}w\left(\begin{bmatrix} A+B & 0 \\ 0 & -(A+B) \end{bmatrix} + \begin{bmatrix} 0 & A-B \\ -(A-B) & 0 \end{bmatrix}\right)$$

$$\leq \frac{1}{2}\left(w\left(\begin{bmatrix} A+B & 0 \\ 0 & -(A+B) \end{bmatrix}\right) + w\left(\begin{bmatrix} 0 & A-B \\ -(A-B) & 0 \end{bmatrix}\right)\right)$$

$$= \frac{w(A+B) + w(A-B)}{2}.$$

$\square$

Corollary 2.7.10 *Let $T = A + iB$ be the Cartesian decomposition of $T \in \mathbb{B}(\mathcal{H})$. Then, for any $\theta \in \mathbb{R}$*

$$\frac{w(T)}{2} \leq w\left(\begin{bmatrix} 0 & A \\ e^{i\theta}B & 0 \end{bmatrix}\right) \leq w(T).$$

Proof Replacing B with iB in (2.7.9), we arrive at

$$\frac{\max\{w(A+iB), w(A-iB)\}}{2} \leq w\left(\begin{bmatrix} 0 & A \\ iB & 0 \end{bmatrix}\right) \leq \frac{w(A+iB)+w(A-iB)}{2}.$$

It follows from Theorem 2.7.7 (v) that

$$\frac{w(T)}{2} = \frac{\max\{w(T), w(T^*)\}}{2} \leq w\left(\begin{bmatrix} 0 & A \\ e^{i\theta}B & 0 \end{bmatrix}\right) \leq \frac{w(T)+w(T^*)}{2} = w(T).$$

$\square$

Corollary 2.7.11 *Let $A, B, X \in \mathbb{B}(\mathcal{H})$ such that A and B are self-adjoint. Then*

$$\|X\| \leq w(X+A) + w(X+iB). \tag{2.7.12}$$

Proof Employing Theorems 2.7.9 and 2.7.7(iv) to arbitrary operators $S, R \in \mathbb{B}(\mathcal{H})$, we get

$$\frac{\|S+R^*\|}{2} \leq w\left(\begin{bmatrix} 0 & S \\ R & 0 \end{bmatrix}\right) \leq \frac{w(S+R)+w(S-R)}{2},$$

whence

$$\|S+R^*\| \leq w(S+R) + w(S-R).$$

Letting $S = X + (A+iB)/2$ and $R = (-A+iB)/2$, we reach the desired inequality. $\square$

2.8 Some Inequalities Involving Eigenvalues and Singular Values

The arithmetic-geometric (A-G) mean inequality asserts that

$$\frac{a+b}{2} \geq \sqrt{ab}$$

for any $a, b > 0$.

A noncommutative version of this inequality for $n \times n$ positive definite matrices A and B is

$$A \nabla B \geq A \sharp B, \tag{2.8.1}$$

where $A \nabla B := (A+B)/2$ and $A \sharp B := A^{\frac{1}{2}}(A^{-\frac{1}{2}}BA^{-\frac{1}{2}})^{\frac{1}{2}}A^{\frac{1}{2}}$ are called the *arithmetic mean* and the *geometric mean* of A and B, respectively. Applying the functional

2.8 Some Inequalities Involving Eigenvalues and Singular Values

calculus for the positive semidefinite matrix $B^{-1/2}AB$ and the numerical arithmetic-geometric mean inequality $t^{1/2} \leq \frac{1+t}{2}$ for $t > 0$, we arrive at

$$(A^{-1/2}BA^{-1/2})^{1/2} \leq \frac{1 + A^{-1/2}BA^{-1/2}}{2}.$$

Multiplying both sides of the inequality by $A^{1/2}$, we get

$$A \sharp B = A^{1/2}(A^{-1/2}BA^{-1/2})^{1/2}A^{1/2} \leq \frac{A+B}{2} = A\nabla B.$$

The singular value form of the arithmetic-geometric inequality (2.8.1) for matrices $A, B \in \mathbb{M}_n$ is

$$2s_j(AB^*) \leq s_j(A^*A + B^*B) \quad (j = 1, \ldots, n), \tag{2.8.2}$$

as shown in [24]. In addition, if $S := A \sharp B$, then we have $B = SA^{-1}S$. Therefore, a variant of (2.8.1) is as follows

$$A + SA^{-1}S \geq 2S, \tag{2.8.3}$$

where A and S are $n \times n$ positive definite matrices. It follows from the *Weyl monotonicity principle* (1.3.6) that (2.8.1) ensures that

$$\lambda_j(A + B) \geq 2\lambda_j(A \sharp B), \quad j = 1, \ldots, n.$$

Bhatia and Kittaneh [27] proved that if A and B are $n \times n$ positive semidefinite matrices, then

$$\lambda_j(A + B) \geq 2\sqrt{\lambda_j(AB)} = 2s_j(A^{\frac{1}{2}}B^{\frac{1}{2}}) \tag{2.8.4}$$

for $j = 1, \ldots, n$.
They asked "is it true"

$$\lambda_j(A + B) \geq 2\sqrt{s_j(AB)}, \quad j = 1, \ldots, n?$$

This question was affirmatively answered by Drury [67]. Lin [154] revisit Drury's solution by simplifying a proof.

Theorem 2.8.1 ([67]) *If A and B are $n \times n$ positive semidefinite matrices, then*

$$\lambda_j(A + B) \geq 2\sqrt{s_j(AB)}, \quad j = 1, \ldots, n. \tag{2.8.5}$$

Without loss of generality, let us assume that A and B are positive definite (the general case can be proved using a standard perturbation argument). Let us fix r in the range $1 \leq r \leq n$. By normalization, we can assume that $s_r(AB) = 1$. We aim to show that $\lambda_r(A+B) \geq 2$.

Note that $s_r(AB) = 1$ is the same as $\lambda_r(AB^2A) = 1$. Consider the spectral representation

$$AB^2A = \sum_{k=1}^{n} \lambda_k(AB^2A) P_k,$$

where $P_1, P_2, \ldots, P_n$ are pairwise orthogonal projections and $\sum_{k=1}^{n} P_k = I$. Then, $\lambda_k(AB^2A) \geq 1$ for $k = 1, \ldots, r$. Define a positive semidefinite matrix

$$B_1 := \left(A^{-1} \left(\sum_{k=1}^{r} P_k \right) A^{-1} \right)^{1/2}.$$

It is easy to see (indeed, from $B^2 \geq B_1^2$) that

$$B = \left(A^{-1} \left(\sum_{k=1}^{r} \lambda_k(AB^2A) P_k \right) A^{-1} \right)^{1/2} \geq B_1.$$

So we are done if we can show

$$\lambda_r(A + B_1) \geq 2. \qquad (2.8.6)$$

As B_1 has rank r, we can split the underlying space as the direct sum of the image and the kernel of B_1. We can then partition B_1 and A conformally in the following form

$$B_1 = \begin{bmatrix} X & 0 \\ 0 & 0 \end{bmatrix}, \quad A = \begin{bmatrix} A_{11} & A_{12} \\ A_{12}^* & A_{22} \end{bmatrix}.$$

Note that AB_1^2A is a projection of rank r, and the same is true for $B_1A^2B_1$. Therefore, we have

$$B_1A^2B_1 = \begin{bmatrix} X(A_{11}^2 + A_{12}A_{12}^*)X & 0 \\ 0 & 0 \end{bmatrix} \text{ implies } X(A_{11}^2 + A_{12}A_{12}^*)X = I_r,$$

where I_r is the $r \times r$ identity matrix.

Finally, observe that

$$A \geq A_1 := \begin{bmatrix} A_{11} & A_{12} \\ A_{12}^* & A_{12}^* A_{11}^{-1} A_{12} \end{bmatrix}.$$

Therefore, (2.8.6) follows from

$$\lambda_r(A_1 + B_1) \geq 2. \qquad (2.8.7)$$

Thus, the remaining effort is made to show (2.8.7), which we formulate as a proposition.

2.8 Some Inequalities Involving Eigenvalues and Singular Values

Proposition 2.8.2 *Let A_{11} and X be $r \times r$ positive definite matrices and let A_{12} be an $(n-r) \times (n-r)$ matrix such that $X(A_{11}^2 + A_{12}A_{12}^*)X = I_r$. Then,*

$$\lambda_r \begin{bmatrix} A_{11} + X & A_{12} \\ A_{12}^* & A_{12}^* A_{11}^{-1} A_{12} \end{bmatrix} \geq 2. \tag{2.8.8}$$

To prove this proposition, we need the following lemma.

Lemma 2.8.3 *Let X be an $r \times r$ positive definite matrix, and let S be an $r \times r$ invertible matrix. Then*

$$\lambda_r \begin{bmatrix} SX^{-1}S^* & (S^{-1})^* \\ S^{-1} & X \end{bmatrix} \geq 2.$$

Proof Consider the polar decomposition $S = U|S|$ of S, where U is unitary and $|S| = (S^*S)^{\frac{1}{2}}$. The matrix $\begin{bmatrix} SX^{-1}S^* & (S^{-1})^* \\ S^{-1} & X \end{bmatrix}$ is unitarily similar to

$$\begin{bmatrix} U^*SX^{-1}S^*U & U^*(S^{-1})^* \\ S^{-1}U & X \end{bmatrix} = \begin{bmatrix} |S|X^{-1}|S| & |S|^{-1} \\ |S|^{-1} & X \end{bmatrix}.$$

As $P := \frac{1}{\sqrt{2}} \begin{bmatrix} I_r & 0 \\ I_r & 0 \end{bmatrix}$ is a partial isometry, we have

$$\lambda_r \begin{bmatrix} SX^{-1}S^* & (S^{-1})^* \\ S^{-1} & X \end{bmatrix} = \lambda_r \begin{bmatrix} |S|X^{-1}|S| & |S|^{-1} \\ |S|^{-1} & X \end{bmatrix}$$

$$\geq \lambda_r \left(P^* \begin{bmatrix} |S|X^{-1}|S| & |S|^{-1} \\ |S|^{-1} & X \end{bmatrix} P \right)$$

$$= \lambda_r \left(\frac{X + |S|X^{-1}|S|}{2} + |S|^{-1} \right)$$

$$\geq \lambda_r(|S| + |S|^{-1}) \geq 2. \quad \text{by (2.8.3)}$$

Thus, the required result is obtained. □

We are ready to prove Proposition 2.8.2.

Proof of Proposition 2.8.2

Consider the factorization

$$\begin{bmatrix} A_{11} + X & A_{12} \\ A_{12}^* & A_{12}^* A_{11}^{-1} A_{12} \end{bmatrix} = \begin{bmatrix} A_{11}^{\frac{1}{2}} & X^{\frac{1}{2}} \\ A_{12}^* A_{11}^{-\frac{1}{2}} & 0 \end{bmatrix} \begin{bmatrix} A_{11}^{\frac{1}{2}} & X^{\frac{1}{2}} \\ A_{12}^* A_{11}^{-\frac{1}{2}} & 0 \end{bmatrix}^*.$$

It is easy to see that $\begin{bmatrix} A_{11} + X & A_{12} \\ A_{12}^* & A_{12}^* A_{11}^{-1} A_{12} \end{bmatrix}$ is unitarily similar to

$$\begin{bmatrix} A_{11}^{\frac{1}{2}} & X^{\frac{1}{2}} \\ A_{12}^* A_{11}^{-\frac{1}{2}} & 0 \end{bmatrix}^* \begin{bmatrix} A_{11}^{\frac{1}{2}} & X^{\frac{1}{2}} \\ A_{12}^* A_{11}^{-\frac{1}{2}} & 0 \end{bmatrix} = \begin{bmatrix} A_{11} + A_{11}^{-\frac{1}{2}} A_{12} A_{12}^* A_{11}^{-\frac{1}{2}} & A_{11}^{\frac{1}{2}} X^{\frac{1}{2}} \\ X^{\frac{1}{2}} A_{11}^{\frac{1}{2}} & X \end{bmatrix}$$

$$= \begin{bmatrix} A_{11}^{-\frac{1}{2}} X^{-2} A_{11}^{-\frac{1}{2}} & A_{11}^{\frac{1}{2}} X^{\frac{1}{2}} \\ X^{\frac{1}{2}} A_{11}^{\frac{1}{2}} & X \end{bmatrix}.$$

Now, setting $S = A_{11}^{-\frac{1}{2}} X^{-\frac{1}{2}}$ in Lemma 2.8.3 yields the desired result. □

The *Hua matrix* is given by

$$H = \begin{bmatrix} (I - A^*A)^{-1} & (I - B^*A)^{-1} \\ (I - A^*B)^{-1} & (I - B^*B)^{-1} \end{bmatrix},$$

where A and B are strictly contractions in the sense that $A^*A < I$ and $B^*B < I$. This block matrix first appeared in Hua's study of the theory of functions of several complex variables; see [126]. The positivity of the Hua matrix immediately leads to

$$|\det(I - A^*B)|^2 \geq \det(I - A^*A) \det(I - B^*B), \qquad (2.8.9)$$

which is known as *Hua's determinantal inequality* in the literature.

For any $X \in \mathbb{M}_n$, it is evident that $|\det X| = \prod_{j=1}^n s_j(X)$. Hua's determinantal inequality (2.8.9) has the following variants

$$\prod_{j=1}^n s_j^2(I - A^*B) \geq \prod_{j=1}^n \lambda_j\Big((I - A^*A)(I - B^*B)\Big)$$

$$= \prod_{j=1}^n s_j\Big((I - A^*A)(I - B^*B)\Big)$$

and

$$\prod_{j=1}^n s_j(I - A^*B) \geq \prod_{j=1}^n \lambda_j\Big((I - A^*A)\sharp(I - B^*B)\Big)$$

$$= \prod_{j=1}^n \lambda_j\Big((I - A^*A)^{1/2}(I - B^*B)^{1/2}\Big).$$

In [152, Proposition 3.1], Lin proved the following strengthening of Hua's determinantal inequality.

2.8 Some Inequalities Involving Eigenvalues and Singular Values

Proposition 2.8.4 *Let $A, B \in \mathbb{M}_n$ be contractions. Then,*

$$s_j^2(I - A^*B) \geq s_j\Big((I - A^*A)(I - B^*B)\Big)$$

for $j = 1, \ldots, n$

It is natural to ask whether other variants of Hua's determinantal inequality have similar strengthening. To affirmatively answer this question, we need the following result.

Theorem 2.8.5 ([153]) *Let $\begin{bmatrix} M_{11} & M_{12} \\ M_{12}^* & M_{22} \end{bmatrix} \in \mathbb{M}_{2n}$, with each block invertible, be positive semidefinite. If $M_{11}^{-1} + M_{22}^{-1} \leq M_{12}^{-1} + (M_{12}^*)^{-1}$, then for each $j = 1, \ldots, n$*

$$s_j^2(M_{12}) \leq \lambda_j(M_{11} M_{22}), \qquad (2.8.10)$$
$$s_j^2(M_{12}) \leq s_j(M_{11} M_{22}), \qquad (2.8.11)$$
$$s_j(M_{12}) \leq \lambda_j(M_{11} \sharp M_{22}), \qquad (2.8.12)$$
$$s_j(M_{12}) \leq \lambda_j(M_{11}^{1/2} M_{22}^{1/2}). \qquad (2.8.13)$$

Proof Based on a result of Fan and Hoffman [19, p. 73], we know that

$$2s_j(M_{12}^{-1}) \geq \lambda_j(M_{12}^{-1} + (M_{12}^*)^{-1}). \qquad (2.8.14)$$

On the other hand, according to a result by Bhatia and Kittaneh [19, p. 262], we have

$$\lambda_j(M_{11}^{-1} + M_{22}^{-1}) \geq 2s_j(M_{11}^{-1/2} M_{22}^{-1/2}) = 2\sqrt{\lambda_j(M_{11}^{-1} M_{22}^{-1})}. \qquad (2.8.15)$$

Combining (2.8.14) and (2.8.15) with $M_{11}^{-1} + M_{22}^{-1} \leq M_{12}^{-1} + (M_{12}^*)^{-1}$ gives

$$s_j^2(M_{12}^{-1}) \geq \lambda_j(M_{11}^{-1} M_{22}^{-1}).$$

This inequality is equivalent to (2.8.10) by utilizing the fact that

$$s_j(X^{-1}) = \frac{1}{s_{n-j+1}(X)}, \quad j = 1, \ldots, n,$$

where $X \in \mathbb{M}_n$ is invertible and also the fact that

$$\lambda_j(X^{-1}) = \frac{1}{\lambda_{n-j+1}(X)}, \quad j = 1, \ldots, n,$$

where $X \in \mathbb{M}_n$ is invertible and all its eigenvalues are real.

The proof of (2.8.11) is similar except that (2.8.15) should be replaced with

$$\lambda_j(M_{11}^{-1}+M_{22}^{-1}) \geq 2\sqrt{s_j(M_{11}^{-1}M_{22}^{-1})},$$

which was established by Drury [67]. This result was previously conjectured in [26]. As $M_{11}^{-1}+M_{22}^{-1} \geq 2M_{11}^{-1}\sharp M_{22}^{-1}$ implies

$$\lambda_j(M_{11}^{-1}+M_{22}^{-1}) \geq 2\lambda_j(M_{11}^{-1}\sharp M_{22}^{-1}),$$

inequality (2.8.12) can be similarly proved.

The proof of (2.8.13) is similar except that (2.8.15) should be replaced with

$$\lambda_j(M_{11}^{-1}+M_{22}^{-1}) \geq 2\lambda_j(M_{11}^{-1/2}M_{22}^{-1/2}),$$

which is given by [26, Eq. (3.12)]. □

This immediately yields

Corollary 2.8.6 *Let $A, B \in \mathbb{M}_n$ be contractions. Then for $j = 1, \ldots, n$*

$$s_j^2(I - A^*B) \geq \lambda_j\Big((I - A^*A)(I - B^*B)\Big),$$
$$s_j(I - A^*B) \geq \lambda_j\Big((I - A^*A)\sharp(I - B^*B)\Big),$$
$$s_j(I - A^*B) \geq \lambda_j\Big((I - A^*A)^{1/2}(I - B^*B)^{1/2}\Big).$$

We remark that none of (2.8.10), (2.8.11), (2.8.12), nor (2.8.13) is stronger than the other.

The following two results are due to Tao [218]. The first one incorporates certain ideas from [238], while the second one utilizes [237].

Theorem 2.8.7 *If $M = \begin{bmatrix} A & X \\ X^* & C \end{bmatrix} \in \mathbb{M}_2(\mathbb{M}_n)$ is positive, then*

$$2s_j(X) \leq s_j(M) \quad \text{for all } j = 1, \ldots, n.$$

Proof Let $D = \begin{bmatrix} 0 & X \\ X^* & 0 \end{bmatrix}$. We have

$$0 \leq \begin{bmatrix} I & 0 \\ 0 & -I \end{bmatrix}\begin{bmatrix} A & X \\ X^* & C \end{bmatrix}\begin{bmatrix} I & 0 \\ 0 & -I \end{bmatrix} = \begin{bmatrix} A & -X \\ -X^* & C \end{bmatrix} = \begin{bmatrix} A & X \\ X^* & C \end{bmatrix} - 2D.$$

Therefore, $2D \leq \begin{bmatrix} A & X \\ X^* & C \end{bmatrix}$. It follows from the Weyl monotonicity principle (1.3.6) that $2\lambda_j(D) \leq \lambda_j\begin{bmatrix} A & X \\ X^* & C \end{bmatrix}$ for all $1 \leq j \leq 2n$. Since the eigenvalues of D in decreasing order are

2.8 Some Inequalities Involving Eigenvalues and Singular Values

$$(s_1(X), \ldots, s_n(X), -s_n(X), \ldots, -s_1(X)),$$

we get the desired inequality. □

Theorem 2.8.8 *The following assertions are equivalent.*
(i) *The inequality [236]*

$$s_j(A - B) \leq s_j(A \oplus B) \quad \text{for all } j = 1, \ldots, n. \tag{2.8.16}$$

holds for all positive semidefinite matrices $A, B \in \mathbb{M}_n$.
(ii) *The arithmetic-geometric mean inequality due to Bhatia and Kittaneh [24]*

$$2s_j(AB^*) \leq s_j(A^*A + B^*B) \quad \text{for all } j = 1, \ldots, n. \tag{2.8.17}$$

holds for all $A, B \in \mathbb{M}_n$.
(iii) *If $M = \begin{bmatrix} A & X \\ X^* & C \end{bmatrix} \in \mathbb{M}_2(\mathbb{M}_n)$ is positive, then*

$$2s_j(X) \leq s_j(M) \quad \text{for all } j = 1, \ldots, n. \tag{2.8.18}$$

Proof (i) $\Longrightarrow$ (ii). Let $X = \begin{bmatrix} A \\ B \end{bmatrix}$ and $Y = \begin{bmatrix} A \\ -B \end{bmatrix}$. It follows from (i) that

$$2s_j \begin{bmatrix} BA^* & 0 \\ 0 & AB^* \end{bmatrix} = 2s_j \begin{bmatrix} 0 & AB^* \\ BA^* & 0 \end{bmatrix} = s_j(XX^* - YY^*)$$

$$\leq s_j \begin{bmatrix} XX^* & 0 \\ 0 & YY^* \end{bmatrix} = s_j \begin{bmatrix} X^*X & 0 \\ 0 & Y^*Y \end{bmatrix}$$

$$= s_j \begin{bmatrix} A^*A + B^*B & 0 \\ 0 & A^*A + B^*B \end{bmatrix}.$$

Hence, $2s_j(AB^*) \leq s_j(A^*A + B^*B)$ for each $1 \leq j \leq n$.

(ii) $\Longrightarrow$ (iii). Let $\begin{bmatrix} A & X \\ X^* & C \end{bmatrix} \geq 0$. From Theorem 2.4.15, there are matrices $S, T \in \mathbb{M}_{2n,2}$ such that $\begin{bmatrix} A & X \\ X^* & C \end{bmatrix} = \begin{bmatrix} S^*S & S^*T \\ T^*S & T^*T \end{bmatrix}$. The inequality (ii) holds also for rectangular matrices. Hence,

$$2s_j(X) = 2s_j(S^*T) \leq s_j(SS^* + TT^*)$$

$$= s_j \begin{bmatrix} SS^* + TT^* & 0 \\ 0 & 0 \end{bmatrix}$$

$$= s_j \left(\begin{bmatrix} S & T \\ 0 & 0 \end{bmatrix} \begin{bmatrix} S & T \\ 0 & 0 \end{bmatrix}^* \right) = s_j \left(\begin{bmatrix} S & T \\ 0 & 0 \end{bmatrix}^* \begin{bmatrix} S & T \\ 0 & 0 \end{bmatrix} \right)$$

$$= s_j \begin{bmatrix} S^*S & S^*T \\ T^*S & T^*T \end{bmatrix} = s_j \begin{bmatrix} A & X \\ X^* & C \end{bmatrix}.$$

(iii) $\Longrightarrow$ (i). Let $A, B \in \mathbb{M}_n$ be positive semidefinite matrices. We have

$$\frac{1}{\sqrt{2}} \begin{bmatrix} I & I \\ -I & I \end{bmatrix} \begin{bmatrix} \frac{A+B}{2} & \frac{A-B}{2} \\ \frac{A-B}{2} & \frac{A+B}{2} \end{bmatrix} \frac{1}{\sqrt{2}} \begin{bmatrix} I & -I \\ I & I \end{bmatrix} = \begin{bmatrix} A & 0 \\ 0 & B \end{bmatrix} \geq 0.$$

From (iii), we have

$$s_j(A - B) \leq s_j \begin{bmatrix} \frac{A+B}{2} & \frac{A-B}{2} \\ \frac{A-B}{2} & \frac{A+B}{2} \end{bmatrix} = s_j \begin{bmatrix} A & 0 \\ 0 & B \end{bmatrix}$$

for all $1 \leq j \leq n$. □

2.9 Exercises and Problems

Exercise 2.9.1 If A, B, and $B - A$ are nonnegative matrices, then $\|A\| \leq \|B\|$.
Hint: Prove $A^*A \leq B^*B$.

Exercise 2.9.2 Show that if A is an invertible accretive operator with the Cartesian decomposition $A = B + iC$, then A^{-1} is accretive.
Hint: Show that $\text{Re}(A^{-1}) = (B + CB^{-1}C)^{-1}$.

Exercise 2.9.3 [107, Problem 70] Let $A, B, C, D \in \mathbb{B}(\mathcal{H})$ be pairwise commuting operators. Show that the operator matrix

$$\begin{bmatrix} A & B \\ C & D \end{bmatrix}$$

is invertible if and only if $AD - BC$ is invertible.
Hint: Refer to Sect. 2.2.

Exercise 2.9.4 [107, Problem 71] Let $A, B, C, D \in \mathbb{B}(\mathcal{H})$ be such that C and D commute and D is invertible. Prove that the operator matrix

$$\begin{bmatrix} A & B \\ C & D \end{bmatrix}$$

2.9 Exercises and Problems

is invertible if and only if $AD - BC$ is invertible.

Hint: See Sect. 2.2.

Exercise 2.9.5 Suppose that the matrix A is invertible. Show that the block matrix

$$\begin{bmatrix} A & B \\ C & D \end{bmatrix}$$

is invertible if and only if the Schur complement $D - CA^{-1}B$ of A is invertible.

Hint: Refer to Sect. 2.2.

Exercise 2.9.6 Suppose that the matrix

$$\begin{bmatrix} A & B \\ B^* & C \end{bmatrix}$$

is invertible and positive. Show that

$$\begin{bmatrix} A & B \\ B^* & C \end{bmatrix}^{-1} = \begin{bmatrix} (A - BC^{-1}B^*)^{-1} & A^{-1}B(B^*A^{-1}B - C)^{-1} \\ (B^*A^{-1}B - C)^{-1}B^*A^{-1} & (C - B^*A^{-1}B)^{-1} \end{bmatrix}.$$

Hint: Use the fact that the inverse of a positive definite matrix and its principal submatrices are positive definite; see also [129].

Exercise 2.9.7 Given matrices $A, B, D \in \mathbb{M}_n$, find all matrices $C \in \mathbb{M}_n$ such that

$$\begin{bmatrix} A & B \\ C & D \end{bmatrix}$$

is invertible.

Hint: Refer to Sect. 2.2.

Exercise 2.9.8 Prove that the block matrix $M = \begin{bmatrix} A & B \\ 0 & 0 \end{bmatrix}$ is Moore–Penrose invertible if and only if ran A + ran B is closed. If this is the case, then

$$M^\dagger = \begin{bmatrix} A^*(AA^* + BB^*)^\dagger & 0 \\ B^*(AA^* + BB^*)^\dagger & 0 \end{bmatrix}$$

Hint: First show that ran M = (ran A + ran B) $\oplus 0$ = ran$(AA^* + BB^*)^{1/2} \oplus 0$ and ran$(AA^* + BB^*)^{1/2}$ = ran$(AA^* + BB^*)$. For more details, refer to [60, Lemma 3].

Exercise 2.9.9 (i) Find an explicit expression for the Moore–Penrose inverse of the block matrix $M = \begin{bmatrix} A & B \\ C & D \end{bmatrix}$ in terms of $A, B, C, D \in \mathbb{M}_n$.

(ii) Find conditions under which M is Moore–Penrose invertible when $A, B, C, D \in \mathbb{B}(\mathcal{H})$ and then find a formula for $M^\dagger$.

It is remarkable that Part (i) with Drazin inverse instead of Moore–Penrose inverse has been a longstanding open problem, see [52, Chap. 5].

Hint: For (i), see [165]. For (ii) see [60].

Exercise 2.9.10 Under some suitable assumptions, find a formula for the Drazin inverse of the block matrix $\begin{bmatrix} A & B \\ C & D \end{bmatrix}$ in terms of $A, B, C, D \in \mathbb{M}_n$.

Hint: See [51, 63, 165].

Exercise 2.9.11 An operator $T \in \mathbb{B}(\mathcal{H})$ is called *Fredholm* if both ker T and ker T^* are finite dimensional. Find necessary and sufficient conditions on A, B, and X in order for the block matrix $\begin{bmatrix} A & B \\ X & C \end{bmatrix} \in \mathbb{M}_2(\mathbb{B}(\mathcal{H}))$ to be Fredholm.

Hint: See [105].

Exercise 2.9.12 An operator $T \in \mathbb{B}(\mathcal{H})$ is called *Weyl* if it is Fredholm and dim ker T = dim ker T^*. Find necessary and sufficient conditions on A, B, and X in order for the block matrix $\begin{bmatrix} A & B \\ X & C \end{bmatrix} \in \mathbb{M}_2(\mathbb{B}(\mathcal{H}))$ to be Weyl.

Hint: See [231].

Exercise 2.9.13 Prove that if T is not unitary, then the space $\mathcal{K}$ in Sz.-Nagy's dilation theorem is necessarily infinite dimensional.

Hint: See [145] and Theorem 2.2.5.

Exercise 2.9.14 Let $[A_{ij}]_{1 \le i,j \le 3}$ be a positive operator matrix. Show that $[\operatorname{tr} |A_{ij}|] \in \mathbb{M}_3$ is positive semidefinite.

Hint: See [68, 147].

Exercise 2.9.15 Let $[A_{ij}]_{1 \le i,j \le 2}$ be a positive operator matrix. Show that $[\operatorname{tr}(A_{ij}^2)] \in \mathbb{M}_2$ is positive semidefinite.

Hint: See [160] as well as [140] for a nice generalization.

Exercise 2.9.16 Show that the Hua matrix

$$H = \begin{bmatrix} (I - A^*A)^{-1} & (I - B^*A)^{-1} \\ (I - A^*B)^{-1} & (I - B^*B)^{-1} \end{bmatrix}$$

2.9 Exercises and Problems

is positive, where A and B are strictly contractions in $\mathbb{M}_n$.

Hint: Refer to Theorem 2.4.1.

Exercise 2.9.17 Let $A \geq 0$ and X be elements in $\mathbb{B}(\mathcal{H})$. Prove that $\begin{bmatrix} 0 & X \\ X^* & A \end{bmatrix} \geq 0$ if and only if $X = 0$.

Hint: Since $\left\langle \begin{bmatrix} 0 & X \\ X^* & A \end{bmatrix} \begin{bmatrix} x \\ y \end{bmatrix}, \begin{bmatrix} x \\ y \end{bmatrix} \right\rangle \geq 0$ for any $x, y \in \mathcal{H}$, we get a contradiction if $X \neq 0$.

Exercise 2.9.18 Show that if $\begin{bmatrix} A & B \\ B^* & C \end{bmatrix}$ is positive, then so is $\begin{bmatrix} C & B^* \\ B & A \end{bmatrix}$.

Hint: Refer to Remark 2.4.2.

Exercise 2.9.19 Let $A, B \in \mathbb{M}_n$ be positive matrices. Prove that

$$w\left(\begin{bmatrix} 0 & A \\ B & 0 \end{bmatrix}\right) = \frac{1}{2}\|A + B\|.$$

Hint: See [1].

Exercise 2.9.20 If $A, B \in \mathbb{M}_n$, then show that

$$\left\|\begin{bmatrix} 0 & A \\ B & 0 \end{bmatrix}\right\| = \max(\|A\|, \|B\|).$$

Hint: Let $W = \begin{bmatrix} 0 & 1 \\ 1 & 0 \end{bmatrix}$. Then consider $\left\|W \begin{bmatrix} 0 & A \\ B & 0 \end{bmatrix}\right\|$.

Exercise 2.9.21 Show that if $M = \begin{bmatrix} A & B \\ B & C \end{bmatrix} \in \mathbb{M}_2(\mathbb{M}_n)$ is positive, then $\|M\| \leq \|A + C\|$ for all unitarily invariant norms $\|\cdot\|$. Furthermore, prove or disprove $2\|B\| \leq \|A + C\|$.

Hint: This is a simple case of the Hiroshima theorem [118].

Exercise 2.9.22 Let $\begin{bmatrix} A & B \\ B^* & C \end{bmatrix} \in \mathbb{M}_2(\mathbb{M}_n)$ is PPT. Show that

$$\prod_{j=1}^k s_j(B) \leq \prod_{j=1}^k s_j(A^{1/2}C^{1/2}), \quad k = 1, 2, \ldots, n.$$

Hint: Use Theorem 2.4.12 to present X as $X = A^{1/2} K B^{1/2}$ for some contraction K.

Exercise 2.9.23 Prove that if $\begin{bmatrix} A & B \\ B^* & C \end{bmatrix} \in \mathbb{M}_2(\mathbb{M}_n)$ is PPT, then $2\,\|\!|B|\!\| \leq \|\!|A+C|\!\|$ for all unitarily invariant norms $\|\!|\cdot|\!\|$ on $\mathbb{M}_n$.

Hint: See [155].

Exercise 2.9.24 Let $a, b, c, d \in \mathbb{C}$. Show that

$$\left\|\begin{bmatrix} a & b \\ c & d \end{bmatrix}\right\|_p \geq \left\|\begin{bmatrix} |a| & |b| \\ |c| & |d| \end{bmatrix}\right\|_p$$

for $1 \leq p \leq 2$; and the reverse inequality holds for $p \geq 2$.

Hint: See [135].

Problem 2.9.25 Let $A, B, C \in \mathbb{M}_n$ and $\begin{bmatrix} A & B \\ B^* & C \end{bmatrix}$ is positive. Prove that

$$\left\|\begin{bmatrix} A & B \\ B^* & C \end{bmatrix}\right\|_p \geq \left\|\begin{bmatrix} \|A\|_p & \|B\|_p \\ \|B^*\|_p & \|C\| \end{bmatrix}\right\|_p$$

for $1 \leq p \leq 2$; and the direction of inequality is reversed for $p \geq 2$.

Hint: See [14].

Problem 2.9.26 (Thompson determinant compression theorem). Let $A = [A_{ij}]_{n \times n}$ be a positive operator matrix with $A_{ij} \in \mathbb{M}_n$. Then $[\mathrm{tr}(A_{ij})]$ is a positive semidefinite matrix in $\mathbb{M}_n$. Prove that

$$\det(A) \leq \det[\det(A_{ij})].$$

Moreover, show that if A is invertible, then equality holds if and only if all $A_{ij} = 0$ for $i \neq j$.

Hint: See [220].

Problem 2.9.27 Let $M = \begin{bmatrix} A & B \\ B^* & C \end{bmatrix} \in \mathbb{M}_2(\mathbb{M}_n)$ be positive and $A + C = kI$ for some $k > 0$. Show that the following assertions are equivalent:

(i) M is PPT;
(ii) $\|\!|M|\!\| \leq \|\!|A + C|\!\|$ for all unitarily invariant norms;
(iii) $\|M\| \leq \|A + C\| = k$.

Hint: See [101].

Problem 2.9.28 Given $A, B \in \mathbb{B}(\mathscr{H})$, consider the upper triangular operator matrix $M_X = \begin{bmatrix} A & X \\ 0 & B \end{bmatrix}$.

2.9 Exercises and Problems

(i) Is $\mathrm{sp}(M_X) \subseteq \mathrm{sp}(A) \cup \mathrm{sp}(B)$? (see [70])

(ii) What is
$$\bigcap_{X \in \mathbb{B}(\mathcal{H})} \mathrm{sp}(M_X)?$$

(iii) Is there any operator $X_0 \in \mathbb{B}(\mathcal{H})$ such that
$$\mathrm{sp}(M_{X_0}) = \bigcap_{X \in \mathbb{B}(\mathcal{H})} \mathrm{sp}(M_X)?$$

(iv) Answer to the above questions when the usual spectrum is replaced with the *essential spectrum* $\mathrm{sp}_e(A) := \{\lambda \in \mathbb{C} : A - \lambda I \text{ is not Fredholm}\}$, the *Weyl spectrum* $\mathrm{sp}_w(A) := \{\lambda \in \mathbb{C} : A - \lambda I \text{ is not Weyl}\}$, the *point spectrum* $\mathrm{sp}_p e(A) := \{\lambda \in \mathbb{C} : A - \lambda I \text{ is not injective}\}$, the *approximate point spectrum* $\mathrm{sp}_{ap} e(A) := \{\lambda \in \mathbb{C} : A - \lambda I \text{ is not bounded below}\}$ and other types of spectrum; see [39, 62, 63]

(v) For which A and B does the operator matrix M_X have a certain property, such as invertibility, Fredholmness, and Weylness, for all $X \in \mathbb{B}(\mathcal{H})$?

Hint: See [53].

Problem 2.9.29 Prove that the following conditions are equivalent:

(i) $w(A) \leq 1$.

(ii) The $(n+1) \times (n+1)$ block matrix

$$\Delta_n(A) = \begin{bmatrix} 2I & A^* & 0 & \cdots & 0 \\ A & 2I & A^* & \cdots & 0 \\ 0 & A & 2I & \cdots & 0 \\ \cdot & \cdot & \cdot & \cdots & \cdot \\ 0 & 0 & 0 & \cdots & 2I \end{bmatrix} \quad (2.9.1)$$

is positive for every n.

(iii) The $(n+1) \times (n+1)$ block matrix

$$\Gamma_n(A) = \begin{bmatrix} 2I & A^* & A^{*2} & \cdots & A^{*n} \\ A & 2I & A^* & \cdots & A^{*n-1} \\ \vdots & \vdots & \vdots & \vdots & \vdots \\ A^n & A^{n-1} & A^{n-2} & \cdots & 2I \end{bmatrix} \quad (2.9.2)$$

is positive for every n.

Hint: See [23].

Problem 2.9.30 Show that the following conditions are equivalent:
 (i) $w(A) \leq 1$.
 (ii) There exist operators X and Y such that

$$X^*X + Y^*Y = I \text{ and } A = 2X^*Y. \tag{2.9.3}$$

 (iii) There exists an operator C such that

$$\|C\| \leq 1 \text{ and } A = 2\left(I - C^*C\right)^{1/2} C. \tag{2.9.4}$$

 (iv) There exists a Hermitian operator H such that the 2×2 block matrix

$$\begin{bmatrix} I + H & A^* \\ A & I - H \end{bmatrix} \tag{2.9.5}$$

is positive.

Hint: See [23].

Problem 2.9.31 Let A_{ij} be $n \times n$ matrices given for $j - i \leq k$. Find the matrices A_{ij}, where $j - i > k$ such that the block matrix $[A_{ij}]$ has norm less than 1. Show that a solution exists if and only if

$$\left\| \begin{bmatrix} A_{i1} & \cdots & A_{i,i+k} \\ \vdots & & \vdots \\ A_{n1} & \cdots & A_{n,i+k} \end{bmatrix} \right\| < 1$$

for $i = 1, 2, \ldots, n - k$; see [225].
 Hint: See [13].

Problem 2.9.32 Let A_{ij} be given $n \times n$ matrices for $|j - i| \leq k$. Find the matrices A_{ij}, $|j - i| > k$ such that the block matrix $[A_{ij}]$ is strictly positive. Show that a sufficient and necessary condition for the existence of a solution is

$$\begin{bmatrix} A_{ii} & \cdots & A_{i,i+k} \\ \vdots & & \vdots \\ A_{i+k,i} & \cdots & A_{i+k,i+k} \end{bmatrix} > 0$$

for $i = 1, 2, \ldots, n - k$.
 Hint: See [71, 225].

Operator Monotone Functions and Positive Maps

This chapter is devoted to a comprehensive study of operator monotone and operator convex functions. We establish connections between operator monotone functions and operator means by employing the Kubo–Ando theory. Furthermore, we extensively explore positive, n-positive, weakly n-positive, and completely positive maps, which include well-known findings from various prominent contributors in this field.

3.1 Operator Monotone and Operator Convex Functions

We first introduce the concepts of operator monotone and operator convex functions. Throughout this section, J denotes an interval on the real line $\mathbb{R}$. We also denote $\mathbb{B}_{sa}(\mathscr{H})(J)$ as the set of all self-adjoint operators in $\mathbb{B}(\mathscr{H})(J)$ with spectra contained in J.

Definition 3.1.1 A continuous function $f : J \to \mathbb{R}$ is called

(i) *operator monotone* if $A, B \in \mathbb{B}_{sa}(\mathscr{H})(J)$ and $A \leq B$ entails that $f(A) \leq f(B)$.
(ii) *operator convex* if

$$f(\lambda A + (1-\lambda) B) \leq \lambda f(A) + (1-\lambda) f(B)$$

for all operators $A, B \in \mathbb{B}_{sa}(\mathscr{H})(J)$ and all $\lambda \in [0, 1]$.
(iii) A function f is said to be *operator decreasing* (*operator concave*, respectively) if $-f$ is operator monotone (operator convex, respectively).

It is clear that operator monotone functions and operator convex functions are monotone and convex in the usual sense, respectively.

Example 3.1.2 (i) For $A = \begin{bmatrix} 1 & -1 \\ -1 & 2 \end{bmatrix} \geq 0$ and $B = \begin{bmatrix} 0 & 0 \\ 0 & 1 \end{bmatrix} \geq 0$, we have $A \geq B \geq 0$, but not $A^2 \geq B^2$. So $f(t) = t^2$ is not operator monotone on $[0, +\infty)$.

(ii) The *Löwner–Heinz theorem* states that the function $f(t) = t^\alpha$ is operator monotone for every $0 < \alpha < 1$.

(iii) The function $\log t$ is operator monotone on $(0, \infty)$, since if $0 < A \leq B$, then the Löwner–Heinz theorem implies that $0 < A^\alpha \leq B^\alpha$ for all $\alpha \in (0, 1]$. Therefore

$$\frac{A^\alpha - I}{\alpha} \leq \frac{B^\alpha - I}{\alpha}.$$

Letting $\alpha \to +0$, we arrive at $\log A \leq \log B$.

(iv) The function $f(t) = \tan t$ is not operator concave on $(-\pi/2, \pi/2)$ [113, Example 4.3], but is operator monotone.

(v) The function t^3 is convex on $[0, \infty)$ but it is not operator convex. This can be shown by considering the matrices $A = \begin{bmatrix} 2 & 1 \\ 1 & 1 \end{bmatrix}$ and $B = \begin{bmatrix} 1 & 0 \\ 0 & A \end{bmatrix}$. In fact, we observe that

$$\frac{A^3 + B^3}{2} - \left(\frac{A+B}{2}\right)^3 = \begin{bmatrix} 11/4 & 9/4 \\ 9/4 & 7/4 \end{bmatrix} \not\geq 0.$$

(vi) The function $f(t) = t^{-1}$ is operator convex on $(0, \infty)$ [7, Exercise V.2.11]. From the standard argument we have only considered the case of $\lambda = \frac{1}{2}$. Let $A, B > 0$. It follows from Theorem 2.4.1 that

$$\begin{bmatrix} A & I \\ I & A^{-1} \end{bmatrix} \geq 0, \quad \begin{bmatrix} B & I \\ I & B^{-1} \end{bmatrix} \geq 0 \quad \text{and} \quad \begin{bmatrix} A+B & 2I \\ 2I & A^{-1}+B^{-1} \end{bmatrix} \geq 0.$$

This implies that $A^{-1} + B^{-1} \geq 4(A^{-1} + B^{-1})^{-1}$. Therefore, $\frac{A^{-1}+B^{-1}}{2} \geq \left(\frac{A^{-1}+B^{-1}}{2}\right)^{-1}$.

The following is a characterization of operator convexity due to Hansen and Pedersen [109, 111], see also [87]. To prove this characterization, we need two lemmas.

Lemma 3.1.3 *If $A \in \mathbb{B}_{\mathrm{sa}}(\mathcal{H})$ and $f \in C(\mathrm{sp}(A))$, then $f(U^*AU) = U^*f(A)U$ for all unitaries U.*

Proof By the Stone–Weierstrass theorem, there exists a sequence $(p_n(t))$ of polynomials converging to f in $C(\mathrm{sp}(A))$. Hence

3.1 Operator Monotone and Operator Convex Functions

$$f(U^*AU) = \lim_n p_n(U^*AU) = \lim_n U^*p_n(A)U = U^*(\lim_n p_n(A))U^* = U^*f(A)U.$$

Note that $\mathrm{sp}(U^*AU) = \mathrm{sp}(A)$, since if $\lambda \notin \mathrm{sp}(A)$, then there exists an operator $B \in \mathbb{B}(\mathscr{H})$ such that $B(A - \lambda I) = (A - \lambda I)B = I$. Therefore,

$$U^*BU(U^*AU - \lambda I) = (U^*AU - \lambda I)U^*BU = I.$$

Hence, $\lambda \notin \mathrm{sp}(U^*AU)$ and vice versa. Furthermore, we used the functional calculus for U^*AU in the first equality and the functional calculus for A in the third equality. □

Lemma 3.1.4 *If $A \in \mathbb{B}_{\mathrm{sa}}(\mathscr{H})$ and $f \in C(\mathrm{sp}(A^*A))$, then $Af(A^*A) = f(AA^*)A$.*

Proof By the Stone–Weierstrass theorem, there is a sequence $(p_n(t))$ of polynomials converging to f in $C(\mathrm{sp}(A^*A))$ (note that $\mathrm{sp}(A^*A) \subseteq [0, \|A\|^2]$). Utilizing the functional calculus for A^*A and the fact that $Ap_n(A^*A) = p_n(AA^*)A$, we get

$$Af(A^*A)) = \lim_n Ap_n(A^*A) = \lim_n p_n(AA^*)A = f(AA^*)A.$$

□

The following theorem is a combination of the results of Davis [59] and Hansen and Pedersen [111].

Theorem 3.1.5 *Let f be a real-valued continuous function on an interval J. Then the following assertions are equivalent:*

(i) *f is operator convex on J;*
(ii) *$f(C^*AC) \leq C^*f(A)C$ for each Hilbert space $\mathscr{H}$, each $A \in \mathbb{B}_{\mathrm{sa}}(\mathscr{H})(J)$ and each isometry $C \in \mathbb{B}(\mathscr{H})$;*
(iii) *$f(\sum_{j=1}^k C_j^*A_jC_j) \leq \sum_{j=1}^k C_j^*f(A_j)C_j$ for each Hilbert space $\mathscr{H}$, each $A_j \in \mathbb{B}_{\mathrm{sa}}(\mathscr{H})(J)$ and each $C_j \in \mathbb{B}(\mathscr{H})$ $(j = 1, \ldots, k)$ with $\sum_{j=1}^k C_j^*C_j = I$;*
(iv) *$f(\sum_{j=1}^k P_jA_jP_j) \leq \sum_{j=1}^k P_jf(A_j)P_j$ for each Hilbert space $\mathscr{H}$, each $A_j \in \mathbb{B}_{\mathrm{sa}}(\mathscr{H})(J)$ and each projection $P_j \in \mathbb{B}(\mathscr{H})$ $(j = 1, \ldots, k)$ with $\sum_{j=1}^k P_j = I$.*

Proof (i) $\Longrightarrow$ (ii). Let $r \in J$ and $X = \begin{bmatrix} A & 0 \\ 0 & rI \end{bmatrix}$. Straightforward computations based on Lemma 3.1.4 show that the operators $U = \begin{bmatrix} C & D \\ 0 & -C^* \end{bmatrix}$ and $V = \begin{bmatrix} C & -D \\ 0 & C^* \end{bmatrix}$, where $D = (I - CC^*)^{1/2}$ are unitaries; see also Proposition 2.2.2. We have

$$U^*XU = \begin{bmatrix} C^*AC & C^*AD \\ DAC & DAD + rCC^* \end{bmatrix}, V^*XV = \begin{bmatrix} C^*AC & -C^*AD \\ -DAC & DAD + rCC^* \end{bmatrix}$$

and

$$\begin{bmatrix} C^*AC & 0 \\ 0 & D^*AD + rCC^* \end{bmatrix} = \frac{U^*XU + V^*XV}{2}.$$

Hence

$$\begin{bmatrix} f(C^*AC) & 0 \\ 0 & f(D^*AD + rCC^*) \end{bmatrix}$$

$$= f \begin{bmatrix} C^*AC & 0 \\ 0 & D^*AD + rCC^* \end{bmatrix} \qquad \text{(by (1.1.7))}$$

$$= f \left(\frac{U^*XU + V^*XV}{2} \right)$$

$$\leq \frac{f(U^*XU) + f(V^*XV)}{2} \qquad \text{(by the operator convexity of } f\text{)}$$

$$= \frac{U^*f(X)U + V^*f(X)V}{2} \qquad \text{(by Lemma 3.1.3)}$$

$$= \begin{bmatrix} C^*f(A)C & 0 \\ 0 & D^*f(A)D + f(r)CC^* \end{bmatrix}.$$

Considering the (1, 1)-components, we reach $f(C^*AC) \leq C^*f(A)C$.

(ii) $\Longrightarrow$ (iii). Let $\mathcal{K}$ be the direct sum of countably many copies of $\mathcal{H}$. Put

$$\tilde{A} = \begin{bmatrix} A_1 & 0 & & & & \cdots \\ 0 & A_2 & & & & \cdots \\ & & \ddots & & & \\ & & & A_k & & \\ & & & & 0 & \\ & & & & & 0 \\ \vdots & \vdots & & & & \ddots \end{bmatrix} \in \mathbb{B}_{sa}(\mathcal{K})$$

and

$$\tilde{C} = \begin{bmatrix} C_1 & 0 & 0 & \cdots \\ \vdots & 0 & 0 & \cdots \\ C_k & 0 & 0 & \cdots \\ \hline 0 & I & 0 & \cdots \\ 0 & 0 & I & \cdots \\ \vdots & \vdots & \vdots & \ddots \end{bmatrix} \in \mathbb{B}(\mathcal{K}).$$

3.1 Operator Monotone and Operator Convex Functions

Then $\tilde{C}$ is an isometry. Hence, for each $x \in \mathcal{H}$ we have

$$\left\langle f\left(\sum_{j=1}^{k} C_j^* A_j C_j\right) x, x \right\rangle = \langle f(\tilde{C}^* \tilde{A} \tilde{C}) \tilde{x}, \tilde{x} \rangle$$

$$\leq \langle \tilde{C}^* f(\tilde{A}) \tilde{C} \tilde{x}, \tilde{x} \rangle$$

$$= \left\langle \sum_{j=1}^{k} C_j^* f(A_j) C_j x, x \right\rangle,$$

where $\tilde{x} = (x, 0, 0, \ldots) \in \mathcal{K}$. Thus, we get (iii).

(iii) $\Longrightarrow$ (iv). Consider $C_j = P_j$.

(iv) $\Longrightarrow$ (i). Let $A, B \in \mathbb{B}_{\mathrm{sa}}(\mathcal{H})(J)$, and let $0 \leq \lambda \leq 1$. Put $X = \begin{bmatrix} A & 0 \\ 0 & B \end{bmatrix}$, $P = \begin{bmatrix} I & 0 \\ 0 & 0 \end{bmatrix}$

and $U = \begin{bmatrix} \sqrt{1-\lambda} & -\sqrt{\lambda} \\ \sqrt{\lambda} & \sqrt{1-\lambda} \end{bmatrix}$. Then U is unitary, P is a projection and

$$\begin{bmatrix} f((1-\lambda)A + \lambda B) & 0 \\ 0 & f(\lambda A + (1-\lambda)B) \end{bmatrix}$$

$$= f(PU^* XUP + (I_{H \oplus H} - P)U^* XU(I_{H \oplus H} - P))$$

$$\leq P\big(f(U^* XU)P + (I_{H \oplus H} - P)f(U^* XU)(I_{H \oplus H} - P)$$

$$= PU^* f(X)UP + (I_{H \oplus H} - P)U^* f(X)U(I_{H \oplus H} - P)$$

$$= \begin{bmatrix} (1-\lambda)f(A) + \lambda f(B) & 0 \\ 0 & \lambda f(A) + (1-\lambda)f(B) \end{bmatrix}.$$

Considering the $(1, 1)$-components, we conclude that f is operator convex. $\square$

Now, we present the *Hansen–Pedersen–Jensen inequality*; see [109, 111]. Inequality in (ii) is called *Hansen's inequality* [108].

Theorem 3.1.6 (The Hansen–Pedersen–Jensen inequality) *Let f be a real-valued continuous function on an interval J containing 0. Then the following assertions are equivalent:*

(i) f is operator convex on J and $f(0) \leq 0$;
(ii) $f(C^* AC) \leq C^* f(A)C$ for each Hilbert space $\mathcal{H}$, each $A \in \mathbb{B}_{\mathrm{sa}}(\mathcal{H})(J)$ and each contraction $C \in \mathbb{B}(\mathcal{H})$;
(iii) $f(\sum_{j=1}^{k} C_j^* A_j C_j) \leq \sum_{j=1}^{k} C_j^* f(A_j) C_j$ for each Hilbert space $\mathcal{H}$, each $A_j \in \mathbb{B}_{\mathrm{sa}}(\mathcal{H})(J)$ and each $C_j \in \mathbb{B}(\mathcal{H})$ $(j = 1, \ldots, k)$ with $\sum_{j=1}^{k} C_j^* C_j \leq I$;
(iv) $f(PAP) \leq Pf(A)P$ for each Hilbert space $\mathcal{H}$, each $A \in \mathbb{B}_{\mathrm{sa}}(\mathcal{H})(J)$ and each projection $P \in \mathbb{B}(\mathcal{H})$.

Proof (i) $\Longrightarrow$ (ii). Let C be a contraction and let put $D = (I - C^*C)^{1/2}$. Then $C^*C + D^*D = I$. From (iii) of Theorem 3.1.5 we get

$$f(C^*AC) = f(C^*AC + D^*0D) \leq C^*f(A)C + D^*f(0)D \leq C^*f(A)C,$$

where the last inequality holds because $f(0) \leq 0$.

(ii) $\Longrightarrow$ (iii). Use the same reasoning as in the proof of (ii) $\Longrightarrow$ (iii) in Theorem 3.1.5.

(iii) $\Longrightarrow$ (iv). Use the same reasoning as in the proof of (iii) $\Longrightarrow$ (iv) in Theorem 3.1.5.

(iv) $\Longrightarrow$ (i). Under the same notation as in the proof of (iv) $\Longrightarrow$ (i) in Theorem 3.1.5, we have

$$\begin{bmatrix} f((1-\lambda)A + \lambda B) & 0 \\ 0 & f(0) \end{bmatrix} = f(PU^*XUP)$$
$$\leq PU^*f(X)UP$$
$$= \begin{bmatrix} (1-\lambda)f(A) + \lambda f(B) & 0 \\ 0 & 0 \end{bmatrix}.$$

Hence, f is operator convex and $f(0) \leq 0$. $\square$

The following theorem establishes a relationship between operator concavity and operator monotonicity on unbounded intervals; see [94, Theorem 1.15].

Theorem 3.1.7 *Suppose that J is an infinite interval, $f : J \to \mathbb{R}$ is a continuous function, $-\infty < \inf J$, and there is $m \in \mathbb{R}$ such that $m \leq f(\lambda)$ for all $\lambda \in J$. Then the following assertions are equivalent:*

(i) f is operator concave on J;
(ii) f is operator monotone on J.

Proof (i) $\Longrightarrow$ (ii). Let $A, B \in \mathbb{B}_{\text{sa}}(\mathcal{H})(J)$ and $A \leq B$. Put $\alpha := \inf\{\lambda : \lambda \in J\}$. Then $\alpha I \leq A \leq B$. Let $0 < \lambda < 1$ be arbitrary. We have

$$\lambda(B - \alpha I) + \alpha I = \lambda A + (1 - \lambda)\left(\frac{\lambda}{1-\lambda}(B - A) + \alpha I\right)$$

and

$$\text{sp}\,(\lambda(B - \alpha I) + \alpha I) \subseteq [\alpha, \infty), \quad \text{sp}\left(\frac{\lambda}{1-\lambda}(B - A) + \alpha I\right) \subseteq [\alpha, \infty).$$

It follows from the operator concavity of f that

3.1 Operator Monotone and Operator Convex Functions

$$f(\lambda(B - \alpha I) + \alpha I) \geq \lambda f(A) + (1 - \lambda)f\left(\frac{\lambda}{1-\lambda}(B-A) + \alpha I\right)$$

$$\geq \lambda f(A) + (1-\lambda)m.$$

Tending λ to 1 we arrive at $f(A) \leq f(B)$.

(ii) $\Longrightarrow$ (i). Assume that C is an isometry and put

$$D := \sqrt{I - CC^*}, \quad X := \begin{bmatrix} A & 0 \\ 0 & \alpha I \end{bmatrix}, \quad \text{and} \quad U := \begin{bmatrix} C & D \\ 0 & -C^* \end{bmatrix}.$$

Then U is a unitary operator. For each $\varepsilon > 0$ there is a number $M > \alpha$ such that

$$U^*XU = \begin{bmatrix} C^*AC & C^*AD \\ D^*AC & D^*AD + \alpha CC^* \end{bmatrix} \leq \begin{bmatrix} C^*AC + \varepsilon I & 0 \\ 0 & MI \end{bmatrix}$$

since by applying Lemma 2.4.1 to the matrix $\begin{bmatrix} \varepsilon I & -C^*AD \\ -D^*AC & MI - (D^*AD + \alpha CC^*) \end{bmatrix}$ we can find a number M such that $(D^*AC)(\varepsilon^{-1}I)(C^*AD) + D^*AD + \alpha CC^* \leq MI$. Such a number actually exists. For example, we may take M=$\|(DAC)(\varepsilon^{-1}I)(C^*AD) + DAD + \alpha CC^*\|$.

By the operator monotonicity of f, we therefore have

$$\begin{bmatrix} C^*f(A)C & C^*f(A)D \\ Df(A)C & Df(A)D + f(\alpha)CC^* \end{bmatrix} = U^*f(X)U = f(U^*XU)$$

$$\leq f\left(\begin{bmatrix} C^*AC + \varepsilon I & 0 \\ 0 & MI \end{bmatrix}\right)$$

$$= \begin{bmatrix} f(C^*AC + \varepsilon I) & 0 \\ 0 & f(MI) \end{bmatrix}.$$

Thus, $C^*f(A)C \leq f(C^*(A + \varepsilon I)C)$. If $\varepsilon \to 0$ and use Theorem 3.1.5, we get the desired inequality. $\square$

Remark 3.1.8 In usual calculus, we have the following natural decreasing sequence

$$C^1(J) \supsetneq \cdots \supsetneq C^{n-1}(J) \supsetneq C^n(J) \supsetneq C^{n+1}(I) \cdots,$$

where $C^n(J)$ represents the set of all functions that are continuously differentiable n-times. This sequence eventually ends at $C^\infty(J)$, which denotes the set of infinitely differentiable functions. In addition, we typically encounter a smaller set called Anal(J) consisting of analytic functions.

We call a continuous real-valued function f defined on an interval $J \subseteq \mathbb{R}$ *matrix monotone of order n* or *n-monotone* for short, if the inequality $f(A) \leq f(B)$ holds for every pair of self-adjoint matrices $A, B \in \mathbb{M}_n$ such that $A \leq B$ and all eigenvalues of A and B are contained in J. Similarly, *matrix convex* (*matrix concave*) functions on J can be defined. We denote the spaces of operator monotone functions and operator convex functions as $P_\infty(J)$ and $K_\infty(J)$, respectively. The spaces for n-monotone functions and n-convex functions are denoted as $P_n(J)$ and $K_n(J)$, respectively. We have then

$$P_1(J) \supseteq \cdots \supseteq P_{n-1}(J) \supseteq P_n(J) \supseteq P_{n+1}(J) \supseteq \cdots \supseteq P_\infty(J)$$
$$K_1(J) \supseteq \cdots \supseteq K_{n-1}(J) \supseteq K_n(J) \supseteq K_{n+1}(J) \supseteq \cdots \supseteq K_\infty(J).$$

Here, we encounter the facts that $\cap_{n=1}^\infty P_n(J) = P_\infty(J)$ and $\cap_{n=1}^\infty K_n(J) = K_\infty(J)$. We regard these two decreasing sequences as noncommutative counterparts of the classical piling class $\{C^n(J), C^\infty(J), \text{Anal}(J)\}$. We can understand that the class of operator monotone functions $P_\infty(J)$ corresponds to the class $\{C^\infty(J), \text{Anal}(J)\}$ through the famous characterization of those functions by Löwner as the restriction of Pick functions.

In these circumstances, it is well recognized that we should not limit our discussions to only those classes $P_\infty(J)$ and $K_\infty(J)$, which refer to the class of operator monotone functions and the class of operator convex functions. These classes $\{P_n(J)\}$ and $\{K_n(J)\}$ are not merely optional alternatives to $P_\infty(J)$ and $K_\infty(J)$. They also play important roles in noncommutative calculus, similar to how the classes $\{C^n(J)\}$ are significant in usual (commutative) calculus.

The first basic question is whether $P_{n+1}(J)$ (respectively, $K_{n+1}(J)$) is strictly contained in $P_n(J)$ (respectively, $K_n(J)$) for every n. This gap problem for arbitrary n has been solved [112, 113, 183].

The following theorem summarizes some of the facts related to matrix monotone functions. Note that the assertions become equivalent when f is operator convex and g is operator monotone due to the piling structure.

Theorem 3.1.9 ([184]) *For $0 < \alpha \leq \infty$, consider the following assertions for a real-valued continuous function f defined on $[0, \alpha)$:*

(1) *f is n-convex and $f(0) \leq 0$,*
(2) *For a matrix $A \in \mathbb{M}_n$ with its spectrum in $[0, \alpha)$ and a contraction $C \in \mathbb{M}_n$,*

$$f(C^*AC) \leq C^* f(A)C,$$

(3) *For two matrices $A, B \in \mathbb{M}_n$ with their spectra in $[0, \alpha)$ and two contractions $C, D \in \mathbb{M}_n$ such that $C^*C + D^*D \leq 1$ the inequality*

$$f(C^*AC + D^*BD) \leq C^* f(A)C + D^* f(B)D$$

3.1 Operator Monotone and Operator Convex Functions

holds,

(4) For a matrix $A \in \mathbb{M}_n$ with its spectrum in $[0, \alpha)$ and a projection $P \in \mathbb{M}_n$ it holds that

$$f(PAP) \le Pf(A)P.$$

(5) The function $g(t) = \frac{f(t)}{t}$ is n-monotone in the open interval $(0, \alpha)$.

Then, the following relations:

$$(1)_{2n} \prec (2)_n \sim (5)_n \preceq (4)_n, (2)_{2n} \prec (3)_n \prec (4)_n, \text{ and } (4)_{2n} \prec (1)_n$$

hold, where it is said that assertions (X) and (Y) are in a relation $m \prec n$ if when (X) holds for the matrix algebra $\mathbb{M}_m$, then (Y) also holds for the matrix algebra $\mathbb{M}_n$. In this case, it is written as $(X)_m \prec (Y)_n$.

Proof Except the implication $(2)_n \sim (5)_n$, claimes follow from the proof of Theorem 3.1.6.

$(2)_n \Rightarrow (5)_n$:

Suppose that $A, B \in \mathbb{M}_n$ with spectrum $(0, \alpha)$ and $A \le C$. Set $C = B^{-1/2}A^{1/2}$. Then, $\|C\| \le 1$. Hence, by the condition $(2)_n$

$$f(A) = f(C^*BC) \le C^*f(B)C = A^{1/2}B^{-1}f(B)A^{1/2}.$$

Therefore, $A^{-1/2}f(A)A^{-1/2} \le B^{-1}f(B)$. This implies that the function $g(t) = \frac{f(t)}{t}$ is n-monotone on $(0, \alpha)$.

$(5)_n \Rightarrow (2)_n$:

Take positive semidefinite matrix $A \in \mathbb{M}_n$ with its spectrum in $[0, \alpha)$ and a contraction $C \in \mathbb{M}_n$. We may assume that A is invertible. Take a positive number $\varepsilon > 0$. From the order relation,

$$A^{1/2}(CC^* + \varepsilon)A^{1/2} \le (1 + \varepsilon)A.$$

We have the inequality

$$f(A^{1/2}(CC^* + \varepsilon)A^{1/2})\left(A^{1/2}(CC^* + \varepsilon)A^{1/2}\right)^{-1} \le f((1+\varepsilon)A)\left((1+\varepsilon)A\right)^{-1}.$$

Hence, multiplying by $A^{1/2}(CC^* + \varepsilon)A^{1/2}$ from both sides and letting $\varepsilon \to 0$ we get the inequality

$$A^{1/2}CC^*A^{1/2}f(A^{1/2}CC^*A^{1/2}) \le A^{1/2}CC^*f(A)CC^*A^{1/2}.$$

Note that here we have the identity,

$$C^*A^{1/2}f(A^{1/2}CC^*A^{1/2}) = f(C^*AC)C^*A^{1/2}.$$

Therefore, the above inequality comes to the form,

$$A^{1/2}Cf(C^*AC)C^*A^{1/2} \leq A^{1/2}CC^*f(A)CC^*A^{1/2}.$$

It follows that
$$Cf(C^*AC)C^* \leq CC^*f(A)CC^*.$$

Hence for a vector x in the underlying space $\mathbb{C}^n$ we have
$$\langle f(C^*AC)C^*x, C^*x\rangle \leq \langle (C^*f(A)C)C^*x, C^*x\rangle.$$

Now consider the orthogonal decomposition of $\mathbb{C}^n$ with respect to the operator C such as $\mathbb{C}^n = \operatorname{ran} C^* \oplus \ker C$ and write $x = x_1 + x_2$. Then,

$$\begin{aligned}
\langle f(C^*AC)x, x\rangle &= \langle f(C^*AC)x_1 + f(0)x_2, x_1 + x_2\rangle \\
&= \langle f(C^*AC)x_1, x_1\rangle + \langle f(C^*AC)x_1, x_2\rangle + f(0)\|x_2\|^2 \\
&= \langle f(C^*AC)x_1, x_1\rangle + f(0)\|x_2\|^2 \\
&\leq \langle f(C^*AC)x_1, x_1\rangle \\
&\leq \langle C^*f(a)cx_1, x_1\rangle \\
&= \langle C^*f(A)Cx, x\rangle.
\end{aligned}$$

Thus, $f(C^*AC) \leq C^*f(A)C$.

In the above computation, we have used the fact that $f(0) \leq 0$, which is derived from the monotonousness of $g(t)$. For, if $g(t)$ is monotone increasing we have the inequality $f(t) \leq \frac{f(t_0)}{t_0}t$ for every $0 < t \leq t_0$. □

Readers may consult Simon's book [208] to study basic facts and other results on monotone functions.

We conclude this section by introducing new directions in the topic inspired by the paper [115]. Two significant orders for matrices can be considered:

- The Löwner order stating that $A \geq B$ if $A - B$ is positive semidefinite,
- Entrywise order, which states that $A = [a_{ij}] \geq B = [b_{ij}]$ if $a_{ij} \geq b_{ij}$ for all $1 \leq i, j \leq n$.

We can also consider two ways to introduce a function of a matrix A. More precisely,

- (Continuous) functional calculus in which a continuous function f, defined on the spectrum of a Hermitian matrix A, acts on it to give $f(A)$.
- entrywise calculus in which $f[A]$ is used to denote the matrix obtained by applying f to each entry of $A = [a_{ij}]$. In other words, $f[A] = [f(a_{ij})]$ if the entries of A are in the domain of f.

Thus, there are four combinations when studying monotonicity or convexity for matrix functions:

3.1 Operator Monotone and Operator Convex Functions

(i) functional calculus and positive semidefiniteness,
(ii) functional calculus and entrywise positivity,
(iii) entrywise calculus and positive semidefiniteness,
(iv) entrywise calculus and entrywise positivity.

Case (i), initiated by Löwner, is very important and has been studied by many mathematicians; see [94]. Case (ii) was investigated by Hansen [110], Case (iii) is treated by Hiai [115], and Case (iv) is trivial.

To understand what happens in Cases (ii) or (iii), we can refer to the following interesting result from [115, Proposition 1.2].

Theorem 3.1.10 *Let f be a real function on $(-\alpha, \alpha)$.*

(1) *If $A \geq B$ implies $f[A] \geq f[B]$ for all symmetric $A, B \in \mathbb{M}_2(\mathbb{R})$ with entries in $(-\alpha, \alpha)$, then f is affine on $(-\alpha, \alpha)$.*
(2) *If $f[\lambda A + (1-\lambda)B] \leq \lambda f[A] + (1-\lambda) f[B]$ for all $0 \leq \lambda \leq 1$ and all symmetric $A, B \geq 0$ in $\mathbb{M}_2(\mathbb{R})$ with entries in $(-\alpha, \alpha)$, then f is affine on $(-\alpha, \alpha)$.*

Proof (1) By taking $f - f(0)$ instead of f, we can assume that $f(0) = 0$. It follows from this assumption that f is non-decreasing on $(-\alpha, \alpha)$. Therefore, $f(x) \geq 0$ for $0 \leq x < \alpha$ and $f(x) \leq 0$ for $-\alpha < x \leq 0$. Let $0 \leq a < \alpha$ and $0 < \lambda < 1$. Since

$$\begin{bmatrix} a & \lambda a \\ \lambda a & a \end{bmatrix} \geq \begin{bmatrix} (1-\lambda)a & 0 \\ 0 & (1-\lambda)a \end{bmatrix} \quad \text{and} \quad \begin{bmatrix} \lambda a & (1-\lambda)a \\ (1-\lambda)a & \lambda a \end{bmatrix} \geq \begin{bmatrix} 0 & a \\ a & 0 \end{bmatrix},$$

we arrive at

$$\begin{bmatrix} f(a) & f(\lambda a) \\ f(\lambda a) & f(a) \end{bmatrix} \geq \begin{bmatrix} f((1-\lambda)a) & 0 \\ 0 & f((1-\lambda)a) \end{bmatrix}$$

and

$$\begin{bmatrix} f(\lambda a) & f((1-\lambda)a) \\ f((1-\lambda)a) & f(\lambda a) \end{bmatrix} \geq \begin{bmatrix} 0 & f(a) \\ f(a) & 0 \end{bmatrix}.$$

Thus, $f(a) = f(\lambda a) + f((1-\lambda)a)$, which means that f is affine on $[0, \alpha)$. Furthermore, since $\begin{bmatrix} a & -a \\ -a & a \end{bmatrix} \geq 0$ and $\begin{bmatrix} -a & a \\ a & -a \end{bmatrix} \leq 0$, we have $\begin{bmatrix} f(a) & f(-a) \\ f(-a) & f(a) \end{bmatrix} \geq 0$ and $\begin{bmatrix} -f(-a) & -f(a) \\ -f(a) & -f(-a) \end{bmatrix} \geq 0$. These ensure that $f(-a) = -f(a)$ for all $a \in [0, \alpha)$. So f is affine on $(-\alpha, \alpha)$.

(2) Let $0 < a < \alpha$ and $s, t \in [-a, a]$. It follows from $\begin{bmatrix} a & s \\ s & a \end{bmatrix}, \begin{bmatrix} a & t \\ t & a \end{bmatrix} \geq 0$, and the assumption that

$$\begin{bmatrix} f(a) & f(\lambda s + (1-\lambda)t) \\ f(\lambda s + (1-\lambda)t) & f(a) \end{bmatrix} \leq \begin{bmatrix} f(a) & \lambda f(s) + (1-\lambda)f(t) \\ \lambda f(s) + (1-\lambda)f(t) & f(a) \end{bmatrix}$$

for every $0 < \lambda < 1$. Hence, $f(\lambda s + (1-\lambda)t) = \lambda f(s) + (1-\lambda)f(t)$. Therefore, f is affine on $(-\alpha, \alpha)$. □

3.2 Operator Means

The study of operator means began with Anderson and Duffin [3], who investigated the arithmetic and harmonic means of positive operators and established the arithmetic-harmonic mean inequality. The geometric mean was introduced by Pusz and Woronowicz [200]. Kubo and Ando [142] then conducted a systematic study of operator means for two strictly positive operators. Later, Ando et al. [9] defined the geometric mean of n strictly positive operators using a symmetric procedure. Drury [69] defined an operator mean as $A \natural B = \left(\frac{2}{\pi} \int_0^\infty (tA + t^{-1}B)^{-1} \frac{dt}{t}\right)^{-1}$. Raïssouli et al. [202] extended this notion to a weighted geometric mean, given by

$$A \natural_\lambda B := \frac{\sin(\lambda \pi)}{\pi} \int_0^\infty t^{\lambda - 1} \left(A^{-1} + tB^{-1}\right)^{-1} dt.$$

In this section, we investigate operator means and their integral representation. We also examine a weighted Cauchy–Schwarz reverse inequality. Furthermore, we explore means of projection transforms on operator means as well as the weights and symmetricity of operator means.

Let OM_+ be the set of all positive operator monotone functions on $(0, \infty)$. The binary operation $(A, B) \mapsto A\sigma B(:= \Phi_{\tilde{f}}(A, B))$ satisfies the following statements:

(i) $A \leq C, B \leq D$ implies that $A\sigma B \leq C\sigma D$,
(ii) $C(A\sigma B)C \leq (CAC)\sigma(CBC)$ for all $C \geq 0$,
(iii) $A_n \searrow A \geq 0$ and $B_n \searrow B \geq 0$ implies that $A_n \sigma B_n \searrow A\sigma A$. Here, $A_n \searrow B$ means that A_n decreasingly tend to A in the strong operator topology,

where Φ_f on $[0, \infty)^2$ is defined by $\Phi_f(x, y) = \begin{cases} yf(x/y) & \text{if } x, y \in (0, \infty) \\ 0, & \text{if } x \cdot y = 0 \end{cases}$ and $\tilde{f}(t) = tf(\frac{1}{t})$ for $t \in (0, \infty)$.

We call such a binary operation σ an *operator connection* and denote the set of operator connections by Σ. In [142], Kubo and Ando show that the aforementioned three statements characterize the class $\{\Phi_{\tilde{f}} : f \in OM_+\}$.

3.2 Operator Means

Theorem 3.2.1 *For every $\sigma \in \Sigma$, there exists a unique $f_\sigma \in OM_+$ such that $A\sigma B = \Phi_{\tilde{f}_\sigma}(A, B)$ for all $A, B \in \mathbb{B}_+(\mathcal{H})$. The map $\sigma \mapsto f_\sigma$ is an affine order isomorphism from Σ onto OM_+.*

To establish this theorem we need some lemmas.

Lemma 3.2.2 *For $C \in \mathbb{B}_{++}(\mathcal{H})$, $C(A\sigma B)C = (CAC)\sigma(CBC)$.*

Proof From the statement (ii), we have

$$A\sigma B = (C^{-1}CACC^{-1})\sigma(C^{-1}CBCC^{-1})$$
$$\geq C^{-1}\left((CAC)\sigma(CBC)\right)C^{-1}$$
$$\geq C^{-1}C(A\sigma B)CC^{-1} = A\sigma B.$$

□

Lemma 3.2.3 *Let $\sigma \in \Sigma$ and $A, B \in \mathbb{B}_+(\mathcal{H})$. If a projection P commutes with A and B, then $((AP)\sigma(BP))P = (A\sigma B)P = P(A\sigma B)$ holds.*

Proof Since the conditions (i) and (ii) hold, we have

$$P(A\sigma B)P \leq (PAP)\sigma(PBP) \leq A\sigma B.$$

This implies

$$\|(A\sigma B)P - P(A\sigma B)P\|^2 = \|((A\sigma B) - P(A\sigma B)P)P\|^2$$
$$\leq \|(A\sigma B) - P(A\sigma B)P\|\|P(A\sigma B - P(A\sigma B)P)P\|$$
$$= 0.$$

Thus, $(A\sigma B)P = P(A\sigma B)P = (P(A\sigma B)P)^* = P(A\sigma B)$. □

Lemma 3.2.4 *For $t \geq 0$, there exists $\alpha_t \geq 0$ such that $I\sigma(tI) = \alpha_t I$.*

Proof Let P be a projection. It follows from the fact that P commutes with I and tI and the preceding lemma that P commutes with $I\sigma(tI)$. Thus, $I\sigma(tI)$ reduces all closed subspaces in $\mathcal{H}$. □

Set $f(t) := I\sigma(tI)$. From the statement (i), we conclude that $f(t)$ is right continuous on $[0, \infty)$. On the other hand, Lemma 3.2.2 shows that $f(t)/t(= ((1/t)I)\sigma I)$ is left continuous on $(0, \infty)$. By combining these results, we can infer that f is continuous on $[0, \infty)$.

Lemma 3.2.5 $f(t)(:= I\sigma(tI))$ *is an operator monotone function on* $[0, \infty)$.

Proof We first show the case where there exist projections $\{P_i\}$ such that $A = \sum_i \alpha_i P_i$, $\sum_i P_i = I$ and $P_i P_j = 0$ $(i \neq j)$. From Lemma 3.2.3,

$$I\sigma A = (I\sigma A)\left(\sum_i P_i\right) = \sum_i (I\sigma A) P_i$$
$$= \sum_i (P_i \sigma(\alpha_i P_i)) = \sum_i (I\sigma(\alpha_i I)) P_i$$
$$= \sum_i f(\alpha_i) P_i = f(A).$$

For a general A, there exists a sequence (A_n) of the above form such that $A_n \searrow A$. Hence, $I\sigma A_n = f(A_n)$ converges to $I\sigma A(= f(A))$ in the strong operator topology. From statement (i), it follows that f is the operator monotone. □

Recall that for a positive operator A, we denote the strictly positive operator $A + \varepsilon I$ by A_ε.

Proof of Theorem 3.2.1 Let $\sigma \in \Sigma$. From the above lemmas, $f(t)(:= I\sigma(tI))$ is operator monotone and $f(A) = I\sigma A$ for all $A \geq 0$. Thus, we have

$$A_\varepsilon \sigma B_\varepsilon = A_\varepsilon^{1/2}(I\sigma(A_\varepsilon^{-1/2} B_\varepsilon A_\varepsilon^{-1/2})) A_\varepsilon^{1/2}$$
$$= A_\varepsilon^{1/2} f(A_\varepsilon^{-1/2} B_\varepsilon A_\varepsilon^{-1/2}) A_\varepsilon^{1/2}$$
$$= \Phi_{\tilde{f}}(A_\varepsilon, B_\varepsilon)$$

for all $A, B \in \mathbb{B}_+(\mathcal{H})$ and $\varepsilon > 0$. Taking the strong limit of each side, we get $A\sigma B = \Phi_{\tilde{f}}(A, B)$, which holds for all $A, B \in \mathbb{B}_+(\mathcal{H})$.

Let us prove the second half of the theorem. The equivalences

$$\sigma = \alpha \sigma_1 + (1-\alpha)\sigma_2 \iff f_\sigma = \alpha f_{\sigma_1} + (1-\alpha) f_{\sigma_2},$$

$$\sigma = 0 \iff f_\sigma = 0$$

and

$$\sigma_1 \leq \sigma_2 \iff f_{\sigma_1} \leq f_{\sigma_2}$$

are obvious. So, it is enough to show that the map $\sigma \mapsto f$ is surjective.

For every $f \in OM_+$, put $A\sigma B := \Phi_{\tilde{f}}(A, B)$. This binary operation σ satisfies the statements (i), (ii), and (iii) from the simple computation and $\sigma \in \Sigma$. □

3.2 Operator Means

Based on the argument above, a normalized positive valued operator monotone function on $(0, \infty)$ can be identified with an operator mean. The binary operation σ satisfying $A\sigma B = A$ (respectively, $A\sigma B = B$) is a trivial example of an operator mean and it is denoted by l (respectively, r).

An operator connection σ that satisfies $I\sigma I = I$ is called an *operator mean*. Hence, there exists an affine order isomorphism $\sigma \longleftrightarrow f_\sigma$ from the set Σ^1 of all operator means onto the set $OM_+^1 := \{f \in OM_+ : f(1) = 1\}$.

It is known that a necessary and sufficient condition for a positive continuous function f on $(0, \infty)$ to be operator monotone is that f has an integral representation as follows:

$$f(t) = \int_{[0,\infty]} \left(\frac{x:t}{x:1}\right) dm(x) \quad (t > 0), \tag{3.2.1}$$

where $x : y = (x^{-1} + y^{-1})^{1/2}$ $(x, y > 0)$ and m is a positive finite Borel measure on $[0, \infty]$ (See [19, V. 53, p. 144]).

Example 3.2.6 We give some examples of the correspondence $m \longleftrightarrow f$.

$$(1-\alpha)\delta_{\{0\}} + \alpha\delta_{\{\infty\}} \longleftrightarrow f_{\nabla_\alpha}(t) := (1-\alpha) + \alpha t \quad (0 \le \alpha \le 1)$$

$$\frac{\sin \alpha \pi}{\pi} \cdot \frac{x^{\alpha-1}}{1+x} dx \longleftrightarrow f_{\#_\alpha}(t) := t^\alpha \quad (0 < \alpha < 1),$$

$$\delta_{\{\alpha/(1-\alpha)\}} \longleftrightarrow f_{!_\alpha}(t) := ((1-\alpha) + \alpha t^{-1})^{-1} \quad (0 < \alpha < 1).$$

For $A, B \ge 0$, since $A \le (A+B)$, using the Douglass majorization Theorem 1.2.4, we can write $A^{1/2} = C^*(A+B)^{1/2}$ for some bounded operator $C : \mathcal{H} \to \ker(A+B)^{1/2}$. That is, $A = (A+B)^{1/2}CC^*(A+B)^{1/2} = (A+B)^{1/2}R(A+B)^{1/2}$ for some positive operator R on $\ker(A+B)^{1/2}$. Similarly, there is a positive operator S on $\ker(A+B)^{1/2}$ such that $B = (A+B)^{1/2}S(A+B)^{1/2}$. This pair (R, S) is used to define the *Pusz–Woronowicz functional calculus* $\Phi(A, B) = (A+B)^{1/2}\Phi(R, S)(A+B)^{1/2}$, where Φ is a real-valued homogeneous continuous function on $[0, \infty)^2$, that is, $\Phi(\lambda r, \lambda s) = \lambda \Phi(r, s)$ for all $\lambda, r, s \ge 0$ (See [187, Definition 3.3]).

Example 3.2.7 We have previously introduced f_{∇_α}, $f_{\#_\alpha}$ and $f_{!_\alpha}$. The corresponding operator means can be written as follows:

$$A\nabla_\alpha B = (1-\alpha)A + \alpha B$$
$$= (A+B)^{1/2}((1-\alpha)R + \alpha S)(A+B)^{1/2},$$

$$A\#_\alpha B = \text{s-}\lim_{\varepsilon \to 0} A_\varepsilon^{1/2}(A_\varepsilon^{-1/2}B_\varepsilon^{1/2}A_\varepsilon^{-1/2})^\alpha A_\varepsilon^{1/2}$$
$$= (A+B)^{1/2}R^{1-\alpha}S^\alpha(A+B)^{1/2}.$$

Moreover,

$$A!_\alpha B = \text{s-}\lim_{\varepsilon\to 0}((1-\alpha)A_\varepsilon^{-1} + \alpha B_\varepsilon^{-1})^{-1}$$

$$= (A+B)^{1/2}\left(\frac{RS}{\alpha R + (1-\alpha)S}\right)(A+B)^{1/2}.$$

In what follows, we write $f_{!_0}(t) := 1$, $f_{\#_0}(t) := 1$, $f_{!_1}(t) := t$, and $f_{\#_1}(t) := t$. The operator means ∇_α, $\#_\alpha$, and $!_\alpha$ are called the *weighted arithmetic mean*, the *weighted geometric mean*, and the *weighted harmonic mean*. Through a straightforward calculation, it can be shown that $f_{!_\alpha} \leq f_{\#_\alpha} \leq f_{\nabla_\alpha}$, which implies $!_\alpha \leq \#_\alpha \leq \nabla_\alpha$ for all $\alpha \in [0,1]$.

In the following, we denote $\nabla := \nabla_{1/2}$, $\# := \#_{1/2}$ and $! := !_{1/2}$.

Example 3.2.8 Since the power functions t^α ($0 \leq \alpha \leq 1$) are in OM_+^1, the integral $\int_0^1 t^\alpha d\alpha = (t-1)/\log t$ is also in OM_+^1. The corresponding operator mean is denoted by λ and is called the *logarithmic mean*. The relation between the logarithmic mean and the operator means stated above is $\# \leq \lambda \leq \nabla$.

We demonstrate that for $A > 0$ and $B \geq 0$, the geometric mean $A\#B = A^{1/2}(A^{-1/2}BA^{-1/2})^{1/2}A^{1/2}$ is the unique positive solution to the Riccati equation $XA^{-1}X = B$.

Proposition 3.2.9 *Let A, B, and X be positive operators and let A is invertible. Then, $X = A\#B$ if and only if $XA^{-1}X = B$.*

Proof

$$X = A\#B \iff A^{-1/2}XA^{-1/2} = (A^{-1/2}BA^{-1/2})^{1/2}$$
$$\iff A^{-1/2}XA^{-1}XA^{-1/2} = (A^{-1/2}XA^{-1/2})^2 = A^{-1/2}BA^{-1/2}$$
$$\iff XA^{-1}X = B.$$

□

In the next result, we explore statements that can be used to characterize the geometric mean.

Proposition 3.2.10 *Let A, B, and X be positive operators in $\mathbb{B}(\mathcal{H})$. Consider the following statements*

(1) $\begin{bmatrix} A & X \\ X & B \end{bmatrix} \geq 0$;

(2) $XA_\varepsilon^{-1}X \leq B$, *for all* $\varepsilon > 0$;

(3) $X \leq A\#B$.

3.2 Operator Means

Then $(1) \iff (2) \implies (3)$ hold.

Proof $(1) \iff (2)$. This is the same as Theorem 2.4.3. However, we present another proof for it. Let $\delta > 0$. Put $S := A_\varepsilon^{-1/2} X B_\delta^{-1/2}$. Since

$$\begin{bmatrix} I & S \\ S^* & I \end{bmatrix} = \begin{bmatrix} A_\varepsilon^{-1/2} & 0 \\ 0 & B_\delta^{-1/2} \end{bmatrix} \begin{bmatrix} A_\varepsilon & X \\ X & B_\delta \end{bmatrix} \begin{bmatrix} A_\varepsilon^{-1/2} & 0 \\ 0 & B_\delta^{-1/2} \end{bmatrix} \geq 0,$$

we have

$$\left\langle \begin{bmatrix} I & -S \\ -S^* & I \end{bmatrix} \begin{bmatrix} x \\ y \end{bmatrix}, \begin{bmatrix} x \\ y \end{bmatrix} \right\rangle = \left\langle \begin{bmatrix} I & S \\ S^* & I \end{bmatrix} \begin{bmatrix} x \\ -y \end{bmatrix}, \begin{bmatrix} x \\ -y \end{bmatrix} \right\rangle \geq 0$$

for $x, y \in \mathcal{H}$. Hence, the inequalities

$$\begin{bmatrix} I & 0 \\ 0 & I \end{bmatrix} \geq \begin{bmatrix} 0 & S \\ S^* & 0 \end{bmatrix} \geq -\begin{bmatrix} I & 0 \\ 0 & I \end{bmatrix}$$

hold. An application of the functional calculus for the self-adjoint matrix $\begin{bmatrix} 0 & S \\ S^* & 0 \end{bmatrix}$ shows that

$$\left\| \begin{bmatrix} 0 & S \\ S^* & 0 \end{bmatrix} \right\| \leq 1,$$

whence $\|A_\varepsilon^{-1/2} X B_\delta^{-1/2}\| = \|S\| \leq 1$ holds, which is equivalent to

$$XA_\varepsilon^{-1}X = B_\delta^{1/2}(B_\delta^{-1/2} X A_\varepsilon^{-1} X B_\delta^{-1/2}) B_\delta^{1/2} \leq B_\delta.$$

$(2) \implies (3)$. From $XA_\varepsilon^{-1}X \leq B$ we conclude that

$$(A_\varepsilon^{-1/2} X A_\varepsilon^{-1/2})^2 = A_\varepsilon^{-1/2} X A_\varepsilon^{-1} X A_\varepsilon^{-1/2} \leq A_\varepsilon^{-1/2} B A_\varepsilon^{-1/2}.$$

Hence, $A_\varepsilon^{-1/2} X A_\varepsilon^{-1/2} \leq (A_\varepsilon^{-1/2} B A_\varepsilon^{-1/2})^{1/2}$, which yields that $X \leq A_\varepsilon \# B$. □

We are ready to prove Ando's characterization of the geometric mean.

Theorem 3.2.11 (Ando's characterization) *Let A and B be positive operators. Then*

$$A \# B = \max \left\{ X \geq 0 : \begin{bmatrix} A & X \\ X & B \end{bmatrix} \geq 0 \right\}.$$

Proof From the above proposition,

$$\left\{ X \geq 0 : \begin{bmatrix} A & X \\ X & B \end{bmatrix} \geq 0 \right\} \subseteq \{ X \geq 0 : X \leq A \# B \}.$$

To show this theorem, it is therefore enough to prove

$$\begin{bmatrix} A & A\#B \\ A\#B & B \end{bmatrix} \geq 0.$$

Thanks to $(A_\varepsilon \# B_\delta) A_\varepsilon^{-1} (A_\varepsilon \# B_\delta) = B_\delta$, we get

$$\begin{bmatrix} A_\varepsilon & A_\varepsilon \# B_\delta \\ A_\varepsilon \# B_\delta & B_\delta \end{bmatrix} \geq 0,$$

which implies the desired result. □

If X is self-adjoint and $\begin{bmatrix} A & X \\ X^* & B \end{bmatrix} \geq 0$, then $\pm X \leq A\#B$. To see this, we use Theorem 2.4.1 to get $A \geq XB^{-1}X$. Then

$$B^{-1/2} A B^{-1/2} \geq B^{-1/2} X B^{-1} X B^{-1/2} = \left(B^{-1/2} X B^{-1/2}\right)^2,$$

whence, $\left(B^{-1/2} A B^{-1/2}\right)^{1/2} \geq B^{-1/2} X B^{-1/2}$. Therefore,

$$A\#B = B^{1/2} \left(B^{-1/2} A B^{-1/2}\right)^{1/2} B^{1/2} \geq X$$

Thus, we get a version of Theorem 3.2.11 as follows (See Proposition 2.4.19).

Theorem 3.2.12 *Let A and B be positive operators. Then*

$$A\#B = \max \left\{ X : X = X^* \text{ and } \begin{bmatrix} A & X \\ X & B \end{bmatrix} \geq 0 \right\}.$$

In [88] Fujii et al. reviewed the Pedersen and Takesaki equation $XBX = C$ [193] as follows. This equation has been studied in the setting of Hilbert C^*-modules in [75].

Theorem 3.2.13 *Let B and C be positive operators with B being invertible. Then the following statements are equivalent:*

1. *The Riccati equation $XBX = C$ has a positive solution.*
2. $(B^{1/2} C B^{1/2})^{1/2} \leq kB$.
3. *There exists the minimum of $\{X \geq 0 : C \leq XBX\}$.*
4. *There exists the minimum of $\left\{ X \geq 0 : \begin{bmatrix} 1 & C^{1/2} \\ C^{1/2} & XBX \end{bmatrix} \geq 0 \right\}$.*

The next result is related to the PPT matrices of operators.

3.2 Operator Means

Theorem 3.2.14 *If* $M = \begin{bmatrix} A & B \\ B^* & C \end{bmatrix}$ *is PPT, then there exists a partial isometry U such that*

$$|B| \le \frac{A\#C + U^*(A\#C)U}{2}.$$

Proof We can assume that $A > 0$. Since M is PPT, we have

$$\begin{bmatrix} C & 0 \\ 0 & C \end{bmatrix} \ge \begin{bmatrix} BA^{-1}B^* & 0 \\ 0 & B^*A^{-1}B \end{bmatrix} = \begin{bmatrix} 0 & -B \\ -B^* & 0 \end{bmatrix} \begin{bmatrix} A & 0 \\ 0 & A \end{bmatrix}^{-1} \begin{bmatrix} 0 & -B \\ -B^* & 0 \end{bmatrix}.$$

It follows from Theorem 2.4.1 that

$$\begin{bmatrix} A & 0 & 0 & -B \\ 0 & A & -B^* & 0 \\ 0 & -B & C & 0 \\ -B^* & 0 & 0 & C \end{bmatrix} \ge 0.$$

From Theorem 3.2.12 we conclude that

$$\begin{bmatrix} 0 & -B \\ -B^* & 0 \end{bmatrix} \le \begin{bmatrix} A & 0 \\ 0 & A \end{bmatrix} \# \begin{bmatrix} C & 0 \\ 0 & C \end{bmatrix} = \begin{bmatrix} A\#C & 0 \\ 0 & A\#C \end{bmatrix},$$

whence

$$\begin{bmatrix} A\#C & B \\ B^* & A\#C \end{bmatrix} \ge 0.$$

Take U to be the partial isometry in the polar decomposition of B. Then

$$\begin{bmatrix} -U^* & I \\ 0 & 0 \end{bmatrix} \begin{bmatrix} A\#C & B^* \\ B & A\#C \end{bmatrix} \begin{bmatrix} -U & 0 \\ I & 0 \end{bmatrix} \ge 0.$$

Hence,

$$\begin{bmatrix} U^*(A\#C)U - B^*U - U^*B + A\#C & 0 \\ 0 & 0 \end{bmatrix} \ge 0.$$

It follows from $U^*B = B^*U = |B|$ that

$$U^*(A\#C)U - |B| - |B| + A\#C \ge 0,$$

which gives the required inequality. □

Since $A\#A^{-1} = I$, we get the next result.

Corollary 3.2.15 *If* $\begin{bmatrix} A & B \\ B^* & A^{-1} \end{bmatrix}$ *is PPT, then* $\|B\| \le 1$.

In the setting of matrices, we use the inverse of A in equation $A\#B = A^{1/2}(A^{-1/2} BA^{-1/2})^{1/2} A^{1/2}$. An interesting question arises when we replace $A^{-1/2}$ with $(A^{1/2})^\dagger$. This leads us to consider the expression:

$$A^{1/2} \left((A^{1/2})^\dagger B (A^{1/2})^\dagger \right)^{1/2} A^{1/2}.$$

Fujii [85] showed that if matrices A and B are positive semidefinite, then

$$A\#B \leq A^{1/2} \left((A^{1/2})^\dagger B (A^{1/2})^\dagger \right)^{1/2} A^{1/2}.$$

Moreover, if the kernel inclusion $\ker A \subseteq \ker B$ is assumed, then the equality holds in the above inequality, as proved by Fujimoto and Seo [92].

To prove their result, they expressed A and B as $A = A_1 \oplus 0$ and $B = B_1 \oplus 0$, respectively, on $\operatorname{ran} A \oplus \ker A$. They observed that $A^\dagger = (A_1)^{-1} \oplus 0$. Fujii [86] presented the next theorem, whose proof requires the following lemma. We denote by $P_A = AA^\dagger$ the range projection of A, that is the projection onto the clusure of the range of A.

Lemma 3.2.16 *If* $\begin{bmatrix} A & X \\ X^* & B \end{bmatrix} \geq 0$, *then* $X = AA^\dagger X = P_A X$ *and* $B \geq X^* A^\dagger X$

Proof By the assumption, we have

$$\begin{bmatrix} (A^{1/2})^\dagger & 0 \\ 0 & I \end{bmatrix} \begin{bmatrix} A & X \\ X^* & B \end{bmatrix} \begin{bmatrix} (A^{1/2})^\dagger & 0 \\ 0 & I \end{bmatrix} = \begin{bmatrix} P_A & (A^{1/2})^\dagger X \\ X^* (A^{1/2})^\dagger & B \end{bmatrix} \geq 0.$$

It follows from

$$0 \leq \begin{bmatrix} I - (A^{1/2})^\dagger X \\ 0 & I \end{bmatrix}^* \begin{bmatrix} P_A & (A^{1/2})^\dagger X \\ X^* (A^{1/2})^\dagger & B \end{bmatrix} \begin{bmatrix} I - (A^{1/2})^\dagger X \\ 0 & I \end{bmatrix}$$

$$= \begin{bmatrix} P_A & 0 \\ 0 & B - X^* A^\dagger X \end{bmatrix}$$

that $B \geq X^* A^\dagger X$.

We now show that $X = P_A X$, which is equivalent to $\ker A \subseteq \ker X^*$. Suppose that $Ax = 0$. Putting $y = -X^* x / \|B\|$, we have

$$0 \leq \left\langle \begin{bmatrix} A & X \\ X^* & B \end{bmatrix} \begin{bmatrix} x \\ y \end{bmatrix}, \begin{bmatrix} x \\ y \end{bmatrix} \right\rangle$$

$$= \langle Xy, x \rangle + \langle X^* x, y \rangle + \langle By, y \rangle$$

$$= \frac{-2}{\|B\|} \|X^* x\|^2 + \frac{1}{\|B\|^2} \langle BX^* x, X^* x \rangle$$

$$\leq -\frac{\|X^* x\|^2}{\|B\|} \leq 0.$$

3.2 Operator Means

Thus, $X^*x = 0$, that is ker $A \subseteq$ ker X^*. □

Theorem 3.2.17 *Suppose that A and B are positive semidefinite matrices. Then*

$$A\#B \leq A^{1/2}\left(\left(A^{1/2}\right)^{\dagger} B \left(A^{1/2}\right)^{\dagger}\right)^{1/2} A^{1/2}.$$

In particular, the equality holds if and only if the projection $P_A = AA^{\dagger}$ onto ran A commutes with B.

Proof For the first half, by employing Theorem 3.2.11, it suffices to show that if $\begin{bmatrix} A & X \\ X & B \end{bmatrix} \geq 0$, then

$$X \leq A^{1/2}\left(\left(A^{1/2}\right)^{\dagger} B \left(A^{1/2}\right)^{\dagger}\right)^{1/2} A^{1/2}.$$

We use the facts that $\left(A^{1/2}\right)^{\dagger} = \left(A^{\dagger}\right)^{1/2}$, and that $\begin{bmatrix} A & X \\ X & B \end{bmatrix} \geq 0$ for the positive semidefinite matrix X. It follows that $X = AA^{\dagger}X = P_A X$ and $B \geq XA^{\dagger}X$, by Lemma 3.2.16.

Now, since $B \geq XA^{\dagger}X$, we have

$$\left(\left(A^{1/2}\right)^{\dagger} B \left(A^{1/2}\right)^{\dagger}\right)^{1/2} \geq \left(A^{1/2}\right)^{\dagger} X \left(A^{1/2}\right)^{\dagger}.$$

The equality $X = P_A X$ yields that

$$A^{1/2}\left(\left(A^{1/2}\right)^{\dagger} B \left(A^{1/2}\right)^{\dagger}\right)^{1/2} A^{1/2} \geq X.$$

Thus, we have $A^{1/2}\left(\left(A^{1/2}\right)^{\dagger} B \left(A^{1/2}\right)^{\dagger}\right)^{1/2} A^{1/2} \geq A\#B$.

Next, suppose that ker $A \subseteq$ ker B. Then we have ran $B \subseteq$ ran A and so

$$A^{1/2}\left(A^{1/2}\right)^{\dagger} B \left(A^{1/2}\right)^{\dagger} A^{1/2} = B.$$

Put $C = \left(A^{1/2}\right)^{\dagger} B \left(A^{1/2}\right)^{\dagger}$. From

$$Y = A^{1/2}\left(\left(A^{1/2}\right)^{\dagger} B \left(A^{1/2}\right)^{\dagger}\right)^{1/2} A^{1/2} = A^{1/2} C^{1/2} A^{1/2}$$

we deduce that

$$\begin{bmatrix} A & Y \\ Y & B \end{bmatrix} = \begin{bmatrix} A^{1/2} & 0 \\ 0 & A^{1/2} \end{bmatrix} \begin{bmatrix} I & C^{1/2} \\ C^{1/2} & C \end{bmatrix} \begin{bmatrix} A^{1/2} & 0 \\ 0 & A^{1/2} \end{bmatrix} \geq 0,$$

which ensures that $Y \leq A\#B$. We have shown $Y = A\#B$.

Now, we show the second half. The notation is as stated above. If $P_A = AA^{\dagger}(= A^{1/2}\left(A^{1/2}\right)^{\dagger})$ commutes with B, we get $P_A B P_A \leq B$. Therefore, we have

$$\begin{bmatrix} A & Y \\ Y & B \end{bmatrix} \geq \begin{bmatrix} A & Y \\ Y & P_A B P_A \end{bmatrix} = \begin{bmatrix} A^{1/2} & 0 \\ 0 & A^{1/2} \end{bmatrix} \begin{bmatrix} I & C^{1/2} \\ C^{1/2} & C \end{bmatrix} \begin{bmatrix} A^{1/2} & 0 \\ 0 & A^{1/2} \end{bmatrix} \geq 0,$$

which entails $Y \leq A \# B$. Thus, $Y = A \# B$.

Conversely, assume that the equality holds. Then $\begin{bmatrix} A & Y \\ Y & B \end{bmatrix} \geq 0$. Hence,

$$B \geq Y A^\dagger Y = A^{1/2} C A^{1/2} = P_A B P_A,$$

which implies that P_A commutes with B. □

Next, we give three other operator versions of the Cauchy–Schwarz inequality due to Jun Ichi Fujii [84].

Theorem 3.2.18 (Operator Cauchy–Schwarz inequality (I)) *For operators $X, Y \in \mathbb{B}(\mathcal{H})$, the inequality*

$$Y^* X^* Y X + X^* Y^* X Y \leq Y^* X^* X Y + X^* Y^* Y X = (XY)^* XY + (YX)^* YX$$

holds with equality only when X commutes with Y.

Proof The desired inequality follows from

$$0 \leq (XY - YX)^*(XY - YX) = (XY)^* XY - Y^* X^* Y X - X^* Y^* X Y + (YX)^* YX.$$

The equality holds only when $(XY - YX)^*(XY - YX) = 0$, and hence $XY - YX = 0$, that is, X commutes with Y. □

Theorem 3.2.19 (Operator Cauchy–Schwarz inequality (II)) *For operators $X, Y \in \mathbb{B}(\mathcal{H})$, the inequality*

$$(X^* Y) \otimes (Y^* X) + (Y^* X) \otimes (X^* Y) \leq (X^* X) \otimes (Y^* Y) + (Y^* Y) \otimes (X^* X)$$

holds with equality only when X and Y are linearly dependent.

Proof The inequality follows from

$$0 \leq (X \otimes Y - Y \otimes X)^*(X \otimes Y - Y \otimes X)$$
$$= (X^* X) \otimes (Y^* Y) - (X^* Y) \otimes (Y^* X) - (Y^* X) \otimes (X^* Y) + (Y^* Y) \otimes (X^* X).$$

Thereby the equality condition is $X \otimes Y = Y \otimes X$, which implies the linear dependence of X and Y. □

3.2 Operator Means

Theorem 3.2.20 (Operator Cauchy–Schwarz inequality (III)) *Let $X, Y \in \mathbb{B}(\mathcal{H})$. Consider the polar decomposition $Y^*X = U|Y^*X|$ of Y^*X, where $U \in \mathbb{B}(\mathcal{H})$ is the partial isometry from $\overline{\mathrm{ran}X^*Y}$ to $\overline{\mathrm{ran}Y^*X}$. Then*

$$|Y^*X| \leq X^*X \# U^*Y^*YU = |X|^2 \# |YU|^2 \quad (\textit{respectively,} \ |X^*Y| \leq |Y|^2 \# |XU^*|^2).$$

If X (respectively, Y) is an invertible operator in $\mathbb{B}(\mathcal{H})$ or operators $\{X, YU\}$ (respectively, $\{Y, XU^*\}$) are linearly dependent, then the equality holds.

Proof Since $X^*Y = U^*|X^*Y|$ is the polar decomposition of X^*Y, we have to only show the former case. By $U^*Y^*X = |Y^*U| = X^*YU$, we have

$$0 \leq \begin{bmatrix} I & 0 \\ 0 & U \end{bmatrix}^* \begin{bmatrix} X^*X & X^*Y \\ Y^*X & Y^*Y \end{bmatrix} \begin{bmatrix} I & 0 \\ 0 & U \end{bmatrix}$$

$$= \begin{bmatrix} X^*X & X^*YU \\ U^*Y^*X & U^*Y^*YU \end{bmatrix} = \begin{bmatrix} X^*X & |Y^*X| \\ |Y^*X| & U^*Y^*YU \end{bmatrix}.$$

It follows from Theorem 3.2.11 that $|Y^*X| \leq X^*X \# U^*Y^*YU$. Suppose $X \in \mathbb{B}(\mathcal{H})$ is invertible. Then

$$X^*X \# U^*Y^*YU = |X|^2 \# U^*Y^*X(X^*X)^{-1}X^*YU$$
$$= |X|(I \# |X|^{-1}|Y^*X|(X^*X)^{-1}|Y^*X||X|^{-1})|X|$$
$$= |X|(I \# (|X|^{-1}|Y^*X||X|^{-1})^2)|X|$$
$$= |X||X|^{-1}|Y^*X||X|^{-1}|X| = |Y^*X|.$$

Also, suppose $sX = rYU$ for some scalars s and r. Then

$$|sr|(X^*X \# U^*Y^*YU) = (|s|^2 X^*X) \# (|r|^2 U^*Y^*YU)$$
$$= (|r|^2 U^*Y^*YU) \# (|r|^2 U^*Y^*YU) = |r|^2 U^*Y^*YU.$$

On the other hand, the relation

$$|sr|^2|Y^*X|^2 = |r|^2(sX)^*Y(UU^*)Y^*(sX) = |r|^4(U^*Y^*YU)^2$$

implies $|sr||Y^*X| = |r|^2 U^*Y^*YU$, which shows the equality. □

In the above theorem, we consider rank-one operators $X = x \otimes \overline{e_1}$ and $Y = y \otimes \overline{e_1}$ in $\mathbb{B}(\mathcal{H})$, where (e_j) denotes the standard basis of a separable Hilbert space $\mathcal{H}$. Moreover, we see that the equality is never satisfied unless x and y are linearly dependent.

Next, we investigate the means of projections. Let P and Q be projections with $P + Q \neq 0$. Let (R, S) be positive operators on $\ker(P + Q)^\perp$ such that $P = (P + Q)^{1/2} R(P + Q)^{1/2}$ and $Q = (P + Q)^{1/2} S(P + Q)^{1/2}$. On the direct sum $\ker(P + Q)^\perp = (\mathrm{ran}(P) \cap$

$\mathrm{ran}(Q)) \oplus (\mathrm{ran}(P) \cap \mathrm{ran}(Q)^\perp) \oplus (\mathrm{ran}(P)^\perp \cap \mathrm{ran}(Q))$, the operators R, S can be denoted as

$$R = \begin{bmatrix} 2^{-1}I & 0 & 0 \\ 0 & I & 0 \\ 0 & 0 & 0 \end{bmatrix} \quad \text{and} \quad S = \begin{bmatrix} 2^{-1}I & 0 & 0 \\ 0 & 0 & 0 \\ 0 & 0 & I \end{bmatrix}.$$

Thus, for $f \in OM_+^1$,

$$\Phi_{\tilde{f}}(P, Q) = (P + Q)^{1/2} \begin{bmatrix} 2^{-1}I & 0 & 0 \\ 0 & \Phi_{\tilde{f}}(I, 0)I & 0 \\ 0 & 0 & \Phi_{\tilde{f}}(0, I)I \end{bmatrix} (P + Q)^{1/2}$$

$$= P \wedge Q + \Phi_{\tilde{f}}(I, 0)(P - P \wedge Q) + \Phi_{\tilde{f}}(0, I)(Q - P \wedge Q).$$

Recall that $P \wedge Q$ is the projection onto $\mathrm{ran}(P) \cap \mathrm{ran}(Q)$.

Proposition 3.2.21 *Let P and Q be projections. Then,*

$$P \sigma Q = P \wedge Q + a(P - P \wedge Q) + b(Q - P \wedge Q),$$

where $a := f_\sigma(0+)$ and $b := \tilde{f}_\sigma(0+)$.

Now, let σ be an operator mean. The following binary operations:

$$A \tilde{\sigma} B := B \sigma A, \quad A \sigma^* B := \text{s-}\lim_{\varepsilon \to 0}(A_\varepsilon^{-1} \sigma B_\varepsilon^{-1})^{-1}, \quad \text{and} \quad A \sigma^\perp B := A \tilde{\sigma}^* B$$

correspond to operator monotone functions

$$\tilde{f}_\sigma(t) = t f_\sigma(1/t), \quad f_\sigma^*(t) = f_\sigma(1/t)^{-1}, \quad \text{and} \quad f_\sigma^\perp(t) = t/f_\sigma(t).$$

These imply the binary operations $\tilde{\sigma}$, σ^*, and $\sigma^\perp$ are in Σ^1. We show some properties of $\tilde{\sigma}$, σ^*, and $\sigma^\perp$ on Σ^1. It follows from $\tilde{\tilde{\sigma}} = \sigma$, $(\sigma^*)^* = \sigma$, and $(\sigma^\perp)^\perp = \sigma$ that these transforms are bijective maps on Σ^1. By a simple calculation, we have the following result.

Proposition 3.2.22 *Let $\sigma_1, \sigma_2 \in \Sigma^1$. If $\sigma_1 \leq \sigma_2$, then*

$$\tilde{\sigma}_1 \leq \tilde{\sigma}_2, \quad \sigma_2^* \leq \sigma_1^*, \quad \text{and} \quad \sigma_2^\perp \leq \sigma_1^\perp.$$

Corollary 3.2.23

$$! = \nabla^* \leq \lambda^* = \lambda^\perp \leq \#^* = \# \leq \lambda \leq \nabla.$$

The injective map $f \mapsto \hat{f} := \frac{1+f}{1+f}$ is called the *Barbour transform* on OM_+. This map plays an important role in the analysis of OM_+ [143, 185]. The Barbour transform has the following properties:

3.2 Operator Means

$$\widehat{OM_+} = OM_+^1\setminus\{1\}, \quad \widehat{OM_+^1} = \{f \in OM_+^1 : f_! \leq f \leq f_\nabla\},$$

$$\widehat{f^\perp} = (\hat{f})^\perp, \quad \widehat{(\tilde{f})} = (\hat{f})^*, \text{ and } \widehat{(f^*)} = \tilde{\hat{f}} \quad (f \in OM_+^1).$$

Next, let f be a positive continuous function on $(0, \infty)$. It is known [19, Theorem V. 25] that f is operator monotone if and only if f is operator concave. Using this, the following is obtained.

Proposition 3.2.24 *Let σ be an operator mean. Then the double inequality $!_\alpha \leq \sigma \leq \nabla_\alpha$ holds, where $\alpha := \frac{df_\sigma}{dx}\big|_{x=1}$.*

Proof Put $f := f_\sigma$. Note that the tangent line of $f(x)$ at $x = 1$ can be written as $y = f'(1)x + (1 - f'(1))$. Since f is an increasing function, we have

$$(f'(1)x + (1 - f'(1))) \geq f(x)$$

for all $x > 0$, which implies $1 - f'(1) \geq f(0+) \geq 0$ and $f'(1) \geq 0$. Hence, $\sigma \leq \nabla_{f'(1)}$. Applying this argument to f^*, we obtain $\sigma^* = \sigma_{f^*} \leq \nabla_{(f^*)'(1)} = \nabla_{f'(1)} = (!_{f'(1)})^*$, which implies $!_{f'(1)} \leq \sigma$. □

We call the positive number $\alpha \left(:= \frac{df_\sigma}{dx}\big|_{x=1}\right)$ the *weight of operator mean σ*.

An operator mean σ having $\tilde{\sigma} = \sigma$ is called a *symmetric operator mean*. Since $(\tilde{f}_\sigma)'(1) = 1 - f'_\sigma(1) = f'_\sigma(1)$, we have $f'_\sigma(1) = \frac{1}{2}$ and $! \leq \sigma \leq \nabla$.

The following result implies that there exists some non-symmetry operator mean σ such that $! \leq \sigma \leq \nabla$.

Proposition 3.2.25

$$\{\sigma \in \Sigma^1 : \tilde{\sigma} = \sigma\} \subsetneq \{\sigma \in \Sigma^1 : ! \leq \sigma \leq \nabla\}.$$

Proof Put $f(t) := \frac{3t+1}{t+3}$. Then the operator mean σ_f which corresponds f is in

$$\{\sigma \in \Sigma^1 : ! \leq \sigma \leq \nabla\}\setminus\{\sigma \in \Sigma^1 : \tilde{\sigma} = \sigma\}.$$

Indeed, from the direct calculation $\tilde{f} \neq f$, that is, $\tilde{\sigma}_f \neq \sigma_f$. Moreover, $\frac{df}{dx}\big|_{x=1} = \frac{1}{2}$. Hence, from Proposition 3.2.24 we have

$$! \leq \sigma_f \leq \nabla.$$

□

Let us consider the Barbour transform $f \mapsto \hat{f} = \frac{t+f}{1+f}$. For $f \in OM_+^1$ with $f^* \neq f$, the operator mean $\sigma_{\hat{f}}$ is not symmetric, but $! \leq \sigma_{\hat{f}} \leq \nabla$.

An operator mean σ having $\sigma^* = \sigma$ is called a *self-adjoint operator mean*. The geometric mean is an easy-to-understand and important example of a self-adjoint operator mean. A nontrivial operator mean σ (that is, $\sigma \neq l$ and $\sigma \neq r$) is self-adjoint if and only if it can be written as the Barbour transform of a symmetric operator connection, namely

$$\{f \in OM_+^1 \setminus \{1\} : f = f^*\} = \{\hat{f} : f = \tilde{f}, f \in OM_+\}.$$

This equality comes from the injectivity of the Barbour transform $\hat{\cdot}$, $\widehat{OM_+} = OM_+^1 \setminus \{1\}$, $\widehat{(\tilde{f})} = (\hat{f})^*$, and $\widehat{(f^*)} = \widetilde{\hat{f}}$ for all $f \in OM_+^1$.

For example, for $r \in [0, 1]$, the function $\widehat{t^{1-r}} = \frac{t^{1-r}+t}{t^{1-r}+1}$ is symmetric. The operator mean which corresponds to this function is called the *Lehmar operator mean*.

Example 3.2.26 For an arbitrary operator mean σ, the operator mean $(\sigma + \tilde{\sigma})/2$ is a symmetric mean. The *Heinz means* σ_{h_α} defined as

$$A\sigma_{h_\alpha} B := (A\#_\alpha B + A\#_{1-\alpha} B)/2 \quad (0 \leq \alpha \leq 1)$$

are typical examples.

It is well-known as Young inequality that for $0 \leq \nu \leq 1$ and $a, b \geq 0$,

$$\nu a + (1-\nu)b \geq a^\nu b^{1-\nu} \geq \frac{a+b-|a-b|}{2}.$$

When $\nu = \frac{1}{2}$, the inequality $a^{\frac{1}{2}} b^{\frac{1}{2}} \geq \frac{a+b-|a-b|}{2}$ is called a *reverse Cauchy–Schwarz inequality*. A natural matrix form of the reverse Cauchy–Schwarz inequality can be written as

$$A^{\frac{1}{2}}(A^{-\frac{1}{2}} B A^{-\frac{1}{2}})^{\frac{1}{2}} A^{\frac{1}{2}} \geq \frac{A+B}{2} - \frac{|A-B|}{2},$$

where A and B are positive semidefinite matrices. Furuichi [93] pointed out that the trace inequality $\operatorname{tr}(A\sigma B) \geq \operatorname{tr}(A\nabla B - \frac{1}{2}|A - B|)$ is not valid in general.

In [123], Hoa et al. showed that the inequality

$$A\sigma B \geq A\nabla B - \frac{1}{2}|A - B| \tag{3.2.2}$$

holds for any operator mean σ and positive semidefinite matrices A and B with $AB + BA \geq 0$. Later, Hoa [122] proved that if σ is symmetric and inequality (3.2.2) holds for any positive semidefinite matrices A and B, then $\sigma = \nabla$.

In this section, we extend the above characterization as follows; see [186].

3.2 Operator Means

Theorem 3.2.27 *Let σ be an operator mean in the sense of Kubo–Ando. Then the following are equivalent:*

1. *There exists $\beta \in [0, 1]$ such that $\sigma = \nabla_\beta$;*
2. *$A\sigma B \geq A\nabla_\alpha B - \max\{\alpha, 1 - \alpha\}|A - B|$ holds for all positive semidefinite matrices A and B;*
3. *$\operatorname{tr}(A\sigma B) \geq \operatorname{tr}(A\nabla_\alpha B - \max\{\alpha, 1 - \alpha\}|A - B|)$ holds for all positive semidefinite matrices A and B.*

We need the following result to prove Theorem 3.2.27.

Proposition 3.2.28 *Let $\alpha \in [0, 1]$, and let $\alpha_0 = \max\{\alpha, 1 - \alpha\}$. If $A\nabla_\alpha B \geq \alpha_0 |A - B|$ for positive semidefinite matrices A and B, then*

$$A\sigma B \geq A\nabla_\alpha B - \alpha_0 |A - B|$$

holds for any operator mean σ.

Proof Note that

$$\begin{aligned} A &= (1 - \alpha)A + \alpha B + \alpha(A - B) \\ &\geq A\nabla_\alpha B - \alpha |A - B| \\ &\geq A\nabla_\alpha B - \alpha_0 |A - B| \end{aligned}$$

and

$$\begin{aligned} B &= (1 - \alpha)A + \alpha B - (1 - \alpha)(A - B) \\ &\geq A\nabla_\alpha B - (1 - \alpha)|A - B| \\ &\geq A\nabla_\alpha B - \alpha_0 |A - B|. \end{aligned}$$

Thus

$$\begin{aligned} A\sigma B &\geq (A\nabla_\alpha B - \alpha_0 |A - B|)\sigma(A\nabla_\alpha B - \alpha_0 |A - B|) \\ &= A\nabla_\alpha B - \alpha_0 |A - B|. \end{aligned}$$

$\square$

Corollary 3.2.29 ([123]) *Let A and B be positive semidefinite matrices such that $AB + BA \geq 0$. Then*

$$A\sigma B \geq A\nabla B - \frac{1}{2}|A - B|$$

holds for all operator means σ.

Proof In Proposition 3.2.28, take $\alpha = \frac{1}{2}$. Since $AB + BA \geq 0$, we have

$$\frac{A+B}{2} = \frac{(A^2 + B^2 + (AB+BA))^{\frac{1}{2}}}{2}$$

$$\geq \frac{(A^2 + B^2 - (AB+BA))^{\frac{1}{2}}}{2}$$

$$= \frac{|A-B|}{2}.$$

□

Lemma 3.2.30 *Let $\alpha \in [0, 1]$ and $\alpha_0 = \max\{\alpha, 1-\alpha\}$, and let σ be an operator mean. If the inequality*

$$\operatorname{tr}(A\nabla_\alpha B - \alpha_0 |A - B|) \leq \operatorname{tr}(A\sigma B)$$

holds for all positive semidefinite matrices A and B, then $\sigma = \nabla_\beta$ for some $\beta \in [0, 1]$.

Proof Let P and Q be projections on a Hilbert space $\mathscr{H}$ with $P \wedge Q = 0$. From the assumption, we have

$$\operatorname{tr}(P\nabla_\alpha Q - \alpha_0 |P - Q|) \leq \operatorname{tr}(P\sigma Q). \tag{3.2.3}$$

Furthermore, $P \sigma Q = aP + bQ$ by Proposition 3.2.21, where $a = \inf_x 1\sigma x$, $b = \lim_{x\to\infty} \frac{1\sigma x}{x}$. Choose two projections

$$P_\theta := \begin{bmatrix} 1 & 0 \\ 0 & 0 \end{bmatrix} \quad \text{and} \quad Q_\theta := \begin{bmatrix} \cos^2\theta & \cos\theta\sin\theta \\ \cos\theta\sin\theta & \sin^2\theta \end{bmatrix} \quad (0 < \theta < \frac{\pi}{2}).$$

Then $P_\theta \wedge Q_\theta = 0$,

$$P_\theta \nabla_\alpha Q_\theta = \begin{bmatrix} (1-\alpha) + \alpha\cos^2\theta & \alpha\cos\theta\sin\theta \\ \alpha\cos\theta\sin\theta & \alpha\sin^2\theta \end{bmatrix}$$

and

$$|P_\theta - Q_\theta| = \begin{bmatrix} \sin\theta & 0 \\ 0 & \sin\theta \end{bmatrix}.$$

Letting $\theta \to +0$ and using (3.2.3), we get

$$\lim_{\theta\to 0} \{\operatorname{tr}(P_\theta \nabla_\alpha Q_\theta - \alpha_0 |P_\theta - Q_\theta|)\} \leq \lim_{\theta\to 0} \operatorname{tr}\left(\begin{bmatrix} a + b\cos^2\theta & b\cos\theta\sin\theta \\ b\cos\theta\sin\theta & b\sin^2\theta \end{bmatrix}\right)$$

or

$$\operatorname{tr}\left(\begin{bmatrix} 1 & 0 \\ 0 & 0 \end{bmatrix}\right) = \operatorname{tr}\left(\begin{bmatrix} (1-\alpha)+\alpha & 0 \\ 0 & 0 \end{bmatrix} - \begin{bmatrix} 0 & 0 \\ 0 & 0 \end{bmatrix}\right) \leq \operatorname{tr}\left(\begin{bmatrix} a+b & 0 \\ 0 & 0 \end{bmatrix}\right).$$

Therefore, we have

$$1 \leq a + b. \tag{3.2.4}$$

Furthermore, since $1\sigma t$ is an operator monotone function, there is a positive Radon measure μ on $[0, \infty]$ such that

$$1\sigma t = a + bt + \int_{(0,\infty)} \frac{(x+1)t}{x+t} d\mu(x).$$

Therefore,

$$1\sigma 1 = a + b + \int_{(0,\infty)} d\mu(t) = 1.$$

Hence, $\mu = 0$ in light of (3.2.4). We have

$$1\sigma t = a + bt, \quad 1 = a + b.$$

Thus, $\sigma = \nabla_\beta$ for some $\beta \in [0, 1]$. $\square$

Proof of Theorem 3.2.27.
$1 \Longrightarrow 2$: Put $\alpha_0 := \max\{\alpha, 1 - \alpha\}$. From the proof of Proposition 3.2.28, we have

$$A \geq A\nabla_\alpha B - \alpha_0 |A - B|$$

and

$$B \geq A\nabla_\alpha B - \alpha_0 |A - B|.$$

Then

$$\begin{aligned} A\sigma B &= A\nabla_\beta B \\ &= (1 - \beta)A + \beta B \\ &\geq (1 - \beta)(A\nabla_\alpha B - \alpha_0 |A - B|) + \beta(A\nabla_\alpha B - \alpha_0 |A - B|) \\ &= A\nabla_\alpha B - \alpha_0 |A - B|. \end{aligned}$$

$2 \Longrightarrow 3$: Immediate.
$3 \Longrightarrow 1$: From Lemm 3.2.30, the operator mean σ should be ∇_β for some $\beta \in [0, 1]$. $\square$
Surprisingly, according to (2) $\Longleftrightarrow$ (3), there is no difference between the operator inequality and its trace evaluation.

3.3 Positive and Completely Positive Linear Maps

Techniques based on 2×2 operator matrices are crucial in transitioning from completely positive maps to completely bounded ones. There is also a canonical way to create an *operator system* (that is, a self-adjoint subspace of a unital C^*-algebra containing the identity) from an *operator space* (that is, a subspace of a C^*-algebra). This method is known as *Paulsen's 2×2 matrix trick*. It relies on the fact that the operator matrix $\begin{bmatrix} A & B \\ B^* & C \end{bmatrix}$ is positive if and only if A and C are positive and $A_\varepsilon^{-1/2} B C_\varepsilon^{-1/2}$ is in the unit ball of $\mathbb{B}(\mathscr{H})$ for all scalars $\varepsilon > 0$. This is a consequence of Theorems 2.4.12 and 2.4.3. For more information see [72, 190, 191], which also cover interesting results related to operator matrices in addition to the methods mentioned in this section.

A (not necessarily linear) map $\Phi : \mathscr{A} \to \mathscr{B}$ between C^*-algebras is said to be a *$*$-map* or a *self-adjoint map* if $\Phi(A^*) = \Phi(A)^*$. It is called a *positive map* if $\Phi(A) \geq 0$ whenever $A \geq 0$.

If $\Phi : \mathscr{A} \to \mathscr{B}$ is a positive linear map, then it maps $\mathscr{A}_{\text{sa}}$ into $\mathscr{B}_{\text{sa}}$ since each self-adjoint element is a difference of two positive elements. By employing the Cartesian decomposition of each element of $\mathscr{A}$ into its real and imaginary parts, we can conclude that Φ is self-adjoint. However, in the framework of real Hilbert spaces, this assertion is not generally valid. For example, assume that $\mathscr{H} = \mathbb{R}^2$. Then the linear functional $\varphi : \mathbb{M}_2(\mathbb{R}) \to \mathbb{R}$ defined by

$$\varphi \begin{bmatrix} \alpha & \beta \\ \gamma & \delta \end{bmatrix} = \alpha + \beta + \delta$$

is positive but not self-adjoint; see [204].

The set of all positive linear maps Φ from $\mathscr{A}$ into $\mathscr{B}$ is denoted by $\mathbb{B}_+(\mathscr{A}, \mathscr{B})$. It is called *strictly positive* if $A > 0$ implies that $\Phi(A) > 0$. We say that Φ is unital or normalized if $\mathscr{A}$ and $\mathscr{B}$ are unital and Φ preserves the unit. In this case, we simply denote both identities of $\mathscr{A}$ and $\mathscr{B}$ by I.

A map $\Phi : \mathscr{A} \to \mathscr{B}$ is called *n-positive* if the map $\Phi_n : \mathbb{M}_n(\mathscr{A}) \to \mathbb{M}_n(\mathscr{B})$ defined by

$$\Phi_n([a_{ij}]_{n \times n}) = [\Phi[a_{ij}]]_{n \times n}$$

is positive. Note that $\Phi \otimes \text{id}_n$ is also used for Φ_n.

For $m \geq n$, we can embed $\mathbb{M}_n(\mathscr{A})$ into $\mathbb{M}_m(\mathscr{A})$ via

$$A \mapsto A' = \begin{bmatrix} A & 0_{n \times (m-n)} \\ 0_{(m-n) \times n} & 0_{(m-n) \times (m-n)} \end{bmatrix},$$

where $0_{p \times q}$ denotes the $p \times q$ zero matrix. We then deduce that if for $m > n$, Φ is m-positive, then it is also n-positive. The map Φ is said to be *completely positive* if it is n-positive for every $n \in \mathbb{N}$. This notion serves as a natural generalization of positive linear functionals. If Φ is completely positive, then so is Φ_n for every $n \geq 1$. Since $\|A\| = \|A'\|$, we can see that $\|\Phi_n\| \leq \|\Phi_m\|$.

It has been proved that a positive linear map $\Phi : \mathscr{A} \to \mathscr{B}$ between unital C^*-algebras is completely positive if either $\mathscr{A}$ or $\mathscr{B}$ is commutative; see [12, 210, 211] and Theorem 3.3.8. In [40], it is demonstrated that every positive linear map $\Phi : \mathscr{A} \to \mathbb{M}_2$ (respectively, $\Phi : \mathbb{M}_2 \to \mathscr{B}$) is 2-positive if and only if $\mathscr{A}$ is commutative (respectively, $\mathscr{B}$ is commutative).

3.3 Positive and Completely Positive Linear Maps

Moreover, a n-positive linear map $\Phi : \mathscr{A} \to \mathscr{B}$ is $(n+1)$-positive if and only if either $\mathscr{A}$ or $\mathscr{B}$ only possesses irreducible representations with dimensions less than n [221].

A map Φ is said to be *completely bounded* if

$$\|\Phi\|_{cb} := \sup_n \|\Phi_n\| < \infty.$$

Note that if Φ is a bounded linear map, then so is Φ_n. In fact, for $A = [A_{ij}]$ we have

$$\|\Phi_n(A)\| \leq \left(\sum_{i,j=1}^n \|\Phi(A_{ij})\|^2 \right)^{1/2}$$

$$\leq \|\Phi\| \left(\sum_{i,j=1}^n \|A_{ij}\|^2 \right)^{1/2} \leq \|\Phi\| \left(\sum_{i,j=1}^n \|A\|^2 \right)^{1/2} = n\|\Phi\|\|A\|.$$

To achieve the next theorem, we first need to establish the following lemma.

Lemma 3.3.1 *If* $\begin{bmatrix} A & B \\ B^* & A \end{bmatrix} \geq 0$, *then* $\|B\| \leq \|A\|$.

Proof From Theorem 2.4.1 we conclude that $B = A^{\frac{1}{2}} K A^{\frac{1}{2}}$ for some contraction K. Hence

$$\|B\| = \|A^{\frac{1}{2}} K A^{\frac{1}{2}}\| \leq \|A^{\frac{1}{2}}\| \|K\| \|A^{\frac{1}{2}}\| \leq \|A^{\frac{1}{2}}\|^2 = \|A\|.$$

$\square$

The first result of this section reads as follows.

Theorem 3.3.2 ([190, Proposition 3.6]) *If $\mathscr{A}$ and $\mathscr{B}$ are unital C^*-algebras and $\Phi : \mathscr{A} \to \mathscr{B}$ is completely positive, then Φ is completely bounded. Moreover,* $\|\Phi\|_{cb} = \|\Phi\|$.

Proof Let $A \in \mathbb{M}_n(\mathscr{A})$ with $\|A\| \leq 1$. It follows from Theorem 2.4.1 that $\begin{bmatrix} I & A \\ A^* & I \end{bmatrix} \in \mathbb{M}_{2n}(\mathscr{A})$ is positive, where we also denote the identity of $\mathbb{M}_n(\mathscr{A})$ by I. By the complete positivity of Φ we have

$$\Phi_{2n}\left(\begin{bmatrix} I & A \\ A^* & I \end{bmatrix} \right) = (\Phi_n)_2 \left(\begin{bmatrix} I & A \\ A^* & I \end{bmatrix} \right) = \begin{bmatrix} \Phi_n(I) & \Phi_n(A) \\ \Phi_n(A)^* & \Phi_n(I) \end{bmatrix} \geq 0.$$

It follows from Lemma 3.3.1 that $\|\Phi_n(A)\| \leq \|\Phi_n(I)\| = \|\Phi(I)\| \leq \|\Phi\|$. Hence, $\|\Phi\|_{cb} \leq \|\Phi\| \leq \|\Phi\|_{cb}$. Therefore, $\|\Phi\|_{cb} = \|\Phi\|$.

$\square$

Example 3.3.3 (i) Any $*$-homomorphism is completely positive.

(ii) The map $\Phi : \mathbb{M}_n \to \mathbb{M}_n$ defined by $\Phi(T) = (n-1)\,\text{tr}(T)I - T$ is $(n-1)$-positive, but not n-positive; see [40].

(iii) Any map of the form $\Phi(A) = \sum_{j=1}^n V_j^* A V_j$ on $\mathbb{B}(\mathcal{H})$ is positive, where V_j's are arbitrary operators. This is because if $A \in \mathbb{B}_+(\mathcal{H})$, then $V_j^* A V_j = (A^{1/2} V_j)^*(A^{1/2} V_j) \in \mathbb{B}_+(\mathcal{H})$. It is unital if and only if $\sum_{j=1}^n V_j^* V_j = I$.

(iv) If $x \in \mathcal{H}$, then $\varphi_x(A) = \langle Ax, x \rangle$ is a positive linear functional. It is a state on $\mathbb{B}(\mathcal{H})$ if and only if x is a unit vector.

(v) Recall that $\mathbb{M}_n$ equipped with the inner product $\langle A, B \rangle = \text{tr}(AB^*)$ forms a Hilbert space. If ρ is a linear functional on $\mathbb{M}_n$, then, by the Riesz representation Theorem 1.1.3, there exists a matrix $X \in \mathbb{M}_n$ such that $\rho(A) = \text{tr}(AX)$ for all $A \in \mathbb{M}_n$. It is evident that ρ is positive if and only if X is positive semidefinite. By considering the standard orthonormal basis $(e_i)_{1 \le i \le n}$ for $\mathbb{C}^n$, we obtain

$$\rho(A) = \text{tr}(X^{1/2} A X^{1/2}) = \sum_{i=1}^n \langle X^{1/2} A X^{1/2} e_i, e_i \rangle = \sum_{i=1}^n e_i^* X^{1/2} A X^{1/2} e_i$$

$$= \sum_{i=1}^n f_i^* A f_i,$$

where $f_i = X^{1/2} e_i$. Using the same reasoning as in part (iii), we conclude that ρ is completely positive; see also the proof $(4) \Rightarrow (1)$ of Theorem 3.5.1.

(vi) The transpose map $\Phi : \mathbb{M}_2 \to \mathbb{M}_2$, defined by $\Phi(A) = A^t = [A_{ji}]$ for $A = [A_{ij}] \in \mathbb{M}_2$, is positive since if $A = B^*B$ is positive, then $A^t = (B^*B)^t = B^t(B^t)^* \ge 0$. It is not 2-positive since

$$\begin{bmatrix} E_{11} & E_{12} \\ E_{21} & E_{22} \end{bmatrix} = \begin{bmatrix} e_1^* e_1 & e_1^* e_2 \\ e_2^* e_1 & e_2^* e_2 \end{bmatrix} = [e_1 \ e_2]^* [e_1 \ e_2] \ge 0$$

but

$$\Phi_2\left(\begin{bmatrix} E_{11} & E_{12} \\ E_{21} & E_{22} \end{bmatrix}\right) = \begin{bmatrix} E_{11} & E_{21} \\ E_{12} & E_{22} \end{bmatrix} = \begin{bmatrix} 1 & 0 & 0 & 0 \\ 0 & 0 & 1 & 0 \\ 0 & 1 & 0 & 0 \\ 0 & 0 & 0 & 1 \end{bmatrix}$$

is not positive since it has a negative eigenvalue. Since the norm of the above matrix is equal to 1 and the norm of $\begin{bmatrix} E_{11} & E_{12} \\ E_{21} & E_{22} \end{bmatrix}$ is 2, we observe that $\|\Phi_2\| \ne \|\Phi\|$. Indeed, $\|\Phi\| = 1 < 2 \le \|\Phi_2\|$.

The next result shows that every positive linear map is continuous.

3.3 Positive and Completely Positive Linear Maps

Theorem 3.3.4 *Every positive linear map $\Phi : \mathscr{A} \to \mathscr{B}$ between C^*-algebras is bounded.*

Proof Since $\mathscr{A}$ is linearly spanned by $S_+ = \{A \in \mathscr{A} : \|A\| \leq 1 \text{ and } A \geq 0\}$, it is enough to show that Φ is bounded on this set. Assume, to the contrary, that Φ is not bounded. Then there would exist a sequence (A_n) in S_+ such that $\|\Phi(A_n)\| \geq n^3$. Since $\sum_{n=1}^{\infty} A_n/n^2$ absolutely converges and $\mathscr{A}$ is a Banach space, $\sum_{n=1}^{\infty} A_n/n^2$ converges to an element $A \in \mathscr{A}$. Since $A \geq A_n/n^2$, we have $\Phi(A) \geq \Phi(A_n)/n^2$. Therefore, $\|\Phi(A)\| \geq \|\Phi(A_n)\|/n^2 \geq n$ for all n, which is a contradiction. $\square$

It is known [176, Corollar 3.3.4] that a bounded linear functional ϕ on a unital C^*-algebra is positive if and only if $\|\phi\| = \phi(I)$. We use this fact and follow a standard argument [191, Proposition 2.11] to prove the following extension, which can be regarded as a reverse of Theorem 3.3.4.

Theorem 3.3.5 *Let $\mathscr{A}$ and $\mathscr{B}$ be unital C^*-algebras and let $\Phi : \mathscr{A} \to \mathscr{B}$ be a unital map with $\|\Phi\| = 1$. Then Φ is positive.*

Proof By using the Gelfand–Naimark–Segal representation, we consider $\mathscr{B}$ as a C^*-subalgebras of $\mathbb{B}(\mathscr{H})$ for some Hilbert space $\mathscr{H}$. Let $x \in \mathscr{H}$ be a unit vector, and set

$$\phi_x(A) = \langle \Phi(A)x, x \rangle.$$

Then, we have

$$|\phi_x(A)| = |\langle \Phi(A)x, x \rangle| \leq \|\Phi(A)\| \|x\| \leq \|\Phi\| \|A\| \leq \|A\|,$$

whence ϕ_x is bounded and

$$\|\phi_x\| \leq 1 = \|\langle \Phi(I)x, x \rangle\| = \|\phi_x(I)\| \leq \|\phi_x\|.$$

Hence, we have $\|\phi_x\| = \phi_x(I)$. Consequently, ϕ_x is positive. Therefore, if A is positive, then $\langle \Phi(A)x, x \rangle \geq 0$ for all unit vectors x. As a result, $\Phi(A)$ is positive. This shows that Φ is a positive map. $\square$

The condition of being unital on Φ is necessary for the above theorem. For example, let Φ be the linear map on $\mathbb{M}_2$ defined as

$$\Phi\left(\begin{bmatrix} A_{11} & A_{12} \\ A_{21} & A_{22} \end{bmatrix}\right) = \begin{bmatrix} A_{11} & A_{12} \\ 0 & 0 \end{bmatrix}.$$

Then, $\|\Phi\| = \|\Phi(I)\| = 1$ and $\Phi(I) \geq 0$, but Φ is not positive. Indeed, since

$$\left\|\begin{bmatrix} A_{11} & A_{12} \\ A_{21} & A_{22} \end{bmatrix}\right\|^2 = \left\|\begin{bmatrix} A_{11} & A_{12} \\ A_{21} & A_{22} \end{bmatrix} \begin{bmatrix} A_{11} & A_{12} \\ A_{21} & A_{22} \end{bmatrix}^*\right\|$$

$$\geq \sup_{x\in\mathbb{C},|x|=1} \left|\left\langle \begin{bmatrix} |A_{11}|^2+|A_{12}|^2 & * \\ * & * \end{bmatrix} \begin{bmatrix} x \\ 0 \end{bmatrix}, \begin{bmatrix} x \\ 0 \end{bmatrix} \right\rangle\right| = \left\|\begin{bmatrix} A_{11} & A_{12} \\ 0 & 0 \end{bmatrix}\right\|^2$$

$$= \left\|\Phi\left(\begin{bmatrix} A_{11} & A_{12} \\ A_{21} & A_{22} \end{bmatrix}\right)\right\|^2,$$

we have $\|\Phi\| \leq 1$. On the contrary, $A = \begin{bmatrix} 1 & 1 \\ 1 & 1 \end{bmatrix}$ is positive, but $\Phi(A) = \begin{bmatrix} 1 & 1 \\ 0 & 0 \end{bmatrix}$ is not positive.

The following result presents a Cauchy–Schwarz inequality involving 2-positive maps.

Theorem 3.3.6 *If $\mathcal{A}$ and $\mathcal{B}$ are C^*-algebras and $\Phi : \mathcal{A} \to \mathcal{B}$ is 2-positive, then*

$$\|\Phi(A^*B)\| \leq \|\Phi(A^*A)\|^{1/2} \|\Phi(B^*B)\|^{1/2}.$$

Proof As mentioned in Chap. 1, the Gelfand–Naimark–Segal theorem allows us to consider $\mathcal{A}$ as a C^*-algebra consisting of bounded linear operators acting on a Hilbert space. It follows from Theorem 2.4.15, that $\begin{bmatrix} A^*A & A^*B \\ B^*A & B^*B \end{bmatrix} \geq 0$. Hence, $\begin{bmatrix} \Phi(A^*A) & \Phi(A^*B) \\ \Phi(B^*A) & \Phi(B^*B) \end{bmatrix} \geq 0$. It follows from Theorem 2.4.12 that there exists a contraction K such that $\Phi(A^*B) = \Phi(A^*A)^{1/2} K \Phi(B^*B)^{1/2}$. Therefore,

$$\|\Phi(A^*B)\| \leq \|\Phi(A^*A)^{1/2}\| \|K\| \|\Phi(B^*B)^{1/2}\| \leq \|\Phi(A^*A)\|^{1/2} \|\Phi(B^*B)\|^{1/2}.$$

□

The following theorem plays a crucial role in establishing various results. There are several versions of this theorem in different contexts. Interested readers are referred to [2, 170, 172] and the references therein.

Theorem 3.3.7 (Stinespring theorem) *[210] Suppose that Φ is a completely positive map from a unital C^*-algebra $\mathcal{A}$ with the identity I into $\mathbb{B}(\mathcal{H})$. Then there exist a representation π of $\mathcal{A}$ on some Hilbert space $\mathcal{K}$ and a bounded linear operator V from $\mathcal{H}$ into $\mathcal{K}$ such that $\|\Phi(I)\| = \|V\|$ and $\Phi(A) = V^*\pi(A)V$ for all $A \in \mathcal{A}$. In particular, if Φ is unital, then V is an isometry.*

Proof Let us define a symmetric bilinear map $\langle \cdot, \cdot \rangle$ on the algebraic tensor product $\mathcal{A} \otimes \mathcal{H}$ by

$$\langle A \otimes x, B \otimes y \rangle := \langle \Phi(B^*A)x, y \rangle$$

for $A, B \in \mathcal{A}, x, y \in \mathcal{H}$ and extend it linearly over the entire $\mathcal{A} \otimes \mathcal{H}$.

3.3 Positive and Completely Positive Linear Maps

Since Φ is completely positive, $\langle \cdot, \cdot \rangle$ is positive semidefinite, as demonstrated below:
For any $n \in \mathbb{N}$, $A_1, A_2, \ldots, A_n \in \mathscr{A}$, $x_1, x_2, \ldots, x_n \in \mathscr{H}$ it holds that

$$\left\langle \sum_{j=1}^n A_j \otimes x_j, \sum_{i=1}^n A_i \otimes x_i \right\rangle = \sum_{i,j=1}^n \langle \Phi(A_i^* A_j) x_j, x_i \rangle$$

$$= \left\langle \Phi_n([A_i^* A_j]) \begin{bmatrix} x_1 \\ \vdots \\ x_n \end{bmatrix}, \begin{bmatrix} x_1 \\ \vdots \\ x_n \end{bmatrix} \right\rangle \geq 0,$$

where the inner product on the right side is in the direct sum $\mathscr{H}^{\oplus n}$ of n copies of $\mathscr{H}$ introduced in Sect. 2.1.

By using standard calculations and the Cauchy–Schwartz inequality, we obtain

$$\mathscr{N} := \{u \in \mathscr{A} \otimes \mathscr{H} : \langle u, u \rangle = 0\} = \{u \in \mathscr{A} \otimes \mathscr{H} : \langle u, v \rangle = 0, \text{ for all } v \in \mathscr{A} \otimes \mathscr{H}\},$$

and hence $\mathscr{N}$ becomes a closed subspace of $\mathscr{A} \otimes \mathscr{H}$. Hence, the bilinear form

$$\langle u + \mathscr{N}, v + \mathscr{N} \rangle = \langle u, v \rangle$$

on the quotient space $(\mathscr{A} \otimes \mathscr{H})/\mathscr{N}$ defines an inner product. Let $\mathscr{K}$ denote the Hilbert space obtained from completing the inner product $(\mathscr{A} \otimes \mathscr{H})/\mathscr{N}$. Hence, $\mathscr{K}$ is the closure of $(\mathscr{A} \otimes \mathscr{H})/\mathscr{N}$.

For any $A \in \mathscr{A}$, let us consider the linear map $\pi_0(A) : \mathscr{A} \otimes \mathscr{H} \to \mathscr{A} \otimes \mathscr{H}$ defined as

$$\pi_0(A) \left(\sum_{i=1}^n A_i \otimes x_i \right) = \sum_{i=1}^n AA_i \otimes x_i.$$

Then,

$$\left\langle \pi_0(A) \sum_{i=1}^n A_i \otimes x_i, \pi_0(A) \sum_{i=1}^n A_i \otimes x_i \right\rangle$$

$$= \sum_{i,j=1}^n \langle AA_j \otimes x_j, AA_i \otimes x_i \rangle$$

$$= \left\langle \Phi_n([A_i^* A^* A A_j]) \begin{bmatrix} x_1 \\ \vdots \\ x_n \end{bmatrix}, \begin{bmatrix} x_1 \\ \vdots \\ x_n \end{bmatrix} \right\rangle$$

$$\leq \|A^* A\| \left\langle \Phi_n([A_i^* A_j]) \begin{bmatrix} x_1 \\ \vdots \\ x_n \end{bmatrix}, \begin{bmatrix} x_1 \\ \vdots \\ x_n \end{bmatrix} \right\rangle$$

$$= \|A\|^2 \left\langle \sum_{j=1}^n A_j \otimes x_j, \sum_{i=1}^n A_i \otimes x_i \right\rangle. \tag{3.3.1}$$

Therefore, $\pi_0(A)(\mathcal{N}) \subseteq \mathcal{N}$.

By using Lemma 1.2.3, the map $\pi_0(A)$ can be extended to a map $\pi(A)$ on $(\mathcal{A} \otimes \mathcal{H})/\mathcal{N}$. From (3.3.1), we have

$$\left\| \pi(A) \left(\sum_{i=1}^n A_i \otimes x_i \right) \right\| \leq \|A\| \left\| \sum_{i=1}^n A_i \otimes x_i \right\|.$$

Therefore, the map $\pi(A)$ is bounded and can be extended to a map on the Hilbert space $\mathcal{K}$. We denote this map by the same notation $\pi(A)$.

Direct calculations show that $\pi : \mathcal{A} \to B(\mathcal{K})$ is a bounded linear $*$-representation and $\pi(I) = I_{\mathcal{K}}$, where $I_{\mathcal{K}}$ denotes the identity operator on $\mathcal{K}$.

Next, we define $V : \mathcal{H} \to \mathcal{K}$ as $V(x) = I \otimes x + \mathcal{N}$. From

$$\begin{aligned}
\|V(x)\|^2 &= \langle V(x), V(x) \rangle \\
&= \langle I \otimes x + \mathcal{N}, I \otimes x + \mathcal{N} \rangle \\
&= \langle I \otimes x, I \otimes x \rangle \\
&= \langle \Phi(I)x, x \rangle \\
&\leq \|\Phi(I)\| \|x\|^2,
\end{aligned}$$

we deduce that V is a bounded linear map and

$$\|V\| = \sup\{\langle \Phi(I)x, x \rangle : \|x\| \leq 1\} = \|\Phi(I)\|.$$

In addition,

$$\begin{aligned}
\langle V^*\pi(A)Vx, y \rangle &= \langle \pi(A)Vx, Vy \rangle \\
&= \langle \pi(A)(I \otimes x), I \otimes y \rangle \\
&= \langle \Phi(A)x, y \rangle
\end{aligned}$$

for any $A \in \mathcal{A}$, $x, y \in \mathcal{H}$. Thus, $V^*\pi(A)V = \Phi(A)$ for all $A \in \mathcal{A}$. If Φ is unital, then $V^*V = V^*\pi(I)V = \Phi(I) = I_{\mathcal{H}}$, and so V is an isometry. □

The following theorem provides conditions for a positive map to be completely positive. We identify $\mathbb{M}_n(C(\Omega))$ with $C(\Omega, \mathbb{M}_n)$ by Lemma 2.1.5.

Theorem 3.3.8 *If $\Phi \in \mathbb{B}_+(\mathcal{A}, \mathcal{B})$ and $\mathcal{A}$ is commutative and unital, then Φ is completely positive.*

3.3 Positive and Completely Positive Linear Maps

Proof Since $\mathscr{A}$ is a unital commutative C^*-algebra, there exists a compact Hausdorff space Ω such that $\mathscr{A}$ is isometrically $*$-isomorphic to $C(\Omega)$. Then, for any $n \in \mathbb{N}$ and any positive element $A = [f_{ij}] \in \mathbb{M}_n(C(\Omega))$ we only need to show that $\Phi_n(A)$ is positive.

Let $A \in \mathbb{M}_n(C(\Omega))_+$ with $\|A\| \leq 1$. Note that for each $x \in \Omega$, the matrix $A(x) = [f_{ij}(x)]$ is positive semidefinite. Let $\varepsilon > 0$. For each $x \in \Omega$ the continuity of A ensures that there is a neighborhood $U(x)$ of x such that $\|A(y) - A(x)\| < \varepsilon$ for each $y \in U(x)$. Since Ω is compact, there is a finite covering $U(x_1), \ldots, U(x_l)$ of Ω. A consequence of the Urysohn Lemma, known as the *partition of unity*, provides us with a family of continuous functions $\{g_j\}_{j=1}^l$ on Ω such that $0 \leq g_j \leq 1$, $\sum_{j=1}^l g_j = 1$ and $g_j(x) = 0$ for all $x \notin U(x_j)$.

For each $1 \leq j \leq l$, set $A_j = A(x_j)$ considered as a matrix of constant functions. Then $\sum_{j=1}^l g_j A_j = \sum_{j=1}^l [A_{pq}^j g_j] \in \mathbb{M}_n(C(\Omega))$ and $\sum_{j=1}^l g_j A_j \geq 0$, where $A_j = [A_{pq}^j]$ for $1 \leq j \leq l$. Note that for any $x \in \Omega$, $(\sum_{j=1}^l g_j A_j)(x) = \sum_{j=1}^l g_j(x) A_j \geq 0$.

For each $x \in \Omega$, there is a $j \in \{1, \ldots, l\}$ such that $x \in U(x_j)$. For $x \in \Omega$, we have

$$\left\| A(x) - \sum_{j=1}^l g_j(x) A_j \right\| = \left\| \sum_{j=1}^l g_j(x)(A(x) - A(x_j)) \right\|$$

$$\leq \sum_{j=1}^l \|A(x) - A(x_j)\| g_j(x) < \sum_{j=1}^l \varepsilon g_j(x) = \varepsilon.$$

Hence,

$$\left\| A - \sum_{j=1}^l g_j A_j \right\|_\infty = \sup_{x \in \Omega} \left\| A(x) - \sum_{j=1}^l g_j(x) A_j \right\| \leq \epsilon.$$

Note that

$$\Phi_n \left(\sum_{j=1}^l g_j A_j \right) = \sum_{j=1}^l \Phi_n(g_j A_j) = \sum_{j=1}^l \Phi_n([A_{pq}^j g_j]) = \sum_{j=1}^l [A_{pq}^j \Phi(g_j)]$$

$$= \sum_{j=1}^l \begin{bmatrix} \Phi(g_j)^{1/2} & 0 & \cdots & 0 \\ 0 & \ddots & & \vdots \\ \vdots & & 0 & \\ 0 & \cdots & 0 & \Phi(g_j)^{1/2} \end{bmatrix} [A_{pq}^j] \begin{bmatrix} \Phi(g_j)^{1/2} & 0 & \cdots & 0 \\ 0 & \ddots & & \vdots \\ \vdots & & 0 & \\ 0 & \cdots & 0 & \Phi(g_j)^{1/2} \end{bmatrix} \geq 0.$$

Since

$$\left\| \Phi_n(A) - \Phi_n\left(\sum_{j=1}^l g_j A_j\right) \right\| = \left\| \Phi_n\left(A - \sum_{j=1}^l g_j A_j\right) \right\|$$

$$\leq \|\Phi_n\| \epsilon$$

and $\Phi_n\left(\sum_{j=1}^{l} g_j A_j\right)$ is positive, $\Phi_n(A)$ is positive by the fact that the set of positive elements of any C^*-algebra is closed. □

The next theorem is known as the Choi inequality [44, Lemma 3.1] for unital positive maps and normal operators.

Theorem 3.3.9 (Choi Inequality) *Suppose that* $\Phi \in \mathbb{B}_+(\mathbb{B}(\mathcal{H}), \mathbb{B}(\mathcal{K}))$ *is unital. Then*

$$\Phi(A^*A) \geq \Phi(A)^*\Phi(A)$$

for all normal operators $A \in \mathbb{B}(\mathcal{H})$.

Proof Recall from Chap. 1 that $C^*(A, I)$ is the commutative C^*-algebra generated by A and I. By Theorem 3.3.8, the restriction of Φ to $C^*(A, I)$ is completely positive, so by the Stinespring Theorem 3.3.7, it admits a decomposition of the form $\Phi(X) = V^*\pi(X)V$ ($X \in C^*(A, I)$), where π is a representation of $C^*(A, I)$ on a Hilbert space $\mathcal{L}$ and V is an isometry from $\mathcal{H}$ into $\mathcal{L}$. We have

$$\Phi(A)^*\Phi(A) = V^*\pi(A^*)V\, V^*\pi(A)V \leq V^*\pi(A^*)\pi(A)V = V^*\pi(A^*A)V = \Phi(A^*A),$$

since $V^*V = I$, we have $\|VV^*\| = \|V^*V\| = 1$. Therefore, $VV^* \leq I$. □

In the following theorem, we present the Choi inequality [17, Proposition 3.2.5] for unital 2-positive maps and arbitrary operators.

Theorem 3.3.10 (Choi inequality for unital 2-positive maps) *Suppose $\mathcal{A}$ and $\mathcal{B}$ are unital C^*-algebras and $\Phi : \mathcal{A} \to \mathcal{B}$ is a unital 2-positive map. Then*

$$\Phi(A^*A) \geq \Phi(A)^*\Phi(A) \qquad (3.3.2)$$

for each $A \in \mathcal{A}$.

Proof The 2-positivity of Φ and the fact that

$$\begin{bmatrix} I & A \\ A^* & A^*A \end{bmatrix} = \begin{bmatrix} I & A \\ 0 & 0 \end{bmatrix}^* \begin{bmatrix} I & A \\ 0 & 0 \end{bmatrix} \geq 0$$

imply

$$\begin{bmatrix} I & \Phi(A) \\ \Phi(A)^* & \Phi(A^*A) \end{bmatrix} = \begin{bmatrix} \Phi(I) & \Phi(A) \\ \Phi(A^*) & \Phi(A^*A) \end{bmatrix} \geq 0.$$

Hence, $\Phi(A^*A) \geq \Phi(A)^*\Phi(A)$. □

3.3 Positive and Completely Positive Linear Maps

The norm of a unital positive map is 1 as shown in the following.

Theorem 3.3.11 (Russo–Dye theorem) *Suppose that $\Phi \in \mathbb{B}_+(\mathbb{B}(\mathcal{H}), \mathbb{B}(\mathcal{K}))$ is unital. Then $\|\Phi\| = 1$.*

Proof Let $A \in \mathbb{B}(\mathcal{H})$ is a contraction. By Proposition 2.2.2, the operator

$$U = \begin{bmatrix} A & (I - AA^*)^{\frac{1}{2}} \\ (I - A^*A)^{\frac{1}{2}} & -A^* \end{bmatrix}$$

is unitary. Applying the Choi inequality to U and the unital positive map

$$\Psi\left(\begin{bmatrix} A_{11} & A_{12} \\ A_{21} & A_{22} \end{bmatrix}\right) := \Phi(A_{11}),$$

we conclude that

$$\|\Phi(A)\| = \|\Psi(U)\| = \|\Psi(U)^*\Psi(U)\|^{\frac{1}{2}} \leq \|\Psi(U^*U)\|^{\frac{1}{2}} = \|\Psi(I)\|^{1/2} = 1.$$

Now, $\Phi(I) = I$ completes the proof. □

Note that the real linear map $\Phi : \mathbb{C} \to \mathbb{C}$ defined by $\Phi(a + ib) = a + i3b$ is clearly unital, positive, and $\|\Phi\| = 3$. Thus, the Russo–Dye theorem does not hold in real linear spaces.

Combining the proofs of Theorems 3.3.9 and 3.3.11 gives the following extension of the Choi inequality. Recall that an operator $A \in \mathbb{B}(\mathcal{H})$ is called *subnormal* if there exist a Hilbert space $\mathcal{K} \supseteq \mathcal{H}$ and a normal operator $N \in \mathbb{B}(\mathcal{K})$ whose matrix corresponding to the decomposition $\mathcal{K} = \mathcal{H} \oplus \mathcal{H}^\perp$ is $\begin{bmatrix} A & * \\ 0 & * \end{bmatrix}$, where $*$ stands for unknown operators.

Theorem 3.3.12 (Choi Inequality for subnormal operators) *Suppose that $\Phi \in \mathbb{B}_+(\mathbb{B}(\mathcal{H}), \mathbb{B}(\mathcal{K}))$ is unital. Then*

$$\Phi(A^*A) \geq \Phi(A)^*\Phi(A)$$

for all subnormal operators $A \in \mathbb{B}(\mathcal{H})$.

As another consequence of the Choi inequality, we immediately get Kadison's inequality. However, we provide an alternative proof for it by using the following lemma, which is derived directly from Theorem 2.4.1.

Lemma 3.3.13 *Let $A \in \mathbb{B}_+(\mathcal{H})$ and $B \in \mathbb{B}_{\mathrm{sa}}(\mathcal{H})$. Then if $\begin{bmatrix} A & B \\ B & I \end{bmatrix} \geq 0$, then $A \geq B^2$;*

Theorem 3.3.14 (Kadison inequality) *[131] Suppose that $\Phi \in \mathbb{B}_+(\mathbb{B}(\mathcal{H}), \mathbb{B}(\mathcal{K}))$ is unital. Then*

$$\Phi(A^2) \geq \Phi(A)^2 \qquad (3.3.3)$$

for all self-adjoint operators $A \in \mathbb{B}(\mathcal{H})$.

Proof It follows from the spectral representation for self-adjoint operators (see Chap. 1) that A can be approximated uniformly by operators $A' = \sum_j \lambda_j E_j$ where $\{E_j\}$ is a *decomposition of the identity* I, that is, $\sum_j E_j = I$ and E_j's are pairwise orthogonal projections. Since Φ is unital, we have $\sum_j \Phi(E_j) = I$. To prove the Kadison inequality we employ Lemma 3.3.13. It is sufficient to prove

$$\begin{bmatrix} \Phi(A'^2) & \Phi(A') \\ \Phi(A') & \Phi(I) \end{bmatrix} \geq 0.$$

We have

$$\begin{bmatrix} \Phi(A'^2) & \Phi(A') \\ \Phi(A') & \Phi(I) \end{bmatrix} = \sum_j \begin{bmatrix} \lambda_j^2 \Phi(E_j) & \lambda_j \Phi(E_j) \\ \lambda_j \Phi(E_j) & \Phi(E_j) \end{bmatrix}$$

$$= \sum_j \begin{bmatrix} \lambda_j & 0 \\ 0 & 1 \end{bmatrix} \begin{bmatrix} \Phi(E_j) & \Phi(E_j) \\ \Phi(E_j) & \Phi(E_j) \end{bmatrix} \begin{bmatrix} \lambda_j & 0 \\ 0 & 1 \end{bmatrix} \geq 0$$

since, by Lemma 2.4.20, $\begin{bmatrix} \Phi(E_j) & \Phi(E_j) \\ \Phi(E_j) & \Phi(E_j) \end{bmatrix} \geq 0$. □

The next theorem appeared in [41] as Theorem 3.1.

Theorem 3.3.15 *If $\Phi : \mathscr{A} \to \mathscr{B}$ is a unital 2-positive linear map between unital C^*-algebras, then the set $\{A \in \mathscr{A} : \Phi(A^*A) = \Phi(A^*)\Phi(A)\}$ is a closed subalgebra of $\mathscr{A}$. Indeed, it is the set $\mathscr{A}_\Phi := \{A \in \mathscr{A} : \Phi(XA) = \Phi(X)\Phi(A) \text{ for all } X \in \mathscr{A}\}$.*

Proof It is easy to see that $\mathscr{A}_\Phi$ is a closed subalgebra of $\mathscr{A}$. We prove that if $\Phi(A^*A) = \Phi(A^*)\Phi(A)$, then $\Phi(XA) = \Phi(X)\Phi(A)$ for all $X \in \mathscr{A}$. Let $H \in \mathscr{A}$ is self-adjoint. By the Kadison inequality (3.3.3), we have

$$\Phi_2\left(\begin{bmatrix} 0 & A^* \\ A & H \end{bmatrix}^2\right) \geq \left(\Phi_2\begin{bmatrix} 0 & A^* \\ A & H \end{bmatrix}\right)^2.$$

Therefore,

3.3 Positive and Completely Positive Linear Maps

$$\begin{bmatrix} \Phi(A^*A) & \Phi(A^*H) \\ \Phi(HA) & \Phi(AA^* + H^2) \end{bmatrix} \geq \begin{bmatrix} \Phi(A^*)\Phi(A) & \Phi(A^*)\Phi(H) \\ \Phi(H)\Phi(A) & \Phi(A)\Phi(A^*) + \Phi(H)^2 \end{bmatrix}.$$

Since $\Phi(A^*A) = \Phi(A^*)\Phi(A)$, we have $\Phi(HA) = \Phi(H)\Phi(A)$. Using the Cartesian decomposition of an arbitrary element $X \in \mathscr{A}$ into its real and imaginary parts, we obtain the required result. □

The following result is given in [44, Proposition 4.3].

Theorem 3.3.16 *Let* $\Phi \in \mathbb{B}_+(\mathbb{B}(\mathscr{H}), \mathbb{B}(\mathscr{K}))$ *with* $\Phi(I) > 0$. *Then*

$$\Phi(BA^{-1}B) \geq \Phi(B)\Phi(A)^{-1}\Phi(B)$$

for all $B \in \mathbb{B}_{\mathrm{sa}}(\mathscr{H})$ *and all* $A \in \mathbb{B}_{++}(\mathscr{H})$.

Note that when $\Phi(I) > 0$, for any $A \in \mathbb{B}_{++}(\mathscr{H})$, it holds that $\Phi(A) \in \mathbb{B}_{++}(\mathscr{K})$.

Proof Put

$$\Psi(X) := \Phi(A)^{-1/2} \Phi(A^{1/2} X A^{1/2}) \Phi(A)^{-1/2}.$$

We observe that $\Psi \in \mathbb{B}_+(\mathbb{B}(\mathscr{H}), \mathbb{B}(\mathscr{K}))$ and $\Psi(I) = \Phi(A)^{-1/2}\Phi(A)\Phi(A)^{-1/2} = I$. Applying Kadison's inequality (3.3.3) to Ψ and self-adjoint operator $X = A^{-1/2}BA^{-1/2}$ we get

$$\Phi(A)^{-1/2}\Phi(BA^{-1}B)\Phi(A)^{-1/2} = \Psi(X^2) \geq \Psi(X)^2 = \Phi(A)^{-1/2}\Phi(B)\Phi(A)^{-1}\Phi(B)\Phi(A)^{-1/2},$$

whence $\Phi(BA^{-1}B) \geq \Phi(B)\Phi(A)^{-1}\Phi(B)$. □

Putting $B = I$ in Theorem 3.3.16, we get the next result.

Corollary 3.3.17 *Let* $\Phi \in \mathbb{B}_+(\mathbb{B}(\mathscr{H}), \mathbb{B}(\mathscr{K}))$ *be unital. Then*

$$\Phi(A^{-1}) \geq \Phi(A)^{-1}$$

for all $A \in \mathbb{B}_{++}(\mathscr{H})$.

Now, we employ the Russo–Dye Theorem 3.3.11 to establish the next theorem [44, Proposition 4.2].

Theorem 3.3.18 *Let* $\Phi \in \mathbb{B}_+(\mathbb{B}(\mathscr{H}), \mathbb{B}(\mathscr{K}))$. *If* $A \in \mathbb{B}_{++}(\mathscr{H})$, $B \in \mathbb{B}(\mathscr{H})$ *and* $A \geq B^*A^{-1}B$ *with* $\Phi(I) > 0$, *then*

$$\Phi(A) \geq \Phi(B^*)\Phi(A)^{-1}\Phi(B).$$

Proof Set
$$\Psi(X) := \Phi(A)^{-1/2}\Phi(A^{1/2}XA^{1/2})\Phi(A)^{-1/2}.$$

Then $\Psi \in \mathbb{B}_+(\mathbb{B}(\mathcal{H}), \mathbb{B}(\mathcal{K}))$ is unital.

With $X = A^{-1/2}BA^{-1/2}$, it follows from $A \geq B^*A^{-1}B$ that $X^*X \leq I$, which yields $\|X\| \leq 1$. Therefore, from the Russo–Dye Theorem 3.3.11, we get $\|\Psi(X)\| \leq 1$, which gives $\Psi(X)^*\Psi(X) \leq I$.

Replacing X with $A^{-1/2}BA^{-1/2}$ yields $\Phi(A)^{-1/2}\Phi(B^*)\Phi(A)^{-1}\Phi(B)\Phi(A)^{-1/2} \leq I$, which is the desired inequality. □

The following result provides a 2-positivity property for positive linear maps and specific block matrices of operators.

We note that we can restate Theorems 3.3.16 and 3.3.18 without assuming $\Phi(I) > 0$ as follows.

Corollary 3.3.19 *[44, Corollary 4.4] Let* $\Phi \in \mathbb{B}_+(\mathbb{B}(\mathcal{H}), \mathbb{B}(\mathcal{K}))$.

(i) If $\begin{bmatrix} A & B \\ B & C \end{bmatrix} \geq 0$, then $\begin{bmatrix} \Phi(A) & \Phi(B) \\ \Phi(B) & \Phi(C) \end{bmatrix} \geq 0$;

(ii) If $\begin{bmatrix} A & B \\ B^* & A \end{bmatrix} \geq 0$, then $\begin{bmatrix} \Phi(A) & \Phi(B) \\ \Phi(B^*) & \Phi(A) \end{bmatrix} \geq 0$.

Proof First, assume that $\Phi(I) > 0$ and $A > 0$. (i) and (ii) follow from Theorem 2.4.1 as well as Theorems 3.3.16 and 3.3.18, respectively. The general case follows from a limit argument as n tends to infinity with the map $\Phi_n(A) = \Phi(A) + \frac{\langle Ax,x\rangle}{n}I$, where $x \in \mathcal{H}$ is a unit vector. □

Now, we introduce one of the elegant theorems in this topic.

Theorem 3.3.20 (Choi–Davis–Jensen theorem) *Suppose that f is a real-valued continuous function defined on an interval J. Then the following assertions are equivalent:*

(i) *f is operator convex on J;*
(ii) *The Choi–Davis–Jensen inequality*

$$f(\Phi(A)) \leq \Phi(f(A)) \tag{3.3.4}$$

holds for every unital $\Phi \in \mathbb{B}_+(\mathbb{B}(\mathcal{H}), \mathbb{B}(\mathcal{K}))$ and every $A \in \mathbb{B}_{sa}(\mathcal{H})(J)$.

Proof (i) $\Longrightarrow$ (ii). Utilizing the spectral theorem for self-adjoint operators, A can be approximated uniformly by operators of the form $A' = \sum_j t_j E_j$ where $\{E_j\}$ is a decomposition

3.3 Positive and Completely Positive Linear Maps

of the identity I. Since Φ is unital, $\sum_j \Phi(E_j) = I$. It follows from Theorem 3.1.5 (iii) with $C_j = \Phi(E_j)^{1/2}$ that

$$f(\Phi(A')) = f\left(\sum_j t_j \Phi(E_j)\right) = f\left(\sum_j C_j^* t_j C_j\right) \leq \sum_j C_j f(t_j) C_j$$

$$= \sum_j f(t_j) \Phi(E_j) = \Phi\left(\sum_j f(t_j) E_j\right) = \Phi(f(A')).$$

Now, the continuity of Φ shows that (ii) holds.

(ii) $\Longrightarrow$ (i). Given an isometry C, set $\Phi(X) = C^* X C$. Then $\Phi \in \mathbb{B}_+(\mathbb{B}(\mathcal{H}), \mathbb{B}(\mathcal{K}))$ and we get (i) from (ii) by utilizing (ii) $\Longrightarrow$ (i) of Theorem 3.1.5. $\square$

When Φ is a nonunital positive map and $\Phi(I) > 0$, the Choi–Davis–Jensen inequality (3.3.4) holds. This is because we can define $\Psi(A) = \Phi(I)^{-1/2} \Phi(A) \Phi(I)^{-1/2}$, which is a unital positive linear map. Therefore, we have

$$f(\Phi(A)) = f\left(\Phi(I)^{1/2} \Psi(A) \Phi(I)^{1/2}\right) \leq \Phi(I)^{1/2} f(\Psi(A)) \Phi(I)^{1/2}$$

$$\leq \Phi(I)^{1/2} \Psi(f(A)) \Phi(I)^{1/2} = \Phi(f(A)).$$

As a consequence, we get an extension of the Hölder–McCarthy inequality.

Corollary 3.3.21 *Let $\Phi \in \mathbb{B}_+(\mathbb{B}(\mathcal{H}), \mathbb{B}(\mathcal{K}))$ and $A \in \mathbb{B}_{++}(\mathcal{H})$. Then*

(i) $\Phi(A^r) \leq \Phi(A)^r$ *for* $0 \leq r \leq 1$;
(ii) $\Phi(A) \leq \Phi(A^r)^{\frac{1}{r}}$ *for* $1 \leq r < \infty$.

Proof The operator convexity of $f(t) = -t^r$, the linearity of Φ and Theorem 3.3.20 give us (i). Applying (i) to $0 < 1/r \leq 1$ we get $\Phi(A^{1/r}) \leq \Phi(A)^{1/r}$. By replacing A with A^r we reach (ii). $\square$

Now, we present a variant of the Russo–Dye theorem where $\mathbb{B}(\mathcal{H})$ is replaced with a unital commutative C^*-algebra; see [191, Theorem 2.4].

Theorem 3.3.22 *Let X be a compact Hausdorff space and $\Phi \in \mathbb{B}_+(C(\Omega), \mathbb{B}(\mathcal{K}))$. Then* $\|\Phi\| = \|\Phi(1)\|$.

Proof Let $f \in C(\Omega)$ with $\|f\| \leq 1$. Let ε be given. For each $x \in \Omega$ the continuity of f ensures that there is a neighborhood $U(x)$ of x such that $|f(y) - f(x)| < \varepsilon$ for each $y \in U(x)$. Since Ω is compact, there is a finite covering $U(x_1), \ldots, U(x_n)$ of Ω. A consequence of the Urysohn Lemma, known as the *partition of unity*, provides us with a family of continuous functions $\{g_j\}_{j=1}^n$ on Ω such that $0 \leq g_j \leq 1$, $\sum_{j=1}^n g_j = 1$ and $g_j(x) = 0$ for all $x \notin U(x_j)$. Set $\alpha_j = f(x_j)$. For each $x \in \Omega$, we have

$$\left| f(x) - \sum_{j=1}^n \alpha_j g_j(x) \right| = \left| \sum_{j=1}^n (f(x) - \alpha_j) g_j(x) \right|$$
$$\leq \sum_{j=1}^n |f(x) - \alpha_j| g_j(x) < \sum_{j=1}^n \varepsilon g_j(x) = \varepsilon.$$

To reach the second inequality note that if $g_j(x) \neq 0$ for some j, then $x \in U(x_j)$, and hence $|f(x) - \alpha_j| < \varepsilon$. Thus

$$\left\| f - \sum_{j=1}^n \alpha_j g_j \right\| < \varepsilon.$$

Next

$$\left\| \sum_{j=1}^n \alpha_j \Phi(g_j) \right\|$$

$$= \left\| \begin{bmatrix} \sum_{j=1}^n \alpha_j \Phi(g_j) & 0 & \cdots & 0 \\ 0 & 0 & \cdots & 0 \\ \vdots & \vdots & & \vdots \\ 0 & 0 & \cdots & 0 \end{bmatrix} \right\|$$

$$= \left\| \begin{bmatrix} \Phi(g_1)^{1/2} & \cdots & \Phi(g_n)^{1/2} \\ 0 & \cdots & 0 \\ \vdots & & \vdots \\ 0 & \cdots & 0 \end{bmatrix} \begin{bmatrix} \alpha_1 & 0 & \cdots & 0 \\ 0 & \alpha_2 & \cdots & 0 \\ \vdots & \vdots & & \vdots \\ 0 & 0 & \cdots & \alpha_n \end{bmatrix} \begin{bmatrix} \Phi(g_1)^{1/2} & 0 & \cdots & 0 \\ \vdots & \vdots & & \vdots \\ \Phi(g_n)^{1/2} & 0 & \cdots & 0 \end{bmatrix} \right\|$$

$$\leq \left\| \begin{bmatrix} \Phi(g_1)^{1/2} & \cdots & \Phi(g_n)^{1/2} \\ 0 & \cdots & 0 \\ \vdots & & \vdots \\ 0 & \cdots & 0 \end{bmatrix} \right\| \max_{1 \leq j \leq n} |\alpha_j| \left\| \begin{bmatrix} \Phi(g_1)^{1/2} & 0 & \cdots & 0 \\ \vdots & \vdots & & \vdots \\ \Phi(g_n)^{1/2} & 0 & \cdots & 0 \end{bmatrix} \right\|$$

$$= \max_{1 \leq j \leq n} |\alpha_j| \left\| \sum_{j=1}^n \Phi(g_j) \right\| \leq \left\| \Phi\left(\sum_{j=1}^n g_j \right) \right\| = \|\Phi(1)\|,$$

where 1 denotes the constant function with value 1. Thus

$$\|\Phi(f)\| \leq \left\| \Phi\left(f - \sum_{j=1}^n \alpha_j g_j \right) \right\| + \left\| \sum_{j=1}^n \alpha_j \Phi(g_j) \right\| < \varepsilon \|\Phi\| + \|\Phi(1)\|.$$

3.3 Positive and Completely Positive Linear Maps

Letting $\varepsilon \to 0$, we get $\|\Phi\| \le \|\Phi(1)\|$. □

Recall that an *approximate identity* for a C*-algebra $\mathscr{A}$ is defined as an increasing net (F_i) of positive elements in the unit ball of $\mathscr{A}$ such that for every element $A \in \mathscr{A}$, we have $A = \lim_i F_i A = \lim_i A F_i$, with the limits taken in the norm topology of $\mathscr{A}$.

Theorem 3.3.23 *Let* $\Phi \in \mathbb{B}(\mathscr{A}, \mathscr{B})$ *is 2-positive. Then* $\|\Phi\| = \sup_i \|\Phi(F_i)\|$ *for all approximate identities* (F_i) *of* $\mathscr{A}$.

Proof Put $M = \sup_i \|\Phi(F_i)\|$. It follows from Theorem 3.3.4 that Φ is bounded. Therefore,

$$M \le \|\Phi\| \sup_i \|F_i\| \le \|\Phi\|.$$

This is because the F_i's are in the unit ball of $\mathscr{A}$. Let $A \in \mathscr{A}$. Since

$$\begin{bmatrix} F_i^2 & F_i A \\ A^* F_i & A^* A \end{bmatrix} = \begin{bmatrix} F_i & A \\ 0 & 0 \end{bmatrix}^* \begin{bmatrix} F_i & A \\ 0 & 0 \end{bmatrix} \ge 0$$

and Φ is 2-positive, we have

$$\begin{bmatrix} \Phi(F_i^2) & \Phi(F_i A) \\ \Phi(A^* F_i) & \Phi(A^* A) \end{bmatrix} \ge 0.$$

It follows from (2.4.1) and the fact that $F_i^2 \le F_i$ (which is valid since $0 \le F_i \le I$) that

$$\Phi(A^* F_i)\Phi(F_i A) \le \|\Phi(F_i^2)\|\Phi(A^* A) \le \|\Phi(F_i)\|\Phi(A^* A) \le M\Phi(A^* A).$$

Taking the limits and using the continuity of Φ, we get

$$\Phi(A^*)\Phi(A) \le M\Phi(A^* A). \tag{3.3.5}$$

Let $\|A\| \le 1$. Then

$$\|\Phi(A)\|^2 = \|\Phi(A)^*\Phi(A)\| \le M\|\Phi(A^* A)\| \le M\|\Phi\|,$$

whence $\|\Phi\|^2 \le M\|\Phi\|$, so $\|\Phi\| \le M$. □

Corollary 3.3.24 *If* $\Phi \in \mathbb{B}(\mathscr{A}, \mathscr{B})$ *is completely positive, then*

$$[\Phi(A_i^*)\Phi(A_j)] \le \|\Phi\|[\Phi(A_i^* A_j)]$$

in $\mathbb{M}_n(\mathscr{A})$ *for all* $A_1, \ldots, A_n \in \mathscr{A}$. *In particular,*

$$\Phi(A^*)\Phi(A) \le \Phi(A^* A) \quad (A \in \mathscr{A})$$

when Φ is unital.

Proof The map Φ_n is 2-positive and $F_i^n := F_i \otimes I_n$ is an approximate identity for $\mathscr{A} \otimes \mathbb{M}_n$ identified with $\mathbb{M}_n(\mathscr{A})$. Moreover, we have $\|\Phi_n(F_i^n)\| = \|\Phi(F_i)\|$. We therefore infer from (3.3.5) that
$$\Phi_n(C^*)\Phi_n(C) \le \|\Phi\|\Phi_n(C^*C)$$
for all $C \in \mathbb{M}_n(\mathscr{A})$. If we take
$$C = \begin{bmatrix} A_1 & A_2 & \cdots & A_n \\ 0 & 0 & \cdots & 0 \\ \vdots & \vdots & \ddots & \vdots \\ 0 & 0 & \cdots & 0 \end{bmatrix}$$
we obtain the desired inequality. □

The following result [46] shows a special role playing by 2-positivity of a map.

Theorem 3.3.25 *If $\Phi : \mathscr{A} \to \mathscr{B}$ is a unital $*$-linear map such that Φ_2 preserves invertibility, then Φ is a $*$- homomorphism.*

Proof Let $A \in \mathscr{A}$ is self-adjoint. It follows from $\begin{bmatrix} A & -I \\ A^2 & -A \end{bmatrix}^2 = 0$ that $\mathrm{sp}\left(\begin{bmatrix} A & -I \\ A^2 & -A \end{bmatrix}\right) = \{0\}$. Since Φ_2 preserves invertibility, for each operator matrix X, we have
$$\mathrm{sp}(\Phi_2(X)) \subseteq \mathrm{sp}(X). \qquad (3.3.6)$$

Hence, $\mathrm{sp}\left(\begin{bmatrix} \Phi(A) & -I \\ \Phi(A^2) & -\Phi(A) \end{bmatrix}\right) = \{0\}$. Since,
$$\begin{bmatrix} \Phi(A)^2 - \Phi(A^2) & 0 \\ \Phi(A^2)\Phi(A) - \Phi(A)\Phi(A^2) & -\Phi(A^2) + \Phi(A)^2 \end{bmatrix} = \begin{bmatrix} \Phi(A) & -I \\ \Phi(A^2) & -\Phi(A) \end{bmatrix}^2,$$
we conclude that the left-hand side of the above equality is a quasi-nilpotent matrix. Hence, $\Phi(A)^2 - \Phi(A^2)$ is quasi-nilpotent.

Due to $\Phi(A)^2 - \Phi(A^2)$ being self-adjoint, we conclude from the spectral theorem that $\Phi(A)^2 - \Phi(A^2) = 0$. Thus, Φ is *Jordan homomorphism* meaning that $\Phi(A)^2 = \Phi(A^2)$. Since Φ_2 is a $*$-map, it follows from (3.3.6) that Φ_2 is positive. It follows from Theorem 3.3.15 that $\mathscr{A}_\Phi$ is a subalgebra of $\mathscr{A}$ containing all self-adjoint operators. Hence, $\mathscr{A}_\Phi = \mathscr{A}$. Therefore, Φ is a $*$-homomorphism. □

3.4 Weakly 2-Positive Maps

In this section, we present some significant results from [169]. A map $\Phi : \mathbb{B}(\mathscr{H}) \to \mathbb{B}(\mathscr{H})$ is called *weakly 2-positive* if Φ_2 preserves the positivity of each operator matrix of the form $\begin{bmatrix} A & C \\ C & B \end{bmatrix}$. Corollary 3.3.19 shows that a positive linear map is weakly 2-positive. On the other hand, the Moore–Penrose inverse † on the matrix algebra $\mathcal{M}_n$ gives a map $\Phi^\dagger$ defined by $\Phi^\dagger(A) = A^\dagger$, which is a nonlinear positive map but it is not weakly 2-positive (and therefore not 2-positive). Since $\Phi^\dagger$ assigns inverses to invertible matrices, we have

$$\begin{bmatrix} 2I & I \\ I & 2I \end{bmatrix} = \begin{bmatrix} 2 & 1 \\ 1 & 2 \end{bmatrix} \otimes I \geq 0 \text{ while } \begin{bmatrix} \Phi^\dagger(2I) & \Phi^\dagger(I) \\ \Phi^\dagger(I) & \Phi^\dagger(2I) \end{bmatrix} = \begin{bmatrix} \frac{1}{2} & 1 \\ 1 & \frac{1}{2} \end{bmatrix} \otimes I \not\geq 0.$$

Next, we give a nontrivial example of a weakly 2-positive map that is not 2-positive. First, let us note that the nonlinear map $X \mapsto (\det X)I$ on $\mathbb{M}_n$ is 2-positive. In fact, if we have the condition $\begin{bmatrix} A & C \\ C^* & B \end{bmatrix} \geq 0$, it implies that $C = A^{\frac{1}{2}} W B^{\frac{1}{2}}$ for some contraction W, employing Theorem 2.4.12. Then, we have $|\det W| \leq 1$ and $\det C = \sqrt{\det A} \det W \sqrt{\det B}$. Using the above criterion again, we conclude that Φ is 2-positive. However, the map $\Phi_\alpha(X) = X^* + \alpha(\det X)I$ for $\alpha \geq 0$ is neither linear nor conjugate linear. It is weakly 2-positive. Moreover, let

$$A = \begin{bmatrix} 1 & 0 \\ 0 & 0 \end{bmatrix}, \quad B = \begin{bmatrix} 2 & 2 \\ 2 & 2 \end{bmatrix}, \quad C = \begin{bmatrix} 1 & 1 \\ 0 & 0 \end{bmatrix}.$$

Then $A^{\frac{1}{2}} = A$, $B^{\frac{1}{2}} = B/2$ and $C = A^{\frac{1}{2}} I B^{\frac{1}{2}}$, so that $\begin{bmatrix} A & C \\ C^* & B \end{bmatrix} \geq 0$. Noting that $\det A = \det B = \det C = 0$, we have

$$\begin{bmatrix} \Phi_\alpha(A) & \Phi_\alpha(C) \\ \Phi_\alpha(C^*) & \Phi_\alpha(B) \end{bmatrix} = \begin{bmatrix} A & C^* \\ C & B \end{bmatrix} + \alpha \begin{bmatrix} 0 & 0 \\ 0 & 0 \end{bmatrix} = \begin{bmatrix} 1 & 0 & 1 & 0 \\ 0 & 0 & 1 & 0 \\ 1 & 1 & 2 & 2 \\ 0 & 0 & 2 & 2 \end{bmatrix},$$

which is not positive since its determinant is negative. Therefore Φ_α is not 2-positive for any $\alpha \geq 0$. Furthermore, these matrices A, B, C can be used to show that the transpose map $\Phi(A) = A^t$ on $\mathbb{M}_2$ is a weakly 2-positive linear map that is not 2-positive (See Example 3.3.3(vi)). Note that the weakly 2-positivity of the transpose map follows from the norm inequality in Proposition 2.4.17 and the isometry property of the transpose map.

Theorem 3.4.1 *Let $A, B, X, Y \in \mathbb{B}(\mathscr{H})$ be arbitrary operators.*

(i) *If Φ is a weakly 2-positive map, then $\Phi(|X^*A^*Y|) \leq \Phi(V^*X^*|A|XV) \# \Phi(Y^*|A^*|Y)$, in which $X^*A^*Y = V|X^*A^*Y|$ denotes the polar decomposition of X^*A^*Y.*

(ii) If Φ is a 2-positive $*$-map, then $|\Phi(X^*A^*Y)| \leq U^*\Phi(X^*|A|X)U \# \Phi(Y^*|A^*|Y)$, in which $\Phi(X^*A^*Y) = U|\Phi(X^*A^*Y)|$ denotes the polar decomposition.

Proof (i) First, note that

$$\begin{bmatrix} |A| & A^* \\ A & |A^*| \end{bmatrix} = \begin{bmatrix} I & 0 \\ 0 & W \end{bmatrix} \begin{bmatrix} |A|^{1/2} & 0 \\ |A|^{1/2} & 0 \end{bmatrix} \begin{bmatrix} |A|^{1/2} & |A|^{1/2} \\ 0 & 0 \end{bmatrix} \begin{bmatrix} I & 0 \\ 0 & W \end{bmatrix}^* \geq 0,$$

where we apply the polar decomposition $A = W|A|$ of A. Hence

$$\begin{bmatrix} X^*|A|X & X^*A^*Y \\ Y^*AX & Y^*|A^*|Y \end{bmatrix} = \begin{bmatrix} X^* & 0 \\ 0 & Y^* \end{bmatrix} \begin{bmatrix} |A| & A^* \\ A & |A^*| \end{bmatrix} \begin{bmatrix} X & 0 \\ 0 & Y \end{bmatrix} \geq 0. \quad (3.4.1)$$

Utilizing the polar decomposition $X^*A^*Y = V|X^*A^*Y|$ of X^*A^*Y, we obtain

$$\begin{bmatrix} V^*(X^*|A|X)V & |X^*A^*Y| \\ |X^*A^*Y| & Y^*|A^*|Y \end{bmatrix} = \begin{bmatrix} V^*(X^*|A|X)V & V^*(X^*A^*Y) \\ (Y^*AX)V & Y^*|A^*|Y \end{bmatrix}$$

$$= \begin{bmatrix} V & 0 \\ 0 & I \end{bmatrix}^* \begin{bmatrix} X^*|A|X & X^*A^*Y \\ Y^*AX & Y^*|A^*|Y \end{bmatrix} \begin{bmatrix} V & 0 \\ 0 & I \end{bmatrix}$$

$$\geq 0.$$

Due to the weak 2-positivity of Φ, we get

$$\begin{bmatrix} \Phi(V^*(X^*|A|X)V) & \Phi(|X^*A^*Y|) \\ \Phi(|X^*A^*Y|) & \Phi(Y^*|A^*|Y) \end{bmatrix} \geq 0.$$

Thus, from Theorem 3.2.11 we obtain that

$$\Phi(|X^*A^*Y|) \leq \Phi(V^*X^*|A|XV) \# \Phi(Y^*|A^*|Y).$$

(ii) It follows from (3.4.1) and 2-positivity of Φ that

$$\begin{bmatrix} \Phi(X^*|A|X) & \Phi(X^*A^*Y) \\ \Phi(Y^*AX) & \Phi(Y^*|A^*|Y) \end{bmatrix} \geq 0,$$

whence, by using the polar decomposition $\Phi(X^*A^*Y) = U|\Phi(X^*A^*Y)|$ of $\Phi(X^*A^*Y)$, we get

$$\begin{bmatrix} U^*\Phi(X^*|A|X)U & |\Phi(X^*A^*Y)| \\ |\Phi(X^*A^*Y)| & \Phi(Y^*|A^*|Y) \end{bmatrix} = \begin{bmatrix} U^*\Phi(X^*|A|X)U & U^*\Phi(X^*A^*Y) \\ \Phi(Y^*AX)U & \Phi(Y^*|A^*|Y) \end{bmatrix}$$

$$= \begin{bmatrix} U & 0 \\ 0 & I \end{bmatrix}^* \begin{bmatrix} \Phi(X^*|A|X) & \Phi(X^*A^*Y) \\ \Phi(Y^*AX) & \Phi(Y^*|A^*|Y) \end{bmatrix} \begin{bmatrix} U & 0 \\ 0 & I \end{bmatrix}$$

$$\geq 0,$$

3.4 Weakly 2-Positive Maps

which gives the desired inequality. $\square$

In light of [92],

1. Theorem 3.4.1 (ii) can be extended as follows: Let Φ be a 2-positive *-map. For any $A, X, Y \in \mathbb{B}(\mathcal{H})$ and any $\alpha, \beta \in [0, 1]$ with $\alpha + \beta = 1$,

$$|\Phi(X^*A^*Y)| \leq U^*\Phi(X^*|A|^{2\alpha}X)U \# \Phi(Y^*|A^*|^{2\beta}Y).$$

2. Let $A = I$ and U be the partial isometry in the polar decomposition of $\Phi(Y^*X) = U|\Phi(Y^*X)|$ of $\Phi(Y^*X)$. Then

$$|\Phi(Y^*X)| \leq \Phi(X^*X) \# U^*\Phi(Y^*Y)U.$$

Therefore,

$$\ker \Phi(X^*X) \subseteq \ker |\Phi(Y^*X)| = \ker \Phi(Y^*X).$$

Hence, $\Phi(Y^*X)\Phi(X^*X)^\dagger \Phi(X^*X) = \Phi(Y^*X)$. Since

$$\begin{bmatrix} X^*X & X^*Y \\ Y^*X & Y^*Y \end{bmatrix} = \begin{bmatrix} X^* & 0 \\ Y^* & 0 \end{bmatrix} \begin{bmatrix} X & Y \\ 0 & 0 \end{bmatrix} \geq 0,$$

it follows from the 2-positivity of Φ that

$$\begin{bmatrix} \Phi(X^*X) & \Phi(X^*Y) \\ \Phi(Y^*X) & \Phi(Y^*Y) \end{bmatrix} \geq 0.$$

Therefore,

$$\begin{bmatrix} I & 0 \\ -\Phi(Y^*X)\Phi(X^*X)^\dagger & I \end{bmatrix} \begin{bmatrix} \Phi(X^*X) & \Phi(X^*Y) \\ \Phi(Y^*X) & \Phi(Y^*Y) \end{bmatrix} \begin{bmatrix} I & -\Phi(X^*X)^\dagger\Phi(X^*Y) \\ 0 & I \end{bmatrix}$$

$$= \begin{bmatrix} \Phi(X^*X) & 0 \\ 0 & \Phi(Y^*Y) - \Phi(Y^*X)\Phi(X^*X)^\dagger\Phi(X^*Y) \end{bmatrix} \geq 0.$$

Hence, we have $\Phi(Y^*Y) \geq \Phi(Y^*X)\Phi(X^*X)^\dagger\Phi(X^*Y)$, whence

$$U^*\Phi(Y^*Y)U \geq |\Phi(Y^*X)|\Phi(X^*X)^\dagger|\Phi(Y^*X)|.$$

3. Moreover, the equality in the inequality in (2) holds if and only if $U^*\Phi(Y^*Y)U = |\Phi(Y^*X)|\Phi(X^*X)^\dagger|\Phi(Y^*X)|$.

Now, consider the separable Hilbert space $\mathcal{H} = \ell_2$. Take the 2-positive map

$$\Phi(A) = \langle Ae, e \rangle,$$

where $A \in \mathbb{B}(\mathcal{H})$, $e = (1, 0, 0, \ldots)$, $X = x \otimes \overline{e}$, and $Y = y \otimes \overline{e}$. Then, from Theorem 3.4.1(ii), we conclude the following Cauchy–Schwarz inequality in the framework of Hilbert spaces:

Corollary 3.4.2 *Let $A \in \mathbb{B}(\mathcal{H})$ and $x, y \in \mathcal{H}$. Then*

$$|\langle Ax, y\rangle|^2 \leq \langle |A|x, x\rangle \langle |A^*|y, y\rangle.$$

Considering the trace functional $\mathrm{tr}(\cdot)$ on $\mathbb{M}_n$, it follows from Theorem 3.4.1(i) that

Corollary 3.4.3 *Let $A, X, Y \in \mathbb{M}_n$. Then*

$$\mathrm{tr}(|X^*A^*Y|)^2 \leq \mathrm{tr}(X^*|A|X)\mathrm{tr}(Y^*|A^*|Y).$$

The next consequence reads as follows.

Corollary 3.4.4 *Let $X \in \mathbb{B}(\mathcal{H})$.*

(i) *If Φ is a weakly 2-positive map, then $\Phi(|X|) \leq \Phi(V^*|X^*|V) \# \Phi(|X|)$, where $X = V|X|$ is the polar decomposition of X.*
(ii) *If Φ is a 2-positive $*$-map, then $|\Phi(X)| \leq U^*\Phi(|X^*|^{1/2})U \# \Phi(|X|^{3/2})$, where $\Phi(X) = U|\Phi(X)|$ denotes the polar decomposition of $\Phi(X)$.*

Proof Let $X = V|X|$ be the polar decomposition of X. It follows from Theorem 3.4.1 (i) that

(i) $\Phi(|X|) = \Phi(|V^*XI|) \leq \Phi(IV^*|X^*|VI) \# \Phi(|X|) = \Phi(V^*|X^*|V) \# \Phi(|X|)$.
(ii) Utilizing Theorem 3.4.1 (ii), we arrive at

$$\begin{aligned} |\Phi(X)| &= |\Phi(V|X|^{1/2}|X|^{1/2})| \\ &\leq U^*\Phi(V|X|^{1/2}V^*)U \# \Phi(|X|^{1/2}|X|^{1/2}|X|^{1/2}) \\ &= U^*\Phi(|X^*|^{1/2})U \# \Phi(|X|^{3/2}). \end{aligned}$$

□

3.5 Positive Linear Maps on Matrix Algebras

We denote $P_k[\mathscr{A}, \mathscr{B}]$ as the convex cone of all k-positive linear maps from $\mathscr{A}$ to $\mathscr{B}$. Similarly, $P_\infty(\mathscr{A}, \mathscr{B})$ denotes the convex cone of all completely positive maps.

In the case when $k = 2$ we observe that the transpose map on $\mathbb{M}_2$ is not 2-positive (See Example 3.3.3(vi)). However, if we introduce the concept of k-copositivity, we know that it is completely copositive.

A linear map $\Phi \colon \mathscr{A} \to \mathscr{B}$ is considered to be *k-copositive* if the map $\Phi_k^c \colon \mathbb{M}_k(\mathscr{A}) \to \mathbb{M}_k(\mathscr{B})$ defined by $(\Phi_n^c)([A_{ij}]) = [\Phi(A_{ji})]$ is positive.

The convex cone of all k-copositive maps from $\mathscr{A}$ into $\mathscr{B}$ is denoted by $P^k[\mathscr{A}, \mathscr{B}]$. If Φ is k-copositive for all $k \in \mathbb{N}$ we say that Φ is *completely copositive*. We denote the convex cone of all completely copositive maps by $P^\infty(\mathscr{A}, \mathscr{B})$.

A positive linear map is said to be *decomposable* if it can be expressed as the sum of a completely positive map and a completely copositive map. Otherwise, it is called *indecomposable*. The convex cone of all decomposable maps from $\mathscr{A}$ into $\mathscr{B}$ is denoted by $D[\mathscr{A}, \mathscr{B}]$.

Let $\mathscr{H} = \mathbb{C}^n$ and $\mathscr{A} = \mathbb{M}_n$. From the definition, we have

$$P[\mathbb{M}_m, \mathbb{M}_n] \supset P_2[\mathbb{M}_m, \mathbb{M}_n] \supset \cdots \supset P_k[\mathbb{M}_m, \mathbb{M}_n] \supset \cdots \supset P_\infty[\mathbb{M}_m, \mathbb{M}_n]$$

$$P[\mathbb{M}_m, \mathbb{M}_n] \supset P^2[\mathbb{M}_m, \mathbb{M}_n] \supset \cdots \supset P^k[\mathbb{M}_m, \mathbb{M}_n] \supset \cdots \supset P^\infty[\mathbb{M}_m, \mathbb{M}_n].$$

Each inclusion is proper, $P_\infty[\mathbb{M}_m, \mathbb{M}_n] = P_{\min\{m,n\}}[\mathbb{M}_m, \mathbb{M}_n]$, and $P^\infty[\mathbb{M}_m, \mathbb{M}_n] = P^{\min\{n,m\}}[\mathbb{M}_m, \mathbb{M}_n]$. Indeed, Choi gave the following characterization. In this section, we use the standard system of matrix units $(E_{ij})_{1 \leq i,j \leq k}$ for the matrix algebra $\mathbb{M}_k$.

Theorem 3.5.1 ([42]) *For a linear map* $\Phi \colon \mathbb{M}_m \to \mathbb{M}_n$, *the following assertions are equivalent.*

(1) Φ is completely positive,
(2) Φ is $\min\{m, n\}$-positive,
(3) The matrix $C_\Phi = \sum_{i,j=1}^m E_{ij} \otimes \Phi(E_{ij}) \in \mathbb{M}_m \otimes \mathbb{M}_n$ is positive,

Moreover, the above is equivalent to (4)

(4) (*Choi–Kraus representation*[141]) There exists a linearly independent finite family $\mathcal{V}$ of $m \times n$ matrices such that $\Phi(X) = \sum_{V \in \mathcal{V}} V^* X V$.

The correspondence $\Phi \longleftrightarrow C_\Phi$ from the space of $\mathcal{L}(\mathbb{M}_m, \mathbb{M}_n)$ of all linear maps onto the space $\mathbb{M}_m \otimes \mathbb{M}_n$ is called the *Jamiolkowski–Choi isomorphism*.

We give the sketch of the proof of Theorem 3.5.1. We can define the adjoint linear map $\Phi^* \colon \mathbb{M}_n \to \mathbb{M}_m$ by $\operatorname{tr}(\Phi(A)B) = \operatorname{tr}(A\Phi^*(B))$ for $A \in \mathbb{M}_m$ and $B \in \mathbb{M}_n$. For any $k \in \mathbb{N}$, the map Φ is k-positive if and only if Φ^* is k-positive.

The implications (1) $\Rightarrow$ (2) follow from the definition.

(2) ⇒ (3): Suppose that $m = \min\{m, n\}$. Since $[E_{ij}]_{i,j=1}^m$ is positive and Φ is m-positive, $\Phi_m([E_{ij}]) = [\Phi(E_{ij})] = \sum_{i,j=1}^m E_{ij} \otimes \Phi(E_{ij})$ is positive.

(3) ⇒ (4): Suppose that $[\Phi(E_{ij})]$ is positive. Note that $E_{ij} = e_i e_j^*$, where e_i's are elements of the standard basis. Each $1 \times mn$ matrix v can be considered as a $1 \times m$ block matrix $[x_1, \ldots, x_m]$ with $1 \times n$ matrices x_i as entries. We can also present v as a column matrix

$$v = \begin{bmatrix} x_1 \\ \vdots \\ x_m \end{bmatrix}$$

with $x_i \in \mathbb{C}^n$. One can associate with v the $n \times m$ matrix $V^* = [x_1, \ldots, x_m]$ having x_j as the jth column. We have

$$V^* E_{ij} V = [x_1, \ldots, x_m] e_i e_j^* [x_1, \ldots, x_m]^* = x_i x_j^*$$

and

$$vv^* = [x_i x_j^*] = [V^* E_{ij} V].$$

By the spectral theorem, there exist vectors $v_k \in \mathbb{C}^{mn}$ such that

$$[\Phi(E_{ij})] = \sum_{k=1}^{mn} v_k v_k^* = \sum_{k=1}^{mn} [V_k^* E_{ij} V_k],$$

where $v_k = \begin{bmatrix} x_1^{(k)} \\ \vdots \\ x_m^{(k)} \end{bmatrix}$, matrices V_k^*'s are associated with v_k's, and $x_l^{(k)} \in \mathbb{C}^n$ for $1 \leq l \leq m$.

Hence,

$$\Phi(E_{ij}) = \sum_{k=1}^{mn} V_k^* E_{ij} V_k$$

for all $1 \leq i, j \leq m$. Since matrix units linearly span $\mathbb{M}_m$ and Φ is linear, we have $\Phi(A) = \sum_{k=1}^{mn} V_k^* A V_k$ for all $A \in \mathbb{M}_m$.

(4) ⇒ (1): We may show that for any $m \times n$ matrix V, the map $\Psi(V)(A) = V^* A V$ from $\mathbb{M}_m$ to $\mathbb{M}_n$ is completely positive. For each $k \in \mathbb{N}$ and any positive semidefinite matrix $[A_{ij}] \in \mathbb{M}_k(\mathbb{M}_m)$

$$\Psi(V)_k([A_{ij}]) = [\Psi(V)(A_{ij})] = [V^* A_{ij} V]$$

$$= \begin{bmatrix} V & 0 & \cdots & 0 \\ 0 & \ddots & 0 & 0 \\ \vdots & & \ddots & \vdots \\ 0 & & \ddots & 0 \\ 0 & \cdots & & V \end{bmatrix}^* [A_{ij}] \begin{bmatrix} V & 0 & \cdots & 0 \\ 0 & \ddots & 0 & 0 \\ \vdots & & \ddots & \vdots \\ 0 & & \ddots & 0 \\ 0 & \cdots & & V \end{bmatrix} \geq 0.$$

Hence, $\Psi(V)$ is completely positive. This implies that Φ is completely positive.

3.5 Positive Linear Maps on Matrix Algebras

In particular, it is known that $CP[\mathbb{M}_m, \mathbb{M}_n] \cong (\mathbb{M}_m \otimes \mathbb{M}_n)_+$.

We note that Størmer [213] showed that $P[\mathbb{M}_2, \mathbb{M}_2] = D[\mathbb{M}_2, \mathbb{M}_2]$ and Woronowicz [228] showed that $P[\mathbb{M}_2, \mathbb{M}_n] = D[\mathbb{M}_2, \mathbb{M}_n]$ if and only if $n \leq 3$.

The first example of an indecomposable map was constructed by Choi. He also established that $P[\mathbb{M}_3, \mathbb{M}_3] \neq D[\mathbb{M}_3, \mathbb{M}_3]$.

Example 3.5.2 Let $\Phi: \mathbb{M}_n \to \mathbb{M}_n$ by $\Phi(X) = (n-1)\operatorname{tr}(X)I_n - X$. Then Φ is a $(n-1)$-positive map, but not n-positive. For example, when $n = 3$,

$$C_\Phi = \begin{bmatrix} 1 & 0 & 0 & 0 & -1 & 0 & 0 & 0 & -1 \\ 0 & 0 & 0 & 0 & 0 & 0 & 0 & 0 & 0 \\ 0 & 0 & 0 & 0 & 0 & 0 & 0 & 0 & 0 \\ 0 & 0 & 0 & 0 & 0 & 0 & 0 & 0 & 0 \\ -1 & 0 & 0 & 0 & 1 & 0 & 0 & 0 & -1 \\ 0 & 0 & 0 & 0 & 0 & 0 & 0 & 0 & 0 \\ 0 & 0 & 0 & 0 & 0 & 0 & 0 & 0 & 0 \\ 0 & 0 & 0 & 0 & 0 & 0 & 0 & 0 & 0 \\ -1 & 0 & 0 & 0 & -1 & 0 & 0 & 0 & 1 \end{bmatrix}$$

is not positive, but $[\Phi(F_{kl})]_{k,l=1}^2$ is positive for any subsystem $(F_{kl})_{1\leq k,l\leq 2}$ of $(E_{ij})_{1\leq i,j\leq 3}$. Hence, Φ is 2-positive by the following result.

Proposition 3.5.3 *(Takasaki-Tomiyama [217]) A map $\Phi: \mathbb{M}_n \to \mathbb{M}_n$ is k-positive if and only if $[\Phi(F_{ij})]$ is positive for any subsystem $(F_{ij})_{1\leq i,j\leq k}$ of the standard system of matrix units $(E_{ij})_{1\leq i,j\leq n}$.*

This follows from the following observation.

Lemma 3.5.4 *[217, Proposition 1.1] Let $\Phi: \mathbb{M}_m \to \mathbb{M}_n$ be a positive linear map and $k \in \mathbb{N}$ with $1 \leq k \leq \min\{m, n\}$. Then, Φ is k-positive if and only if $(P \otimes I_n)[\Phi(E_{ij})](P \otimes I_n) \geq 0$ for every projection P of $\mathbb{M}_m$ of rank k, where $(E_{ij})_{1\leq i,j\leq m}$ is the standard system of matrix units.*

Proof Suppose that Φ is k-positive. Then, for any projection P of rank k there is a unitary $U \in \mathbb{M}_m$ such that $UPU^* = \begin{bmatrix} 1 & & & \\ & \ddots & & O_{k,m-k} \\ & & 1 & \\ \hline O_{m-k,k} & & O_{m-k,m-k} \end{bmatrix}$.

Then, $UP\mathbb{M}_m PU^* \cong \mathbb{M}_k$. Since Φ_k is positive,

$$(P \otimes I_n) C_\Phi (P \otimes I_n)$$

$$= (P \otimes I_n)(\sum_{i,j=1}^{m} E_{ij} \otimes \Phi(E_{ij}))(P \otimes I_n)$$

$$= (U^* \otimes I_n)(\sum_{ij\ j=1}^{m} UPE_{ij}PU^* \otimes \Phi(E_{ij}))(U \otimes I_n)$$

$$= (U^* \otimes I_n)(id_k \otimes \Phi)(\sum_{i,,j=1}^{m} UPE_{ij}PU^* \otimes E_{ij})(U \otimes I_n) \geq 0,$$

where $\sum_{i,,j=1}^{m} UPE_{ij}PU^* \otimes E_{ij} = (UP \otimes I_m)(\sum_{i,j=1}^{m} E_{ij} \otimes E_{ij})(PU^* \otimes I_m) \in (\mathbb{M}_k \otimes \mathbb{M}_m)_+$.

Conversely, suppose that for any projection $P \in \mathbb{M}_m$ of rank k, the matrix $(P \otimes I_n) C_\Phi (P \otimes I_n)$ is positive. Then, we have only show that $\Phi_k(x^*x) \geq 0$ for any $x \in \mathbb{C}^k \otimes \mathbb{C}^m$.

Let $x = [x_1^1, \ldots, x_m^1, x_1^2, \ldots, x_m^2, \ldots, x_1^k, \ldots x_m^k]$. Then

$$\Phi_k(x^*x) = \left[\Phi \left(\begin{bmatrix} \overline{x_1^i} x_1^j & \cdots & \overline{x_1^i} x_m^j \\ \vdots & & \vdots \\ \overline{x_m^i} x_1^j & \cdots & \overline{x_m^i} x_m^j \end{bmatrix} \right) \right]_{i,j=1}^{k}$$

$$= [\sum_{r,s=1}^{m} \overline{x_r^i} x_s^j \Phi(E_{ij})]_{i,j=1}^{k}$$

$$= \sum_{r,s=1}^{m} \sum_{i,j=1}^{k} \overline{x_r^i} x_s^j F_{ij} \otimes \Phi(E_{rs}) \quad ((F_{ij})_{1 \leq i,j \leq k} \text{ is the system of matrix units for } \mathbb{M}_k)$$

$$= \sum_{r,s=1}^{m} W^* E_{rs} W \otimes \Phi(E_{rs})$$

$$= (W \otimes I_n)^* (P_W \otimes I_n) C_\Phi (P_W \otimes I_n)(W \otimes I_n)$$

$$\geq 0,$$

where $W = \begin{bmatrix} x_1^1 & \cdots & x_1^k & 0 & \cdots & 0 \\ \vdots & & \vdots & \vdots & & \vdots \\ \vdots & & \vdots & \vdots & & \vdots \\ x_m^1 & \cdots & x_m^k & 0 & \cdots & 0 \end{bmatrix}$ and P_W is the support projection of W of rank k. □

3.5 Positive Linear Maps on Matrix Algebras

Note that $\Phi(U^*XU) = U^*\Phi(X)U$ for any unitary $U \in \mathbb{M}_n$, therefore it is enough to verify the positivity of $[\Phi(E_{ij})]_{i,j=1}^{n-1}$ to prove the $(n-1)$-positivity of Φ. In the above C_Φ it is easy to show that

$$\left[\begin{array}{ccc|ccc} 1 & 0 & 0 & 0 & -1 & 0 \\ 0 & 0 & 0 & 0 & 0 & 0 \\ 0 & 0 & 0 & 0 & 0 & 0 \\ \hline 0 & 0 & 0 & 0 & 0 & 0 \\ -1 & 0 & 0 & 0 & 1 & 0 \\ 0 & 0 & 0 & 0 & 0 & 0 \end{array}\right]$$

is positive. Hence, Φ is 2-positive.

Example 3.5.5 (*Choi indecomposable map*) [43] Let $\Phi_{3,1} \colon \mathbb{M}_3 \to \mathbb{M}_3$ be a positive linear map defined by

$$\Phi_{3,1}(X) = \begin{bmatrix} x_{11} + x_{33} & -x_{12} & -x_{13} \\ -x_{21} & x_{11} + x_{22} & -x_{23} \\ -x_{31} & -x_{32} & x_{22} + x_{33} \end{bmatrix},$$

where $X = \begin{bmatrix} x_{11} & x_{12} & x_{13} \\ x_{21} & x_{22} & x_{23} \\ x_{31} & x_{32} & x_{33} \end{bmatrix}$. Then, $\Phi_{3,1}$ is indecomposable.

To show the indecomposability of $\Phi_{3,1}$ we use the following criterion due to Størmer [214].

Theorem 3.5.6 (Størmer's criterion) *[214] Let $\mathscr{A}$ be a C^*-algebra and $\Phi \colon \mathscr{A} \to \mathbb{B}(\mathscr{H})$ be a linear map. Then, Φ is decomposable if and only if for all $n \in \mathbb{N}$, whenever $[X_{ij}]$ and $[X_{ji}]$ belong to $\mathbb{M}_n(\mathscr{A})_+$, it holds that $[\Phi(X_{ij})] \in \mathbb{M}_n(\mathbb{B}(\mathscr{H}))_+$.*

Corollary 3.5.7 *Let $\mathscr{A}$ be a C^*-algebra and $\Phi \colon \mathscr{A} \to \mathbb{B}(\mathscr{H})$ be a linear map. If there exists $k \in \mathbb{N}$ with $k \geq 2$ such that $[X_{ij}], [X_{ji}] \in \mathbb{M}_k(\mathscr{A})$ and $[\Phi(X_{ij})]$ is not positive, then Φ can not be decomposed as a sum of a k-positive map and k-copositive map.*

Proof If $\Phi = \Phi_1 + \Phi_2$ such that Φ_1 is k-positive and Φ_2 is k-copositive, then $(\Phi_1)_k([X_{ij}])$ is positive and since $[X_{ji}]$ is positive, $(\Phi_2)_k^c([X_{ji}]) = [\Phi_2(X_{ij})] = \Phi_2([X_{ij}])$ is positive. Hence, $\Phi_k([X_{ij}]) = \Phi_1([(X_{ij})]) + \Phi_2([X_{ij}])$ is positive. This is a contradiction to the assumption. Therefore, we get the conclusion. □

Applying this criterion, we can show the indecomposability of $\Phi_{3,1}$ as follows. Let

$$[X_{ij}] := \left[\begin{array}{ccc|ccc|ccc} 1 & 0 & 0 & 0 & 1 & 0 & 0 & 0 & 1 \\ 0 & 2 & 0 & 1 & 0 & 0 & 0 & 0 & 0 \\ 0 & 0 & \frac{1}{2} & 0 & 0 & 0 & 1 & 0 & 0 \\ \hline 0 & 1 & 0 & \frac{1}{2} & 0 & 0 & 0 & 0 & 0 \\ 1 & 0 & 0 & 0 & 1 & 0 & 0 & 0 & 1 \\ 0 & 0 & 0 & 0 & 0 & 2 & 0 & 1 & 0 \\ \hline 0 & 0 & 1 & 0 & 0 & 0 & 2 & 0 & 0 \\ 0 & 0 & 0 & 0 & 0 & 1 & 0 & \frac{1}{2} & 0 \\ 1 & 0 & 0 & 0 & 1 & 0 & 0 & 0 & 1 \end{array}\right].$$

Then, it is easy to verify that $[X_{ij}] = [X_{ji}]$ and $[X_{ij}]$ can be expressed as the form

$$U^* \begin{bmatrix} 1 & 1 & 1 \\ 1 & 1 & 1 \\ 1 & 1 & 1 \end{bmatrix} U + V_1^* \begin{bmatrix} 2 & 1 \\ 1 & \frac{1}{2} \end{bmatrix} V_1 + V_2^* \begin{bmatrix} \frac{1}{2} & 1 \\ 1 & 2 \end{bmatrix} V_2 + V_3^* \begin{bmatrix} 2 & 1 \\ 1 & \frac{1}{2} \end{bmatrix} V_3,$$

where

$$U = \left[\begin{array}{ccc|ccc|ccc} 1 & 0 & 0 & 0 & 0 & 0 & 0 & 0 & 0 \\ 0 & 0 & 0 & 0 & 1 & 0 & 0 & 0 & 0 \\ 0 & 0 & 0 & 0 & 0 & 0 & 0 & 0 & 1 \end{array}\right],$$

$$V_1 = \left[\begin{array}{ccc|ccc|ccc} 0 & 1 & 0 & 0 & 0 & 0 & 0 & 0 & 0 \\ 0 & 0 & 0 & 1 & 0 & 0 & 0 & 0 & 0 \end{array}\right],$$

$$V_2 = \left[\begin{array}{ccc|ccc|ccc} 0 & 0 & 1 & 0 & 0 & 0 & 0 & 0 & 0 \\ 0 & 0 & 0 & 0 & 0 & 0 & 1 & 0 & 0 \end{array}\right],$$

$$V_3 = \left[\begin{array}{ccc|ccc|ccc} 0 & 0 & 0 & 0 & 0 & 1 & 0 & 0 & 0 \\ 0 & 0 & 0 & 0 & 0 & 0 & 0 & 1 & 0 \end{array}\right].$$

Therefore, $[X_{ij}]$ and $[X_{ji}]$ belongs to $\mathbb{M}_3(\mathbb{M}_3)_+$.

$$[\Phi_{3,1}(X_{ij})] = \left[\begin{array}{ccc|ccc|ccc} \frac{3}{2} & 0 & 0 & 0 & -1 & 0 & 0 & 0 & -1 \\ 0 & 3 & 0 & -1 & 0 & 0 & 0 & 0 & 0 \\ 0 & 0 & \frac{5}{2} & 0 & 0 & 0 & -1 & 0 & 0 \\ \hline 0 & -1 & 0 & \frac{5}{2} & 0 & 0 & 0 & 0 & 0 \\ -1 & 0 & 0 & 0 & \frac{3}{2} & 0 & 0 & 0 & -1 \\ 0 & 0 & 0 & 0 & 0 & 3 & 0 & -1 & 0 \\ \hline 0 & 0 & -1 & 0 & 0 & 0 & 3 & 0 & 0 \\ 0 & 0 & 0 & 0 & 0 & -1 & 0 & \frac{5}{2} & 0 \\ -1 & 0 & 0 & 0 & -1 & 0 & 0 & 0 & \frac{3}{2} \end{array}\right].$$

3.5 Positive Linear Maps on Matrix Algebras

Since the principal submatrix $\begin{bmatrix} \frac{3}{2} & -1 & -1 \\ -1 & \frac{3}{2} & -1 \\ -1 & -1 & \frac{3}{2} \end{bmatrix}$ is not positive, the map $\Phi_{3,1}$ is not decomposable.

The following is an example of an extended Choi map.

Example 3.5.8 Let $S = [\delta_{i,i+1}]$. For $n, k \in \mathbb{N}$ with $n \geq 3$ and $1 \leq k \leq n-1$, and for $X \in \mathbb{M}_n$, let us define $\varepsilon([X_{ij}]) = \mathrm{diag}(X_{11}, \ldots, X_{nn})$ We can then define $\Phi_{n,k}(X) = (n-k)\varepsilon(X) + \sum_{i=1}^{k} \varepsilon(S^i X {S^i}^*) - X$ and $\Phi_{n,0}(X) = n\varepsilon(X) - X$. It is worth noting that $\Phi_{n,0}$ is completely positive and $\Phi_{n,n-1}$ is completely copositive.

$\Phi_{n,1}$ is a positive and indecomposable map that cannot be expressed as the sum of n-positive map and 2-copositive map. Specifically, $\Phi_{3,1}$, also known as a Choi map, cannot be decomposed into a 2-positive map and a 2-copositive map. This type of map is called an *atomic map*.

Note that Yamagami [232] showed that all $\Phi_{n,k}$'s are positive. This is further explored in [180, 181] as follows:

Theorem 3.5.9 *(1)* $\Phi_{n,n-2}$ *cannot be decomposed into a sum of a 3-positive and 3-copositive.*
(2) There are $[B_{ij}], [B_{ji}]$ *in* $(\mathbb{M}_3 \otimes \mathbb{M}_m)_+$ *such that* $[B_{ij}]$ *is not in* $\mathbb{M}_{3+} \otimes \mathbb{M}_{m+}$ *for* $m \geq 4$.
(3) $\Phi_{n,1}$ *cannot be decomposed into a sum of a 2-positive and a 2-copositive, that is, it is atomic.*

Note that $\Phi : \mathbb{M}_n \to \mathbb{M}_n$ can be decomposed into a sum of a k-positive map and a k-copositive map under the condition that if $[X_{ij}]$ and $[X_{ji}]$ belong to $\mathbb{M}_k(\mathbb{M}_n)_+$, then $[\Phi(X_{ij})] \in \mathbb{M}_k(\mathbb{M}_n)_+$.

Proof (1) Let $[X_{ij}] \in \mathbb{M}_3(\mathbb{M}_n)$ be the matrix defined by

$$[X_{ij}] = \begin{bmatrix} 1 & 0 & 0 & \cdots & 0 & 0 & 0 & 1 & 0 & \cdots & 0 & 0 & 0 & 0 & 0 & \cdots & 0 & 1 \\ 0 & 2 & 0 & \cdots & 0 & 0 & 1 & 0 & 0 & \cdots & 0 & 0 & 0 & 0 & 0 & \cdots & 0 & 0 \\ 0 & 0 & 0 & \cdots & 0 & 0 & 0 & 0 & 0 & \cdots & 0 & 0 & 0 & 0 & 0 & \cdots & 0 & 0 \\ \vdots & \vdots & \vdots & \ddots & \vdots & \vdots & \vdots & \vdots & \vdots & \ddots & \vdots & \vdots & \vdots & \vdots & \vdots & \ddots & \vdots & \vdots \\ 0 & 0 & 0 & \cdots & 0 & 0 & 0 & 0 & 0 & \cdots & 0 & 0 & 0 & 0 & 0 & \cdots & 0 & 0 \\ 0 & 0 & 0 & \cdots & 0 & 1/2 & 0 & 0 & 0 & \cdots & 0 & 0 & 1 & 0 & 0 & \cdots & 0 & 0 \\ \hline 0 & 1 & 0 & \cdots & 0 & 0 & 1/2 & 0 & 0 & \cdots & 0 & 0 & 0 & 0 & 0 & \cdots & 0 & 0 \\ 1 & 0 & 0 & \cdots & 0 & 0 & 0 & 1 & 0 & \cdots & 0 & 0 & 0 & 0 & 0 & \cdots & 0 & 1 \\ 0 & 0 & 0 & \cdots & 0 & 0 & 0 & 0 & 0 & \cdots & 0 & 0 & 0 & 0 & 0 & \cdots & 0 & 0 \\ \vdots & \vdots & \vdots & \ddots & \vdots & \vdots & \vdots & \vdots & \vdots & \ddots & \vdots & \vdots & \vdots & \vdots & \vdots & \ddots & \vdots & \vdots \\ 0 & 0 & 0 & \cdots & 0 & 0 & 0 & 0 & 0 & \cdots & 0 & 0 & 0 & 0 & 0 & \cdots & 0 & 0 \\ 0 & 0 & 0 & \cdots & 0 & 0 & 0 & 0 & 0 & \cdots & 0 & 2 & 0 & 1 & 0 & \cdots & 0 & 0 \\ \hline 0 & 0 & 0 & \cdots & 0 & 1 & 0 & 0 & 0 & \cdots & 0 & 0 & 2 & 0 & 0 & \cdots & 0 & 0 \\ 0 & 0 & 0 & \cdots & 0 & 0 & 0 & 0 & 0 & \cdots & 0 & 1 & 0 & 1/2 & 0 & \cdots & 0 & 0 \\ 0 & 0 & 0 & \cdots & 0 & 0 & 0 & 0 & 0 & \cdots & 0 & 0 & 0 & 0 & 0 & \cdots & 0 & 0 \\ \vdots & \vdots & \vdots & \ddots & \vdots & \vdots & \vdots & \vdots & \vdots & \ddots & \vdots & \vdots & \vdots & \vdots & \vdots & \ddots & \vdots & \vdots \\ 0 & 0 & 0 & \cdots & 0 & 0 & 0 & 0 & 0 & \cdots & 0 & 0 & 0 & 0 & 0 & \cdots & 0 & 0 \\ 1 & 0 & 0 & \cdots & 0 & 0 & 0 & 1 & 0 & \cdots & 0 & 0 & 0 & 0 & 0 & \cdots & 0 & 1 \end{bmatrix}.$$

The matrix $[X_{ij}] = [X_{ji}]$ can be expressed as the form

$$U^* \begin{bmatrix} 1 & 1 & 1 \\ 1 & 1 & 1 \\ 1 & 1 & 1 \end{bmatrix} U + V_1^* \begin{bmatrix} 2 & 1 \\ 1 & \frac{1}{2} \end{bmatrix} V_1 + V_1^* \begin{bmatrix} \frac{1}{2} & 1 \\ 1 & 2 \end{bmatrix} V_2 + V_3^* \begin{bmatrix} 2 & 1 \\ 1 & \frac{1}{2} \end{bmatrix} V_3,$$

where U is $3 \times 3n$ matrix and V_i are $2 \times 3n$ matrices ($i = 1, 2, 3$). Therefore, $[X_{ij}]$ and $[X_{ji}]$ belong to $\mathbb{M}_3(\mathbb{M}_n)_+$.

We have $\langle [\Phi_{n,n-2}(X_{ij})]x, x \rangle = -\frac{1}{2}$, where

$$x = [\underbrace{2, 0, \cdots, 0}_{n}, \underbrace{0, 1, 0, \cdots, 0}_{n}, \underbrace{0, 0, \cdots, 0, 2}_{n}].$$

Hence, from the above remark we conclude that $\Phi_{n,n-2}$ cannot be decomposed into a sum of a 3-positive map and a 3-copositive map from Corollary 3.5.7.

(2) Suppose that $[X_{ij}] \in \mathbb{M}_{3+} \otimes \mathbb{M}_{n+}$. Hence, $[X_{ij}] = \sum_k X_k \otimes Y_k$ for some $X_k \in \mathbb{M}_{3+}$, $Y_k \in \mathbb{M}_{n+}$. Then, $[\Phi_{n,n-2}(X_{ij})] = \sum_k X_k \otimes \Phi_{n,n-2}(Y_k) \in \mathbb{M}_{3+} \otimes \mathbb{M}_{n+}$. On the contrary, since $\langle [\Phi_{n,n-2}(X_{ij})]x, x \rangle < 0$ in (1), we get a contradiction. Therefore, $[X_{ij}] \notin \mathbb{M}_{3+} \otimes \mathbb{M}_{n+}$.

(3) We present a sketch of the proof. For more explanations refer to [181, Theorem 3].

3.5 Positive Linear Maps on Matrix Algebras

For $X = \begin{bmatrix} 0 & 1 & \cdots & 1 \\ 0 & 0 & \cdots & 0 \\ \vdots & \vdots & \ddots & \vdots \\ 0 & 0 & \cdots & 0 \end{bmatrix} \in \mathbb{M}_n(\mathbb{R})$ and $y = [0, 1, \ldots, 1] \in \mathbb{C}^n$, we have

$$\langle \{(n-1)\Phi_{n,1}(X^*X) - \Phi_{n,1}(X)^*\Phi_{n,1}(X)\} y^t, y^t \rangle = (1-n) < 0.$$

Hence, from (3.3.5) in Theorem 3.3.23 we conclude that $\Phi_{n,1}$ is not 2-positive on $\mathbb{M}_n(\mathbb{R})$.

Suppose that $\Phi_{n,1} = \Psi_1 + \Psi_2$, as a decomposition into the sum of a 2-positive map Ψ_1 and a 2-copositive map Ψ_2. Let $(E_{ij})_{1 \leq i,j \leq n}$ be the standard system of matrix units in $\mathbb{M}_n$. Since $\Phi_{n,1}(E_{ij}) = (n-2)E_{ii} + E_{i+1 i+1}$, we have $\Psi_2(E_{ii})E_{ll} = E_{ll}\Psi_2(E_{ii}) = 0$ for $l = i+2, \ldots, i+n-1 \pmod{n}$. Since Ψ_2 is 2-copositive, for any i, j, we have

$$\begin{bmatrix} \Psi_2(E_{jj}) & \Psi_2(E_{ij}) \\ \Psi_2(E_{ji}) & \Psi_2(E_{ii}) \end{bmatrix} = \Psi_2^c \left(\begin{bmatrix} E_{jj} & E_{ji} \\ E_{ij} & E_{ii} \end{bmatrix} \right) \geq 0.$$

It follows from (2.4.5) in Proposition 2.4.16 that $|\langle \Psi_2(E_{ij})x, y \rangle|^2 \leq \langle \Psi_2(E_{jj})y, y \rangle \langle \Psi_2(E_{ii})x, x \rangle$. We have

$$\Psi_2(E_{ij})E_{ll} = E_{ll}\Psi_2(E_{ji}) = 0 \quad \text{for} \quad l = i+2, \ldots, i+n-1 \pmod{n}.$$

From this observation, we can set $\overline{\Psi_2}([x_{ij}]) = [\overline{\Psi_2(x_{ij})}]$ for $[x_{ij}] \in \mathbb{M}_n$. The map $\tilde{\Psi}_2([x_{ij}]) = \frac{1}{2}(\Psi_2 + \overline{\Psi_2})([x_{ij}])$ is 2-positive on $\mathbb{M}_n(\mathbb{R})$ since for $x = [x_1, \ldots, x_n]$, $y = [y_1, \ldots, y_n]$, we have $\tilde{\Psi}_2(x^t y) = \tilde{\Psi}_2(y^t x)$ and

$$(\tilde{\Psi}_2)_2 \begin{bmatrix} x^t x & x^t y \\ y^t x & y^t y \end{bmatrix} = \begin{bmatrix} \tilde{\Psi}_2(x^t x) & \tilde{\Psi}_2(x^t y) \\ \tilde{\Psi}_2(y^t x) & \tilde{\Psi}_2(y^t y) \end{bmatrix}$$

$$= \begin{bmatrix} \tilde{\Psi}_2(x^t x) & \tilde{\Psi}_2(y^t x) \\ \tilde{\Psi}_2(x^t y) & \tilde{\Psi}_2(y^t y) \end{bmatrix}$$

$$= (\tilde{\Psi}_2)_2^c \begin{bmatrix} x^t x & x^t y \\ y^t x & y^t y \end{bmatrix} \geq 0$$

by the 2-copositivity of $\tilde{\Psi}_2$. Since $\tilde{\Psi}_1$ is 2-positive on $\mathbb{M}_n(\mathbb{R})$ and $\Phi_{n,1} = \tilde{\Psi}_1 + \tilde{\Psi}_2$ on $\mathbb{M}_n(\mathbb{R})$, we infer that $\Phi_{n,1}$ is 2-positive. This is a contradiction to the first observation.

Hence, $\Phi_{n,1}$ cannot be written as a sum of a 2-positive map and a 2-copositive map. □

It is shown in [104] that for $3 \leq n$ and $2 \leq k \leq n-2$ all $\Phi_{n,k}$ are atomic as demonstrated by the duality below; see [144].

Peres [194] and Horodecki's family [124] gave a characterization of quantum entanglement states that are linked to positive linear maps on matrix algebras. They achieved this by utilizing the duality between $\mathbb{B}(\mathbb{M}_m, \mathbb{M}_n)$ and $\mathbb{M}_m \otimes \mathbb{M}_n$, where

$$\langle X, \Phi \rangle = \sum_{i,j=1}^{m} \mathrm{tr}(\Phi(E_{i,j})A_{ij}^t) = \mathrm{tr}(C_\Phi X^t)$$

for $X = \sum_{i,j=1}^{m} E_{ij} \otimes A_{ij} \in \mathbb{M}_m \otimes \mathbb{M}_n$ and $\Phi \in \mathbb{B}(\mathbb{M}_m, \mathbb{M}_n)$.

For vector $z \in \mathbb{C}^m \otimes \mathbb{C}^n$ we say that z is a *k-simple vector* in $\mathbb{C}^m \otimes \mathbb{C}^n$ if $z = \sum_{i=1}^{m} e_i \otimes z_i$ such that the linear span of $\{z_1, z_2, \ldots, z_m\}$ has dimension less than or equal to k. In other words, the *Schmidt number* of z is less than or equal to k.

For a matrix $X = \sum_{i,j=1}^{m} E_{ij} \otimes X_{ij} \in \mathbb{M}_m \otimes \mathbb{M}_n$ the *partial transpose* X^τ of X is defined by

$$X^\tau = \sum_{i,j=1}^{m} E_{ij}^t \otimes X_{ij} = \sum_{i,j=1}^{m} E_{ji} \otimes X_{ij} = \sum_{i,j=1}^{m} E_{ij} \otimes X_{ji}.$$

For each $k = 1, 2, \ldots, \min\{m, n\}$ we define the convex cones $\mathcal{V}_k$ and $\mathcal{V}^k$ in $\mathbb{M}_m \otimes \mathbb{M}_n$ as

$$\mathcal{V}_k = \{zz^* : z \text{ is an k-simple vector in } \mathbb{C}^m \otimes \mathbb{C}^n\}^{\circ\circ}$$

and

$$\mathcal{V}^k = \{(zz^*)^\tau : z \text{ is an k-simple vector in } \mathbb{C}^m \otimes \mathbb{C}^n\}^{\circ\circ},$$

where $\mathcal{C}^\circ = \{y \in \mathcal{Y} : \langle x, y \rangle \geq 0 \text{ for each } x \in \mathcal{C}\}$ is the dual cone of a subset $\mathcal{C}$ of normed space $\mathcal{X}$. Here we assume that $\mathcal{X}$ and $\mathcal{Y}$ are normed spaces, which are dual to each other with respect to a bilinear pairing $\langle \cdot, \cdot \rangle$; see [74, 144] for details.

For an s-simple vector $z = \sum_{i=1}^{m} e_i \otimes z_i \in \mathbb{C}^m \otimes \mathbb{C}^n$ take a generator $\{u_1, u_2, \ldots, u_r\}$ of the linear span of $\{z_1, z_2, \ldots, z_m\}$ in $\mathbb{C}^n$, and denote $a_{ik} \in \mathbb{C}$, $a_k \in \mathbb{C}^m$ by

$$z_i = \sum_{k=1}^{s} a_{ik} u_k \in \mathbb{C}^n, i = 1, 2, \ldots, m$$

and

$$a_k = \sum_{i=1}^{m} a_{ik} e_i \in \mathbb{C}^m, k = 1, 2, \ldots, s.$$

Then, we have

$$zz^* = \sum_{i,j=1}^{m} E_{ij} \otimes z_i z_j^* \in \mathbb{M}_m \otimes \mathbb{M}_n, z_i z_j^* = \sum_{k,l=1}^{s} = a_{ik}\overline{a_{jl}} u_k u_l^* \in \mathbb{M}_n,$$

3.5 Positive Linear Maps on Matrix Algebras

and it follows that

$$\langle zz^*, \Phi \rangle = \sum_{i,j=1}^{m} \langle z_i z_j^*, \Phi(E_{ij}) \rangle$$

$$= \sum_{i,j=1}^{m} \sum_{k,l=1}^{s} a_{ik}\overline{a_{jl}} \langle u_k u_l^*, \Phi(E_{ij}) \rangle = \sum_{i,j=1}^{m} \sum_{k,l=1}^{s} a_{ik}\overline{a_{jl}} \langle \Phi(E_{ij})\overline{u}_l, \overline{u}_k \rangle.$$

Therefore

$$\langle zz^*, \Phi \rangle = \sum_{i,j=1}^{m} \sum_{k,l=1}^{s} a_{ik}\overline{a_{jl}} \langle E_{kl} \otimes \Phi(E_{ij}) u, u \rangle,$$

where

$$u = \sum_{k=1}^{s} e_k \otimes \overline{u}_k \in \mathbb{C}^s \otimes \mathbb{C}^n.$$

If we put

$$w = \sum_{k=1}^{s} e_k \otimes a_k \in \mathbb{C}^s \otimes \mathbb{C}^m,$$

then we have

$$\Phi_s(ww^*) = \sum_{k,l=1}^{s} E_{kl} \otimes \Phi(a_k a_l^*) = \sum_{k,l=1}^{s} \sum_{i,j=1}^{m} a_{ik}\overline{a_{jl}} E_{kl} \otimes \Phi(E_{ij}).$$

Therefore,

$$\langle zz^*, \Phi \rangle = \langle \Phi_s(ww^*) u, u \rangle.$$

From the above argument, we have

$$\Phi \in \mathcal{P}_k \iff \langle X, \Phi \rangle \geq 0 \text{ for each } X \in \mathcal{V}_k,$$
$$X \in \mathcal{V}_k \iff \langle X, \Phi \rangle \geq 0 \text{ for each } \Phi \in \mathcal{P}_k.$$

Here, we use $\mathcal{P}_k = \mathcal{P}_k[\mathbb{M}_m, \mathbb{M}_n]$ and $\mathcal{P}^l = \mathcal{P}^l[\mathbb{M}_m, \mathbb{M}_n]$.

In general, $(\mathcal{V}_k \cap \mathcal{V}^l, \mathcal{P}_k + \mathcal{P}^l)$ forms a dual pair. In particular, in light of Theorem 3.5.1, $\mathcal{D} = \mathcal{P}_{\min\{m,n\}} + \mathcal{P}^{\min\{m,n\}}$ represents the cone of all decomposable maps. The corresponding dual cone is $\mathcal{V}_{\min\{m,n\}} \cap \mathcal{V}^{\min\{m,n\}}$, which is referred to as $\mathcal{PPT} = \{X \in (\mathbb{M}_m \otimes \mathbb{M}_n)_+ : X^\tau \in (\mathbb{M}_m \otimes \mathbb{M}_n)_+\}$. By the same observation as the dual pair $(\mathcal{V}_k, \mathcal{P}_k)$, we have

$$\Phi \in \mathcal{D} \iff \langle X, \Phi \rangle \geq 0 \text{ for each } X \in \mathcal{PPT}$$
$$X \in \mathcal{PPT} \iff \langle X, \Phi \rangle \geq 0 \text{ for each } \Phi \in \mathcal{D}.$$

If a Hilbert space $\mathcal{H}$ represents a quantum system, then a *quantum state* of the system $\mathcal{H}$ is given by a positive trace class operator ρ acting on $\mathcal{H}$ such that $\text{tr}(\rho) = 1$. Let $\mathcal{H}_{AB} = \mathcal{H}_A \otimes \mathcal{H}_B$ represent a bipartite composed quantum system, where $\mathcal{H}_A$ and $\mathcal{H}_B$ are Hilbert spaces with dimensions $\dim(\mathcal{H}_A) = m$ and $\dim(\mathcal{H}_B) = n$, respectively. We identify $\mathcal{H}_A$ and $\mathcal{H}_B$ with $\mathbb{C}^m$ and $\mathbb{C}^n$, respectively.

A quantum state ρ on $\mathbb{C}^m \otimes \mathbb{C}^n$ is called *separable* if it can be written as $\rho = \sum_{i=1}^{N} p_i \rho_A^{(i)} \otimes \rho_B^{(i)}$ for some $n \in \mathbb{N}$, a probability distribution $\{p_i\}_{i=1}^{N}$, and quantum states $\rho_A^{(i)}$ on $\mathbb{C}^m$ and $\rho_B^{(i)}$ on $\mathbb{C}^n$ ($1 \le i \le N$). Any state that is not separable is called *entangled*. Let $m = n = d$ and $\{e_i\}_{i=1}^{d}$ be an arbitrary orthogonal basis in $\mathbb{C}^d$. Then we define $\psi_d^+ = \frac{1}{\sqrt{d}} \sum_{i=1}^{d} e_i \otimes e_i$. The state $\psi_d^+ \psi_d^{+*}$ is called a *maximal entangled state* in $\mathbb{C}^d \otimes \mathbb{C}^d$.

Dualities between cones may be explained in the following diagram together with the inclusion relations between the cones:

$$\begin{array}{ccc} \mathcal{V}_1 \subseteq & \mathcal{PPT} \subseteq & \mathcal{V}_{\min\{m,n\}} = (\mathbb{M}_m \otimes \mathbb{M}_n)_+ \\ \updownarrow & \updownarrow & \updownarrow \\ \mathcal{P}_1 \supseteq & \mathcal{D} \supseteq & \mathcal{P}_{\min\{m,n\}} \cong (\mathbb{M}_m \otimes \mathbb{M}_n)_+, \end{array}$$

where $\cong$ denotes the Jamiolkowski–Choi isomorphism.

Since $(x \otimes y)(x \otimes y)^* = (xx^* \otimes yy^*)$, we can conclude that $\mathcal{V}_1 = \mathbb{M}_{m+} \otimes \mathbb{M}_{n+}$. A positive semidefinite matrix in $(\mathbb{M}_m \otimes \mathbb{M}_n)_+$ is said to be *entanglement* if it is not separable. In other words, it belongs to $(\mathbb{M}_m \otimes \mathbb{M}_n)_+ \setminus \mathbb{M}_{m+} \otimes \mathbb{M}_{n+}$.

The duality relation between $\mathcal{V}_1$ and $\mathcal{P}_1$ gives us a characterization of separability; $X \in (\mathbb{M}_m \otimes \mathbb{M}_n)_+$ is separable if and only if $\langle X, \Phi \rangle \ge 0$ for every $\Phi \in P[\mathbb{M}_m, \mathbb{M}_n] = \mathcal{P}_1$. Equivalently, $X \in (\mathbb{M}_m \otimes \mathbb{M}_n)_+$ is entanglement if and only if there exists a positive linear map Φ such that $\langle X, \Phi \rangle < 0$. This positive map is called *entanglement witness*. It is known that the problem of detecting entanglement is NP-hard [103].

The following is the recipe for constructing indecomposable positive maps on C^*-algebras, which could potentially be used as entanglement witnesses. This is an extended version of Piani and Mora's work on matrix algebras [198].

Theorem 3.5.10 ([18]) *Let $\mathcal{A}$ and $\mathcal{B}$ be unital C^*-algebras and let $\Phi: \mathcal{A} \to \mathcal{B}$ be a linear map. If there exists some $k \ge 2$ such that $\Phi_k: \mathbb{M}_k(\mathcal{A}) \to \mathbb{M}_k(\mathcal{B})$ is decomposable, then Φ is completely positive. In other words, if $\Phi: \mathcal{A} \to \mathcal{B}$ is not completely positive, then for any $k \in \mathbb{N}$ with $k \ge 2$, Φ_k is indecomposable.*

To prove this theorem, we need some results. The following result provides a factorization property for positive maps.

Theorem 3.5.11 *Let $\mathcal{A}, \mathcal{C}$ be C^*-algebras, and let $\mathcal{H}$ be a Hilbert space. Suppose that $\Phi: \mathcal{A} \to \mathcal{C}$ is a linear map and $\Psi: \mathcal{A} \otimes \mathbb{B}(\mathcal{H}) \to \mathcal{C} \otimes \mathbb{B}(\mathcal{H})$ is a positive linear*

3.5 Positive Linear Maps on Matrix Algebras

map. *Suppose that* $\Phi \otimes \text{id} - \Psi$ *is positive. Then* $\Psi = \Theta \otimes \text{id}$ *for some positive linear map* $\Theta \colon \mathscr{A} \to \mathscr{C}$.

Proof For any $A \in \mathscr{A}$ with $A \geq 0$ and $x \in \mathscr{H}$ we have

$$0 \leq \Psi(A \otimes xx^*) \leq \Phi(A) \otimes xx^*.$$

By a standard argument, we conclude that $\Psi(A \otimes xx^*) = \Theta_x(A) \otimes xx^*$ for some $\Theta_x(A) \in \mathscr{C}$. It follows that $\Theta_x(A)$ is independent of x from the linearity of Ψ. Therefore, for any $A \in \mathscr{A}$ and $y \in \mathscr{H}$ with $\|y\| = 1$, we have $\Psi(A \otimes yy^*) = \Theta(A) \otimes yy^*$ for some $\Theta(A)$. The linearity of Θ comes from the linearity of Ψ. Moreover, the positivity of $\Theta(A \otimes yy^*)$ comes from that of $\Psi(A \otimes yy^*)$. Hence, $(id_\mathscr{C} \otimes \text{tr})(\Theta(A) \otimes yy^*) = \Theta(A)$ is positive. □

We still need one more result to prove Theorem 3.5.10.

Lemma 3.5.12 *Let $\mathscr{A}$ and $\mathscr{C}$ be unital C^*-algebras, and let $\Phi \colon \mathscr{A} \to \mathscr{C}$ be a nonzero positive linear map. Then for $k \geq 2$, the map $\Phi \otimes \text{id}_k \colon \mathscr{A} \otimes \mathbb{M}_k \to \mathscr{C} \otimes \mathbb{M}_k$ is not completely copositive.*

Proof Let $(E_{ij})_{1 \leq i,j \leq k}$ be the standard system of matrix units for $\mathbb{M}_k$. Suppose that $\Phi \otimes \text{id}_k$ is completely copositive. Then, since $I \otimes (\sum_{ij=1}^{k} E_{ij} \otimes E_{ij}) = [I \otimes E_{ij}] (= [B_{ij}]) \in (\mathscr{A} \otimes \mathbb{M}_k) \otimes \mathbb{M}_k$ is positive and $\Phi \otimes \text{id}_k$ is completely copositive, $(\Phi \otimes \text{id}_k)^c([B_{ij}]) = [\Phi \otimes \text{id}_k(B_{ji})] = [\Phi(I) \otimes E_{ji}]$ is positive, that is,

$$[\Phi(I) \otimes E_{ji}] = \Phi(I) \otimes (\sum_{ij=1}^{n} E_{ji} \otimes E_{ij})$$

is positive. Nevertheless, $\sum_{ij=1}^{n} E_{ji} \otimes E_{ij}$ is not positive, and we reach a contradiction. □

Proof of Theorem 3.5.10

Suppose that $\Phi_k = \Phi \otimes \text{id}_k$ is decomposable for some k. Then, $\Phi \otimes \text{id}_k = \Psi + \Psi'$ for some Ψ and Ψ', where Ψ is completely positive and Ψ' is completely copositive. In particular, Ψ and Ψ' are positive. Then, by Theorem 3.5.11, there exist positive linear maps $\Psi_1 \colon \mathscr{A} \to \mathscr{C}$ and $\Psi_1' \colon \mathscr{A} \to \mathscr{C}$ such that $\Psi = \Psi_1 \otimes \text{id}_k$ and $\Psi' = \Psi_1' \otimes \text{id}_k$. From Lemma 3.5.12 we know that Ψ' is not completely copositive. Hence, Ψ' should be zero. Hence, $\Phi = \Psi_1$ and we have the conclusion. □

3.6 Exercises and Problems

Exercise 3.6.1 Prove that

(i) The function $f(t) = t^\alpha$ is operator monotone on $[0, \infty)$ if and only if $\alpha \in [0, 1]$.
(ii) The function $f(t) = t^\alpha$ is operator convex on $(0, \infty)$ if $\alpha \in [1, 2] \cup [-1, 0]$. It is operator concave on $(0, \infty)$ if $\alpha \in [0, 1]$.
(iii) The function e^t on $\mathbb{R}$ is neither operator monotone nor operator convex.

Hint: See [19, Theorem V.2.10 and Exercise V.2.11] for (i) and (ii).

Exercise 3.6.2 Show that the geometric mean $A \# B$ is concave on pairs of positive operators, or equivalently, prove that

$$\sum_{i=1}^{n}(A_i \# B_i) \leq \left(\sum_{i=1}^{n} A_i\right) \# \left(\sum_{i=1}^{n} B_i\right).$$

Exercise 3.6.3 Prove that if $\begin{bmatrix} A & B \\ B^* & C \end{bmatrix}$ and $\begin{bmatrix} A' & B \\ B^* & C' \end{bmatrix}$ are positive, then so is $\begin{bmatrix} A \# A' & B \\ B^* & C \# C' \end{bmatrix}$.

Hint: Use Theorem 2.4.3, the condition (ii) in operator means, and $A \natural B = A \natural^* B = s - \lim(A_\epsilon^{-1} \natural B_\epsilon^{-1})^{-1}$.

Exercise 3.6.4 Let A and B be positive semidefinite matrices in $\mathbb{M}_n$.

1. Establish that
$$A \#_p B \leq A^{1/2}((A^{1/2})^\dagger B (A^{1/2})^\dagger)^p A^{1/2}$$
for $p \in [0, 1]$.
2. If $\ker A \subseteq \ker B$, then show the equality holds.
3. Is it true that the equality holds if and only if the proposition $P_A = AA^\dagger$ onto $\operatorname{ran}(A)$ commutes with B?

Hint: See [179].

Exercise 3.6.5 Suppose that Φ is a positive linear functional $\mathbb{M}_n$. Show that there exists a positive semidefinite matrix X such that $\Phi(A) = \operatorname{tr}(AX)$ for all $A \in \mathbb{M}_n$.

Hint: Consider $\mathbb{M}_n$ equipped with the inner product $\langle A, B \rangle = \operatorname{tr}(AB^*)$ and apply the Riesz representation theorem.

Exercise 3.6.6 Let Φ be a positive linear map on a unital C^*-algebra $\mathscr{A}$. Prove that $\|\Phi\| \leq 2\|\Phi(I)\|$.

3.6 Exercises and Problems

Hint: See [191, Proposition 2.1].

Exercise 3.6.7 Let $\Phi \in \mathbb{B}_+(\mathbb{B}(\mathcal{H}), \mathbb{B}(\mathcal{K}))$. If $\begin{bmatrix} A & B \\ B^* & C \end{bmatrix} \geq 0$ and if the set $\{A, B, B^*, C\}$ is linearly dependent in $\mathbb{B}(\mathcal{H})$, then $\begin{bmatrix} \Phi(A) & \Phi(B) \\ \Phi(B^*) & \Phi(C) \end{bmatrix} \geq 0$.

Hint: See [44, Theorem 4.5].

Exercise 3.6.8 Find max $\left\{ X : X^* = X, \begin{bmatrix} A & V & X \\ V & B & W \\ X & W & C \end{bmatrix} \geq 0 \right\}$.

Hint: See [49].

Problem 3.6.9 (i) Is there any example of a 2-positive map from $\mathbb{M}_3$ to $\mathbb{M}_4$ which is not decomposable?
(ii) For $n \geq 4$, is any $(n-1)$-positive map from $\mathbb{M}_n$ to $\mathbb{M}_n$ decomposable?
Note that when $n = 3$, any 2-positive map from $\mathbb{M}_3$ to $\mathbb{M}_3$ is decomposable by [235].
Hint: See [198, 235].

Problem 3.6.10 Let $\mathscr{A}$ be a unital C^*-algebra and Φ be a completely bounded operator from $\mathscr{A}$ into $\mathbb{B}(\mathcal{H})$. Then there exist completely positive operators ψ and θ from $\mathscr{A}$ into $\mathbb{B}(\mathcal{H})$ with $\psi(I) = \theta(I) = \|\Phi\|_{cb}$ such that $\Phi : \mathbb{M}_2(\mathscr{A}) \to \mathbb{M}_2(\mathbb{B}(\mathcal{H}))$ given by
$$\Phi\left(\begin{bmatrix} A & B \\ C & D \end{bmatrix}\right) = \begin{bmatrix} \psi(A) & \Phi(B) \\ \Phi(C)^* & \theta(D) \end{bmatrix}$$ is completely positive.

Hint: See [199, Theorem 7.3].

Problem 3.6.11 (*Arveson's extension theorem*) Let $\mathscr{A}$ be a C^*-algebra, let $\mathscr{S} \subseteq \mathscr{A}$ be an operator system, and let $\Phi : \mathscr{S} \to \mathbb{B}(\mathcal{H})$ be a completely positive map. Prove that there exists a completely positive map $\Psi : \mathscr{A} \to \mathbb{B}(\mathcal{H})$, which extends Φ. Give an example to show that one cannot replace complete positivity with positivity.

Hint: See [12] and [191, Chap. 7].

Problem 3.6.12 Let $\mathscr{A}$ and $\mathscr{B}$ be unital C^*-algebras, let $\mathscr{M}$ be a subspace of $\mathscr{A}$, and let $\Phi : \mathscr{M} \to \mathscr{B}$ be a linear map. Set
$$\mathcal{S} = \left\{ \begin{bmatrix} \lambda I & A \\ B^* & \mu I \end{bmatrix} : \lambda, \mu \in \mathbb{C}, A, B \in \mathscr{A} \right\}$$
and define $\Psi : \mathcal{S} \to \mathbb{M}_2(\mathscr{B})$ by
$$\Psi\left(\begin{bmatrix} \lambda I & A \\ B^* & \mu I \end{bmatrix}\right) = \begin{bmatrix} \lambda I & \Phi(A) \\ \Phi(B)^* & \mu I \end{bmatrix}.$$

Show that if $\|\Phi\|_{cb} \leq 1$, then Ψ is completely positive.

Hint: See [191, Lemma 8.1].

Problem 3.6.13 (*Wittstock's extension theorem*) Let $\mathscr{A}$ be a unital C^*-algebra, let $\mathscr{M}$ be a subspace of $\mathscr{A}$, and let $\Phi : \mathscr{M} \to \mathbb{B}(\mathscr{H})$ be a completely bounded map. Prove that there exists a completely bounded map $\Psi : \mathscr{A} \to \mathbb{B}(\mathscr{H})$ that is an extension of Φ and $\|\Psi\|_{cb} = \|\Phi\|_{cb}$.

Hint: See [227].

Operator Variance and Covariance 4

In this chapter, we expand upon the concepts of variance and covariance, going beyond the boundaries of classical probability theory to operate within a noncommutative framework. Within this setting, we establish upper bounds for unitarily invariant norms associated with the covariance of bounded linear operators and matrices.

4.1 Bounds for the Operator Norm of the Covariance

Let (Ω, μ) be a probability measure space. The expectation of a random variable $f \in L^2(\Omega, \mu)$ is defined as $\mathbb{E}f = \int_\Omega f\, d\mu$. The covariance between two random variables f and g is defined as $\text{cov}(f, g) := \mathbb{E}(\overline{f}g) - \overline{\mathbb{E}f}\,\mathbb{E}g$. When $g = f$, we arrive at the variance of f, so $\text{var}(f) := \text{cov}(f, f) = \mathbb{E}(|f|^2) - |\mathbb{E}f|^2$. In probability theory, it is known that

$$\text{var}(f) \geq 0.$$

Furthermore, $\text{cov}(f, g)$ provides a semi-inner product on $L^2(\Omega, \mu)$. Therefore, by the Cauchy–Schwarz inequality, the *covariance–variance inequality*

$$|\text{cov}(f, g)|^2 \leq \text{var}(f)\,\text{var}(g).$$

holds.

There are several ways to extend the notions of variance and covariance to the operators in $\mathbb{B}(\mathscr{H})$. One way [150] is to consider two fixed vectors $x, y \in \mathscr{H}$, and then define the covariance and variance of operators $S, T \in \mathbb{B}(\mathscr{H})$ as follows:

$$\text{cov}_{x,y}(S, T) = \|y\|^2 \langle Sx, Tx\rangle - \langle Sx, y\rangle\langle y, Tx\rangle \quad \text{and} \quad \text{var}_{x,y}(T) = \text{cov}_{x,y}(T, T).$$

© The Author(s), under exclusive license to Springer Nature Switzerland AG 2024
M. S. Moslehian and H. Osaka, *Advanced Techniques with Block Matrices of Operators*, Frontiers in Mathematics, https://doi.org/10.1007/978-3-031-64546-4_4

In the case where $\|x\| = 1$ and $y = x$ we obtain the notion of covariance of two operators T and S introduced in [89] as

$$\mathrm{cov}_x(S, T) = \langle Sx, Tx\rangle - \langle Sx, x\rangle\langle x, Tx\rangle \quad \text{and} \quad \mathrm{var}_x(T) = \mathrm{cov}_x(T, T)$$

Enomoto, in his work [73], established a strong connection between the operator covariance–variance inequality and the Heisenberg uncertainty principle. He also emphasized that this connection corresponds precisely to the generalized Schrödinger inequality.

Another natural extension is obtained by considering a map $\Phi : \mathbb{B}(\mathcal{H}) \to \mathbb{B}(\mathcal{K})$ and putting

$$\mathrm{cov}(A, B) := \Phi(A^*B) - \Phi(A)^*\Phi(B) \quad \text{and} \quad \mathrm{var}(A) := \mathrm{cov}(A, A) = \Phi(A^*A) - \Phi(A)^*\Phi(A).$$

We define the set of all operators of the form UAU^*, where U is an arbitrary unitary as the *unitary orbit* of an operator A. The diameter of the unitary orbit, denoted as d_A, is given by

$$d_A = \sup\{\|AU - UA\| : U \text{ is unitary}\} = \sup_{\|X\|=1} \|AX - XA\| = 2\Delta(A),$$

where

$$\Delta(A) = \inf_{\lambda \in \mathbb{C}} \|A - \lambda I\|$$

is the $\|\cdot\|$-*distance* of A from the scalar operators [30]. It is known that $\Delta(A) \leq \|A\|$ and $\Delta(A) = c(A)$ for any normal operator A, where $c(A)$ represents the radius of the smallest disk in the complex plane containing the spectrum $\sigma(A)$ of A; see [208].

It is evident that the Choi inequality (3.3.2), applicable to a unital 2-positive map Φ and an arbitrary operator A, implies that $\mathrm{var}(A) \geq 0$. In this section, we aim to establish upper bounds for both $\mathrm{var}(A)$ and its operator norm. These bounds are considered complementary inequalities to the Kadison inequality. We show that if $\mathscr{A}$ and $\mathscr{B}$ are unital C^*-algebras and $\Phi : \mathscr{A} \to \mathscr{B}$ is a unital n-positive linear map for some $n \geq 3$, then

$$\|\mathrm{cov}(A, B)\| \leq \Delta(A)\,\Delta(B)$$

for all operators $A, B \in \mathscr{A}$.

Part (i) of the following result and its next counterexample is due to Mathias [162].

Theorem 4.1.1 *Let $\Phi : \mathscr{A} \to \mathscr{B}$ be a $*$-3-positive (not necessarily linear) map between C^*-algebras such that $\Phi(I) \leq I$. For every operators $A, B \in \mathscr{A}$, it holds that*

(i) *the* noncommutative covariance–variance inequality

$$\begin{bmatrix} \mathrm{var}(A) & \mathrm{cov}(A, B) \\ \mathrm{cov}(A, B)^* & \mathrm{var}(B) \end{bmatrix} \geq 0 \quad (A, B \in \mathscr{A}). \tag{4.1.1}$$

4.1 Bounds for the Operator Norm of the Covariance

(ii)
$$\| \operatorname{cov}(A, B) \|^2 \leq \| \operatorname{var}(A) \| \, \| \operatorname{var}(B) \|. \tag{4.1.2}$$

Proof (i) Let A and B be two operators in $\mathscr{A}$. We have

$$0 \leq \begin{bmatrix} A^* \\ B^* \\ I \end{bmatrix} \begin{bmatrix} A & B & I \end{bmatrix} = \begin{bmatrix} A^*A & A^*B & A^* \\ B^*A & B^*B & B^* \\ A & B & I \end{bmatrix}.$$

Since Φ is 3-positive, we have

$$0 \leq \begin{bmatrix} \Phi(A^*A) & \Phi(A^*B) & \Phi(A)^* \\ \Phi(B^*A) & \Phi(B^*B) & \Phi(B)^* \\ \Phi(A) & \Phi(B) & \Phi(I) \end{bmatrix} \leq \begin{bmatrix} \Phi(A^*A) & \Phi(A^*B) & \Phi(A)^* \\ \Phi(B^*A) & \Phi(B^*B) & \Phi(B)^* \\ \Phi(A) & \Phi(B) & I \end{bmatrix},$$

whence

$$0 \leq \begin{bmatrix} \Phi(A^*A) & \Phi(A^*B) & \Phi(A)^* & 0 \\ \Phi(B^*A) & \Phi(B^*B) & \Phi(B)^* & 0 \\ \Phi(A) & \Phi(B) & I & 0 \\ 0 & 0 & 0 & I \end{bmatrix}. \tag{4.1.3}$$

Employing Theorem 2.4.1 to (4.1.3) we reach

$$\begin{bmatrix} \Phi(A^*A) & \Phi(A^*B) \\ \Phi(B^*A) & \Phi(B^*B) \end{bmatrix} \geq \begin{bmatrix} \Phi(A)^* & 0 \\ \Phi(B)^* & 0 \end{bmatrix} \begin{bmatrix} I & 0 \\ 0 & I \end{bmatrix}^{-1} \begin{bmatrix} \Phi(A) & \Phi(B) \\ 0 & 0 \end{bmatrix}$$
$$= \begin{bmatrix} \Phi(A)^*\Phi(A) & \Phi(A)^*\Phi(B) \\ \Phi(B)^*\Phi(A) & \Phi(B)^*\Phi(B) \end{bmatrix},$$

or, equivalently,

$$\begin{bmatrix} \Phi(A^*A) - \Phi(A)^*\Phi(A) & \Phi(A^*B) - \Phi(A)^*\Phi(B) \\ \Phi(B^*A) - \Phi(B)^*\Phi(A) & \Phi(B^*B) - \Phi(B)^*\Phi(B) \end{bmatrix} \geq 0.$$

Therefore,

$$\begin{bmatrix} \operatorname{var}(A) & \operatorname{cov}(A, B) \\ \operatorname{cov}(A, B)^* & \operatorname{var}(B) \end{bmatrix} \geq 0.$$

(ii) Applying (2.6.1) to (4.1.1) yields that

$$\begin{bmatrix} \|\operatorname{var}(A)\| & \|\operatorname{cov}(A,B)\| \\ \|\operatorname{cov}(A,B)^*\| & \|\operatorname{var}(B)\| \end{bmatrix} \geq 0,$$

from which we conclude (4.1.2).

□

Example 4.1.2 A result of Choi [40, 3. Theorem 1] asserts that the map $\Phi(X) = \frac{2}{3}\operatorname{tr}(X)I - X$ is 2-positive but not 3-positive on $\mathbb{M}_3$. It is straightforward to see that

$$A = \begin{bmatrix} 0 & 0 & 0 \\ 0 & 0 & 0 \\ 1 & 0 & 0 \end{bmatrix} \quad \text{and} \quad B = \begin{bmatrix} 0 & 0 & 0 \\ 0 & 0 & 0 \\ 0 & 1 & 0 \end{bmatrix}$$

does not satisfy the variance–covariance inequality (4.1.1).

Bhatia and Sharma [29] proved that

$$\operatorname{var}(A) \leq \Delta(A)^2. \tag{4.1.4}$$

for every operator A and every unital positive linear map Φ. This result is extended in [55] as follows.

Proposition 4.1.3 *Let $\mathscr{A}$ and $\mathscr{B}$ be two unital C^*-algebras. If $\Phi : \mathscr{A} \to \mathscr{B}$ is a unital $*$-linear map, then*

$$\operatorname{var}(A) \leq \Phi(|A - \lambda I|^2) \tag{4.1.5}$$

for all $A \in \mathscr{A}$ and $\lambda \in \mathbb{C}$.

Proof Since Φ is a unital $*$-linear map, we have

$$\begin{aligned} \Phi(|A - \lambda I|^2) &\geq \Phi\left((A - \lambda I)^*(A - \lambda I)\right) - (\Phi(A - \lambda I))^* \Phi(A - \lambda I) \\ &= \Phi(A^*A) - \lambda \Phi(A^*) - \overline{\lambda}\Phi(A) + \overline{\lambda}\lambda I - \Phi(A)^*\Phi(A) \\ &\quad + \lambda \Phi(A^*) + \overline{\lambda}\Phi(A) - \overline{\lambda}\lambda I \\ &= \Phi(A^*A) - \Phi(A)^*\Phi(A) \end{aligned}$$

for all $\lambda \in \mathbb{C}$ and for all $A \in \mathscr{A}$.

□

If Φ is a unital positive linear map, then

$$\operatorname{var}(A) \leq \Phi(|A - \lambda I|^2) \leq \Phi(\|A - \lambda I\|^2 I) = \|A - \lambda I\|^2 \tag{4.1.6}$$

4.1 Bounds for the Operator Norm of the Covariance

for every $\lambda \in \mathbb{C}$. Hence

$$\mathrm{var}(A) \leq \inf_{\lambda \in \mathbb{C}} \|A - \lambda I\|^2 = \Delta(A)^2,$$

that is (4.1.4).

The following example shows that inequality (4.1.6) is really finer than inequality (4.1.4).

Example 4.1.4 Let $\Phi : \mathbb{M}_2 \to \mathbb{C}$ be defined by $\Phi\left(\begin{bmatrix} a_{11} & a_{12} \\ a_{21} & a_{22} \end{bmatrix}\right) = a_{11}$. It is evident that Φ is a unital positive linear map. Take $A = \begin{bmatrix} 1 & 2 \\ 2 & 4 \end{bmatrix}$. Then, we have $\mathrm{sp}(A) = \{0, 5\}$. Since A is positive semidefinite, we have $\inf_{\lambda \in \mathbb{C}} \|A - \lambda I_{2 \times 2}\|^2 = (2.5)^2 = 6.25$; see [208, Corollary 1]. We have

$$\mathrm{var}(A) = 4 \quad \text{and} \quad \Phi(|A|^2) = 5.$$

Therefore,

$$\mathrm{var}(A) < \Phi(|A|^2) < \inf_{\lambda \in \mathbb{C}} \|A - \lambda I_{2 \times 2}\|^2.$$

Now, we prove Theorem 2.6 of [174] in a much simpler way. In fact, it is deduced from Theorem 4.1.1 (ii) and (4.1.4).

Theorem 4.1.5 *Let $\Phi : \mathscr{A} \to \mathscr{B}$ be a unital 3-positive linear map between C^*-algebras. Then*

$$\|\mathrm{cov}(A, B)\| \leq \Delta(A) \Delta(B) \tag{4.1.7}$$

for all operators $A, B \in \mathscr{A}$.

We note that inequality (4.1.7) may be invalid if Φ is only a unital positive linear map. To illustrate this, consider the transpose map $\Phi : \mathbb{M}_2 \to \mathbb{M}_2$. Example 3.3.3 (vi) shows that this type of map is positive but not 2-positive. Hence, Φ is not a 3-positive map. Consequently, for matrices

$$A = \begin{bmatrix} 1 & 2 \\ 2 & 4 \end{bmatrix} \quad \text{and} \quad B = \begin{bmatrix} 1 & 0 \\ 0 & 4 \end{bmatrix},$$

we have

$$\|\mathrm{cov}(A, B)\| = 6.$$

Since A and B are positive matrices with $\mathrm{sp}(A) = \{0, 5\}$ and $\mathrm{sp}(B) = \{1, 4\}$, we conclude that $\Delta(A) = c(A) = \frac{5}{2}$ and $\Delta(B) = c(B) = \frac{3}{2}$. Therefore,

$$\| \operatorname{cov}(A, B) \| = 6 > \frac{15}{4} = \Delta(A) \Delta(B).$$

Bhatia and Davis [28] proved that for a unital positive linear map Φ and a Hermitian matrix A with eigenvalues in an interval $[m, M]$, the following inequality holds:

$$\operatorname{var}(A) \leq \left(\frac{M - m}{2} \right)^2.$$

In the subsequent result, we present a different version of this theorem that is applicable to unital completely positive maps and more general operators. Let us recall that the ball of diameter $[x, y]$ in a normed space $\mathcal{X}$ is the set of all elements $z \in \mathcal{X}$ such that $\|z - (x + y)/2\| \leq \|(x - y)/2\|$. The geometric property that $A \in \mathscr{A}$ belongs to the ball of diameter $[mI, MI]$ with $m, M \in \mathbb{C}$ is equivalent to the condition that

$$\operatorname{Re}\left((MI - A)^*(A - mI) \right) \geq 0,$$

since

$$\frac{1}{4}|M - m|^2 I - \left| A - \frac{M + m}{2} \right|^2 = \operatorname{Re}\left((MI - A)^*(A - mI) \right).$$

Proposition 4.1.6 *Let $\mathscr{A}$ be a unital C^*-algebra, let $\Phi : \mathscr{A} \to \mathbb{B}(\mathscr{H})$ be a unital positive map, and let $A \in \mathscr{A}$ belongs to the ball of diameter $[mI, MI]$ for some complex numbers m, M. Then*

$$\operatorname{var}(A) \leq \frac{1}{4}|M - m|^2 I.$$

Proof For any complex number $\lambda \in \mathbb{C}$, we have

$$\Phi(|A|^2) - |\Phi(A)|^2 = \Phi(|A - \lambda I|^2) - |\Phi(A - \lambda I)|^2. \tag{4.1.8}$$

The assumption of the proposition implies that

$$\left| A - \frac{M + m}{2} I \right|^2 \leq \frac{1}{4}|M - m|^2 I,$$

whence

$$\Phi\left(\left| A - \frac{M + m}{2} I \right|^2 \right) \leq \frac{1}{4}|M - m|^2 I. \tag{4.1.9}$$

It follows from (4.1.8) and (4.1.9) that

$$\Phi(|A|^2) - |\Phi(A)|^2 \leq \Phi\left(\left| A - \frac{M + m}{2} I \right|^2 \right) \leq \frac{1}{4}|M - m|^2 I.$$

□

4.1 Bounds for the Operator Norm of the Covariance

Renaud [202] established that

$$\left|\text{tr}(XAB) - \text{tr}(XA)\text{tr}(XB)\right| \leq krs \qquad (4.1.10)$$

for $1 \leq k \leq 4$. In this case, $A, B \in \mathbb{B}(\mathcal{H})$ are operators whose numerical ranges lie in the circular discs of radii r and s, respectively, and X is a positive matrix with trace one. Finally, as the last result of this section, we generalize Renaud's result to finite traces on a noncommutative probability space.

A finite trace τ on a von Neumann algebra $\mathscr{A}$ is a positive linear functional on $\mathscr{A}$ such that $\tau(I) < \infty$ and $\tau(AB) = \tau(BA)$ for all $A, B \in \mathscr{A}$. A trace τ is said to be *normal* if $\sup_i \tau(A_i) = \tau(\sup_i A_i)$ for any bounded increasing net (A_i) in $\mathscr{A}_+$. It is faithful if $A \geq 0$ and $\tau(A) = 0$ entail $A = 0$. It is normalized if $\tau(I) = 1$. A noncommutative probability space $(\mathscr{A}, \tau)$ is a von Neumann algebra $\mathscr{A}$ with a normal faithful normalized finite trace τ. In this case, for $1 \leq p < \infty$ and $A \in \mathscr{A}$, $\|A\|_p = (\tau(|A|^p))^{\frac{1}{p}}$ gives rise to a norm on $\mathscr{A}$. By a density operator, we mean a positive element A of $\mathscr{A}$ such that $\tau(A) = 1$. The next result reads as follows.

Theorem 4.1.7 *Let $(\mathscr{A}, \tau)$ be a noncommutative probability space and let $X \in \mathscr{A}$ be a density operator. Then*

(i) $|\tau(XAB) - \tau(XA)\tau(XB)| \leq \|A - \alpha I\|_4 \|B - \beta I\|_4 \|X\|_2$,
(ii) $|\tau(XAB) - \tau(XA)\tau(XB)| \leq \|A - \alpha I\|_2 \|B - \beta I\|_2 \|X\|$,
(iii) $|\tau(XAB) - \tau(XA)\tau(XB)| \leq \|A - \alpha I\| \|B - \beta I\|$,

for all $A, B \in \mathscr{A}$ and $\alpha, \beta \in \mathbb{C}$.

Proof We define the map $\varphi : \mathscr{A} \to \mathbb{C}$ by $\varphi(A) = \tau(XA)$. It is clear that φ is a unital positive linear functional. Therefore, by using Proposition 4.1.3, we have

$$\tau(XAA^*) - \tau(XA)\tau(XA^*) \leq \tau(X|A^* - \bar{\alpha}I|^2) \qquad (4.1.11)$$

for every $A \in \mathscr{A}$ and $\alpha \in \mathbb{C}$. Hence, by using the Cauchy–Schwarz inequality and applying the fact that for all $C \in \mathscr{A}$, $\|C\|_p = \|C^*\|_p$ ($p \geq 1$), we get

$$\tau(XAA^*) - \tau(XA)\tau(XA^*) \leq \left(\tau(|A^* - \bar{\alpha}I|^4)\right)^{\frac{1}{2}} \left(\tau(|X|^2)\right)^{\frac{1}{2}}$$
$$= \|A - \alpha I\|_4^2 \|X\|_2 \qquad (4.1.12)$$

for every $\alpha \in \mathbb{C}$. Similarly

$$\tau(XB^*B) - \tau(XB^*)\tau(XB) \leq \|B - \beta I\|_4^2 \|X\|_2 \qquad (4.1.13)$$

for every $\beta \in \mathbb{C}$. Now, we define the map $(\cdot,\cdot)_\tau : \mathscr{A} \times \mathscr{A} \to \mathbb{C}$ as

$$(A,B)_\tau = \tau(XA^*B) - \tau(XA^*)\tau(XB).$$

Evidently, $(\cdot,\cdot)$ is a semi-inner product on $\mathscr{A}$. Using the Cauchy–Schwarz inequality and inequalities (4.1.12) and (4.1.13), we arrive at

$$\begin{aligned} |\tau(XAB) - \tau(XA)\tau(XB)| &= |(A^*,B)_\tau| \\ &\leq (A^*,A^*)_\tau^{\frac{1}{2}} (B,B)_\tau^{\frac{1}{2}} \\ &\leq \|A - \alpha I\|_4 \|B - \beta I\|_4 \|X\|_2 \end{aligned}$$

for all $A, B \in \mathscr{A}$ and $\alpha, \beta \in \mathbb{C}$. This proves inequality (i).

Note that on the right side of inequality (4.1.11) (and similarly for inequality (4.1.13)), we can write $|A^* - \bar{\alpha}I|X|A^* - \bar{\alpha}I| \leq |A^* - \bar{\alpha}I| \|X\| |A^* - \bar{\alpha}I|$. Hence, $\tau(X|A^* - \bar{\alpha}I|^2) \leq \|A - \alpha I\|_2^2 \|X\|$. Using this fact and applying the Cauchy–Schwarz inequality to $(\cdot,\cdot)_\tau$, we can get inequality (ii).

Finally, we have $X^{\frac{1}{2}}|A^* - \bar{\alpha}I|^2 X^{\frac{1}{2}} \leq X^{\frac{1}{2}} \|A - \alpha I\|^2 X^{\frac{1}{2}}$. So $\tau(X|A^* - \bar{\alpha}I|^2) \leq \|A - \alpha I\|^2 \tau(X)$. Applying this inequality to the right side of (4.1.11) (and similarly for (4.1.13)) and using the Cauchy–Schwarz inequality we obtain inequality (iii). $\square$

4.2 Bounds for Unitarily Invariant Norms of the Covariance

It follows from the Stinespring Theorem 3.3.7 that for any unital completely positive map $\Phi : \mathscr{A} \to \mathbb{B}(\mathscr{H})$ between C^*-algebras there exist a Hilbert space $\mathscr{K}$, an isometry $V : \mathscr{H} \to \mathscr{K}$ and a unital $*$-homomorphism $\pi : \mathscr{A} \to \mathbb{B}(\mathscr{K})$ such that $\Phi(T) = V^*\pi(T)V$ for all $T \in \mathscr{A}$. We assume that $\mathscr{K}$ is the closure of $\pi(\mathscr{A})V\mathscr{H}$, and then we get the minimal Stinespring representation, which is unique up to a unitary equivalence. Moreover, if $\dim(\mathscr{A}) = k$ and $\dim(\mathscr{H}) = n$, then $\dim(\mathscr{K}) \leq k^2 n$. The equality occurs if $\mathscr{A} = \mathbb{M}_m$ for some m, see [20, Theorem 3.1.2]. Therefore, we deduce that

$$\left\|\left|I_{\dim(\mathscr{K})}\right|\right\| \leq \left\|\left|I_{k^2 n}\right|\right\|,$$

since, by the Fan dominance theorem and the Weyl monotonicity principle (1.3.6), a sufficient condition to have $\|\|A\|\| \leq \|\|B\|\|$ is that $A \leq B$.

In Sect. 4.1, we focus on approximating the operator norm of $\text{cov}(A,B)$ when dealing with an n-positive linear map Φ. We also use the results in [161] to estimate its unitarily invariant norm. Although part (i) of the following theorem can be deduced from Proposition 4.1.3, we prove it with a different method.

Theorem 4.2.1 *Let $\mathscr{A}$ be a finite-dimensional C^*-algebra of dimension k and let $\Phi : \mathscr{A} \to \mathbb{M}_n$ be a unital completely positive map. Then*

4.2 Bounds for Unitarily Invariant Norms of the Covariance

(i) $\|\mathrm{var}(A)\| \leq \|I_{k^2n}\| \Delta(A)^2$
for all $A \in \mathscr{A}$.
(ii) $\|\mathrm{cov}(A, B)\| \leq \|I_n\| \|I_{k^2n}\| \Delta(A) \Delta(B)$
for all $A, B \in \mathscr{A}$.

Proof (i) By using the Stinespring Theorem 3.3.7, the positivity of

$$\begin{pmatrix} \Phi(A^*A) - \Phi(A^*)\Phi(A) & \Phi(A^*B) - \Phi(A^*)\Phi(B) \\ \Phi(B^*A) - \Phi(B^*)\Phi(A) & \Phi(B^*B) - \Phi(B^*)\Phi(B) \end{pmatrix} \tag{4.2.1}$$

will follow once we prove the positivity of

$$\begin{pmatrix} V^*\pi(A^*A)V - V^*\pi(A^*)VV^*\pi(A)V & V^*\pi(A^*B)V - V^*\pi(A^*)VV^*\pi(B)V \\ V^*\pi(B^*A)V - V^*\pi(B^*)VV^*\pi(A)V & V^*\pi(B^*B)V - V^*\pi(B^*)VV^*\pi(B)V \end{pmatrix}. \tag{4.2.2}$$

As V is an isometry, we have $VV^* \leq I_{\dim(\mathscr{K})}$. It follows from Lemma 2.4.20 that

$$\begin{bmatrix} VV^* & VV^* \\ VV^* & VV^* \end{bmatrix} \leq \begin{bmatrix} I_{\dim(\mathscr{K})} & I_{\dim(\mathscr{K})} \\ I_{\dim(\mathscr{K})} & I_{\dim(\mathscr{K})} \end{bmatrix}.$$

Hence,

$$\begin{bmatrix} \pi(A)^* & 0 \\ 0 & \pi(B)^* \end{bmatrix} \begin{bmatrix} VV^* & VV^* \\ VV^* & VV^* \end{bmatrix} \begin{bmatrix} \pi(A) & 0 \\ 0 & \pi(B) \end{bmatrix}$$
$$\leq \begin{bmatrix} \pi(A)^* & 0 \\ 0 & \pi(B)^* \end{bmatrix} \begin{bmatrix} I_{\dim(\mathscr{K})} & I_{\dim(\mathscr{K})} \\ I_{\dim(\mathscr{K})} & I_{\dim(\mathscr{K})} \end{bmatrix} \begin{bmatrix} \pi(A) & 0 \\ 0 & \pi(B) \end{bmatrix},$$

whence

$$\begin{bmatrix} \pi(A)^*VV^*\pi(A) & \pi(A)^*VV^*\pi(B) \\ \pi(B)^*VV^*\pi(A) & \pi(B)^*VV^*\pi(B) \end{bmatrix} \leq \begin{bmatrix} \pi(A^*A) & \pi(A^*B) \\ \pi(B^*A) & \pi(B^*B) \end{bmatrix}. \tag{4.2.3}$$

The positivity of (4.2.2) follows by pre-multiplying (4.2.3) with $\begin{bmatrix} V^* & 0 \\ 0 & V^* \end{bmatrix}$ and post-multiplying it with $\begin{bmatrix} V & 0 \\ 0 & V \end{bmatrix}$. Utilizing the Stinespring Theorem 3.3.7, we have

$$\Phi(A^*A) - \Phi(A^*)\Phi(A)$$
$$= V^*\pi(A^*A)V - V^*\pi(A^*)VV^*\pi(A)V$$
$$= V^*\pi\big((A - \lambda I)^*(A - \lambda I)\big)V - V^*\pi(A - \lambda I)^*VV^*\pi(A - \lambda I)V$$
$$= V^*\pi(A - \lambda I)^*(I_{\dim(\mathscr{K})} - VV^*)\pi(A - \lambda I)V$$

for every $\lambda \in \mathbb{C}$.

Note that $I_{\dim(\mathcal{K})} - VV^*$ is a projection and π is a $*$-homomorphism, hence

$$\prod_{j=1}^{k} s_j \Big(\Phi(A^*A) - \Phi(A^*)\Phi(A) \Big)$$

$$= \prod_{j=1}^{k} s_j \Big(V^* \pi(A - \lambda I)^* (I_{\dim(\mathcal{K})} - VV^*) \pi(A - \lambda I) V \Big)$$

$$\leq \prod_{j=1}^{k} \Big[s_j \Big(V^* \pi(A - \lambda I)^* (I_{\dim(\mathcal{K})} - VV^*) \Big) s_j \Big((I_{\dim(\mathcal{K})} - VV^*) \pi(A - \lambda I) V \Big) \Big]$$

(by Lemma 1.3.3)

$$\leq \prod_{j=1}^{k} \Big[s_j \big(\pi(A - \lambda I)^* \big) s_j \big(\pi(A - \lambda I) \big) \Big] \qquad \text{(by Lemma 1.3.1 (i))}$$

$$= \prod_{j=1}^{k} s_j (\pi(|A - \lambda I|^2))$$

(since eigenvalues of matrices XY and YX are the same). (4.2.4)

for all $k = 1, 2, \ldots, n$ and $\lambda \in \mathbb{C}$. Since the weak log-majorization inequality implies the weak majorization inequality (see [238, Theorem 10.15]), we get from (4.2.4) that

$$\sum_{j=1}^{k} s_j \big(\Phi(A^*A) - \Phi(A^*)\Phi(A) \big) \leq \sum_{j=1}^{k} s_j (\pi(|A - \lambda I|^2)) \qquad (k = 1, 2, \ldots, n).$$

Thus, by using Lemma 1.3.1 (ii), we reach

$$\big\| \Phi(A^*A) - \Phi(A^*)\Phi(A) \big\| \leq \big\| \pi(|A - \lambda I|^2) \big\|$$
$$= \big\| \pi(|A - \lambda I|) I_{\dim(\mathcal{K})} \pi(|A - \lambda I|) \big\|$$
$$\leq \big\| \pi(|A - \lambda I|) \big\| \, \big\| I_{\dim(\mathcal{K})} \big\| \, \big\| \pi(|A - \lambda I|) \big\|$$
$$\leq \|A - \lambda I\|^2 \, \big\| I_{k^2 n} \big\| \qquad \text{(since } \pi \text{ is norm decreasing)}.$$

Therefore,

$$\big\| \Phi(A^*A) - \Phi(A^*)\Phi(A) \big\| \leq \big\| I_{k^2 n} \big\| \left(\inf_{\lambda \in \mathbb{C}} \|A - \lambda I\| \right)^2 = \big\| I_{k^2 n} \big\| \, \Delta(A)^2.$$

(ii) Since (4.2.1) is positive, by Lemma 2.4.12, there exists a contraction $K \in \mathbb{M}_n$ such that

$$\Phi(A^*B) - \Phi(A^*)\Phi(B) = \big(\Phi(A^*A) - \Phi(A^*)\Phi(A) \big)^{\frac{1}{2}} K \big(\Phi(B^*B) - \Phi(B^*)\Phi(B) \big)^{\frac{1}{2}}.$$

4.2 Bounds for Unitarily Invariant Norms of the Covariance

It follows that

$$\|\Phi(A^*B) - \Phi(A^*)\Phi(B)\|$$
$$= \left\|\left(\Phi(A^*A) - \Phi(A^*)\Phi(A)\right)^{\frac{1}{2}} K \left(\Phi(B^*B) - \Phi(B^*)\Phi(B)\right)^{\frac{1}{2}}\right\|$$
$$\leq \|\Phi(A^*A) - \Phi(A^*)\Phi(A)\|^{\frac{1}{2}} \|K\| \|\Phi(B^*B) - \Phi(B^*)\Phi(B)\|^{\frac{1}{2}}$$
$$\qquad \text{(by Lemma 1.3.1 (ii))}$$
$$\leq \|\Phi(A^*A) - \Phi(A^*)\Phi(A)\|^{\frac{1}{2}} \|I_n\| \|\Phi(B^*B) - \Phi(B^*)\Phi(B)\|^{\frac{1}{2}}$$
$$\qquad \text{(by Lemma 1.3.2)}$$
$$\leq \|I_n\| \|I_{k^2n}\| \inf_{\lambda \in \mathbb{C}} \|A - \lambda I\| \inf_{\mu \in \mathbb{C}} \|B - \mu I\| \quad \text{(by part (i))}$$
$$= \|I_n\| \|I_{k^2n}\| \Delta(A) \Delta(B).$$

$\square$

As a consequence, we get the following upper bound for the covariance of two operators and for specific unitarily invariant norms.

Corollary 4.2.2 *If* $\Phi : \mathbb{M}_m \to \mathbb{M}_n$ *is a unital completely positive map, then*

$$\|\operatorname{cov}(A,B)\| \leq \frac{1}{4} \max_{\|X\|=1, \|Y\|=1} \|A^*X - XA^*\| \|BY - YB\| = \Delta(A)\Delta(B).$$

and

$$\|\operatorname{cov}(A,B)\|_p \leq (mn)^{2/p} \Delta(A)\Delta(B) \quad (p \geq 1)$$

for all $A, B \in \mathbb{M}_m$.

Proof First, observe that $\dim(\mathbb{M}_m) = m^2$. Second, note that the operator norm $\|\cdot\|$ and the Schatten p-norm $\|\cdot\|_p$, whenever $p \geq 1$, are unitarily invariant norms. Furthermore, $\|I_k\|_p = k^{1/p}$ for every positive integer $k \geq 1$. It is now sufficient to use Theorem 4.2.1. $\square$

If A is self-adjoint with $mI \leq A \leq MI$ for some real numbers m and M, then $\Delta(A) = (M-m)/2$. As a consequence of Theorem 4.2.1, we obtain the following result.

Corollary 4.2.3 *Let* $\Phi : \mathbb{M}_m \to \mathbb{M}_n$ *be a unital completely positive map and* $A, B \in \mathbb{M}_n$ *be Hermitian matrices with* $mI_m \leq A \leq MI_m$, $m'I_m \leq B \leq M'I_m$ *for some constants* $m, m', M,$ *and* M'. *Then*

$$\|\operatorname{cov}(A,B)\| \leq \frac{1}{4}(M-m)(M'-m') \|I_n\| \|I_{m^2n}\|.$$

Theorem 4.2.4 Let $12 \leq \eta$ be a positive integer, and let $\Phi : \mathbb{M}_m \to \mathbb{M}_n$ be a unital η-positive linear map. Then

$$\|\text{cov}(A, B)\| \leq \|I_n\| \, \|I_{m^2n}\| \, \Delta(A) \Delta(B) \tag{4.2.5}$$

for all $A, B \in \mathbb{M}_m$.

Proof First, assume that A and B are Hermitian matrices. By the same reasoning as in Theorem 4.1.1, we have

$$\begin{bmatrix} \Phi(A^*A) - \Phi(A)^*\Phi(A) & \Phi(A^*B) - \Phi(A)^*\Phi(B) \\ \Phi(B^*A) - \Phi(B)^*\Phi(A) & \Phi(B^*B) - \Phi(B)^*\Phi(B) \end{bmatrix} \geq 0. \tag{4.2.6}$$

As A is Hermitian, the unital C^*-algebra $C^*(A, I_m)$ generated by A and the identity I_m is commutative. Hence, the restriction of Φ to $C^*(A, I_m)$ is a unital completely positive map by Theorem 3.3. Thus, Theorem 4.2.1 (i) gives us the inequality

$$\left\|\Phi(A^*A) - \Phi(A^*)\Phi(A)\right\|^{\frac{1}{2}} \leq \sqrt{\|I_{m^2n}\|} \, \|A\|.$$

A similar formula is valid for B instead of A. Now, the same reasoning as in the proof of Theorem 4.2.1 (ii), along with (4.2.6), shows that the inequality

$$\left\|\Phi(A^*B) - \Phi(A^*)\Phi(B)\right\| \leq \|I_n\| \, \|I_{m^2n}\| \, \|A\| \, \|B\| \tag{4.2.7}$$

holds for any Hermitian matrices A and B and any 3-positive map Φ.

Second, let A and B be arbitrary and Hermitian matrices, respectively. Applying inequality (4.2.7) to 3-positive map $\Phi_2 : \mathbb{M}_2(\mathbb{M}_m) \to \mathbb{M}_2(\mathbb{M}_n)$ and Hermitian matrices $\begin{bmatrix} 0 & A \\ A^* & 0 \end{bmatrix}$ and $\begin{bmatrix} 0 & 0 \\ 0 & B \end{bmatrix}$ and using Lemma 1.3.4, we get

$$\left\| \Phi_2\left(\begin{bmatrix} 0 & A \\ A^* & 0 \end{bmatrix} \begin{bmatrix} 0 & 0 \\ 0 & B \end{bmatrix} \right) - \Phi_2\left(\begin{bmatrix} 0 & A \\ A^* & 0 \end{bmatrix} \right) \Phi_2\left(\begin{bmatrix} 0 & 0 \\ 0 & B \end{bmatrix} \right) \right\|$$
$$\leq \|I_n\| \, \|I_{m^2n}\| \, \|A\| \, \|B\|$$

Since

$$\Phi_2\left(\begin{bmatrix} 0 & A \\ A^* & 0 \end{bmatrix} \begin{bmatrix} 0 & 0 \\ 0 & B \end{bmatrix} \right) = \begin{bmatrix} 0 & \Phi(AB) \\ 0 & 0 \end{bmatrix}$$

and

$$\Phi_2\left(\begin{bmatrix} 0 & A \\ A^* & 0 \end{bmatrix} \right) \Phi_2\left(\begin{bmatrix} 0 & 0 \\ 0 & B \end{bmatrix} \right) = \begin{bmatrix} 0 & \Phi(A)\Phi(B) \\ 0 & 0 \end{bmatrix},$$

4.2 Bounds for Unitarily Invariant Norms of the Covariance

we have

$$\left\|\begin{bmatrix} 0 & \Phi(AB) - \Phi(A)\Phi(B) \\ 0 & 0 \end{bmatrix}\right\| \leq \|I_n\| \, \|I_{m^2n}\| \, \|A\| \, \|B\|.$$

Hence

$$\left\|\begin{bmatrix} \Phi(AB) - \Phi(A)\Phi(B) & 0 \\ 0 & 0 \end{bmatrix}\right\|$$

$$= \left\|\begin{bmatrix} I_n & 0 \\ 0 & I_n \end{bmatrix} \begin{bmatrix} 0 & \Phi(AB) - \Phi(A)\Phi(B) \\ 0 & 0 \end{bmatrix} \begin{bmatrix} 0 & I_n \\ I_n & 0 \end{bmatrix}\right\|$$

$$= \left\|\begin{bmatrix} 0 & \Phi(AB) - \Phi(A)\Phi(B) \\ 0 & 0 \end{bmatrix}\right\|$$

$$\leq \|I_n\| \, \|I_{m^2n}\| \, \|A\| \, \|B\|$$

for arbitrary matrix A, Hermitian matrix B, and 6-positive map Φ.

Third, by repeating the same argument as above to the latter inequality for an arbitrary matrix B, we conclude that

$$\left\|\begin{bmatrix} \Phi(AB) - \Phi(A)\Phi(B) & 0 \\ 0 & 0 \end{bmatrix}\right\| \leq \|I_n\| \, \|I_{m^2n}\| \, \|A\| \, \|B\|.$$

In our notation (1.3.5), we can write

$$\|\Phi(AB) - \Phi(A)\Phi(B)\| \leq \|I_n\| \, \|I_{m^2n}\| \, \|A\| \, \|B\|$$

for any arbitrary matrices A and B, and 12-positive map Φ. It follows from the latter inequality that

$$\|\Phi(AB) - \Phi(A)\Phi(B)\|$$
$$= \left\|\Phi\big((A - \lambda I_m)(B - \mu I_m)\big) - \Phi(A - \lambda I_m)\Phi(B - \mu I_m)\right\|$$
$$\leq \|I_n\| \, \|I_{m^2n}\| \, \|A - \lambda I_m\| \, \|B - \lambda I_m\|$$

for all $\lambda, \mu \in \mathbb{C}$. Thus

$$\|\Phi(AB) - \Phi(A)\Phi(B)\| \leq \|I_n\| \, \|I_{m^2n}\| \inf_{\lambda \in \mathbb{C}} \|A - \lambda I_m\| \inf_{\mu \in \mathbb{C}} \|B - \mu I_m\|$$
$$= \|I_n\| \, \|I_{m^2n}\| \, \Delta(A) \, \Delta(B).$$

$\square$

Remark 4.2.5 It is noted that Theorem 4.2.4 is not valid if Φ is supposed to be unital 2-positive linear map. To see this, consider the map $\Phi : \mathbb{M}_3 \to \mathbb{M}_3$ defined as $\Phi(A) = 2\text{tr}(A)I_3 - A$. Then Φ is 2-positive but not 3-positive as we osbserved in Example 3.5.2;

see [44]. Taking $A = \begin{bmatrix} 1 & 0 & 0 \\ 0 & 0 & 1 \\ 0 & 1 & 0 \end{bmatrix}$ and $B = \begin{bmatrix} 0 & 0 & 1 \\ 0 & 0 & 0 \\ 1 & 0 & 1 \end{bmatrix}$, we can easily observe that 2×2 block matrix in (4.2.6) is not positive and (4.2.5) does not hold for the operator norm. The case when $2 < \eta < 12$ remains unsolved.

4.3 Concluding Remarks

Let ρ be a quantum state, also known as a density operator. In the theory of quantum measurement, the classical expectation value of an observable, represented by a self-adjoint operator A, in a quantum state ρ, is given by $\operatorname{tr}(\rho A)$. The *classical variance* of a quantum state ρ and an observable operator A is defined as follows:

$$V_\rho(A) := \operatorname{tr}(\rho A^2) - (\operatorname{tr}(\rho A))^2.$$

The *Heisenberg uncertainty relation* asserts that

$$V_\rho(A) V_\rho(B) \geq \frac{1}{4} |\operatorname{tr}(\rho[A, B])|^2 \tag{4.3.1}$$

for a quantum state ρ. Recall that $[A, B] = AB - BA$ represents the commutator of observables A and B. A stronger result was given by Schrödinger [205], which can be expressed as

$$V_\rho(A) V_\rho(B) - |\operatorname{Re}(\operatorname{Cov}_\rho(A, B))|^2 \geq \frac{1}{4} |\operatorname{tr}(\rho[A, B])|^2. \tag{4.3.2}$$

In this inequality, the *classical covariance* is defined by

$$\operatorname{Cov}_{\rho}(A) := \operatorname{tr}(\rho AB) - \operatorname{tr}(\rho A)\operatorname{tr}(\rho B).$$

Yanagi et al. [233] defined the *one-parameter correlation* and the *one-parameter Wigner–Yanase skew information* (also known as the Wigner–Yanase–Dyson skew information; see [148]) for operators A and B, respectively, as follows

$$\operatorname{Corr}^\alpha_\rho(A, B) := \operatorname{tr}(\rho A^* B) - \operatorname{tr}(\rho^{1-\alpha} A^* \rho^\alpha B) \text{ and } I^\alpha_\rho(A) := \operatorname{Corr}^\alpha_\rho(A, A),$$

where $\alpha \in [0, 1]$. They showed a trace inequality representing the relation between these two quantities as

$$\left|\operatorname{Re}(\operatorname{Corr}^\alpha_\rho(A, B))\right|^2 \leq I^\alpha_\rho(A) I^\alpha_\rho(B). \tag{4.3.3}$$

Tracial positive linear maps are the natural generalizations of the usual trace on $\mathbb{M}_n$. A linear map $\Phi : \mathscr{A} \to \mathscr{B}$ between C^*-algebras is said to be a *tracial map* if

4.3 Concluding Remarks

$$\Phi(AB) = \Phi(BA)$$

for all $A, B \in \mathscr{A}$. A positive operator $\rho \in \mathscr{A}$ is said to be Φ-*density* if $\Phi(\rho) = I$.

Tsui [48] proved that each tracial positive linear map $\Phi : \mathscr{A} \to \mathbb{B}(\mathscr{H})$ admits a factorization as the composition of a tracial positive linear map $\mathscr{A} \to C(\Omega)$ and a positive linear map $C(\Omega) \to \mathbb{B}(\mathscr{H})$. As a consequence, every tracial positive linear map is completely positive. Several examples are presented in [48] as follows.

Example 4.3.1 (i) If $\mathscr{A}$ is a unital C^*-algebra with a unique tracial state τ (in particular, if $\mathscr{A}$ is a finite factor), then every tracial positive linear map $\Phi : \mathscr{A} \to \mathbb{B}(\mathscr{H})$ is of the form $\Phi(A) = \tau(A) I$. Indeed, if Φ is unital, then every unit vector $x \in \mathscr{H}$ induces a tracial state $\langle \Phi(\cdot)x, x \rangle$ on $\mathscr{A}$ and

$$\langle (\Phi(A) - \tau(A)I)x, x \rangle = \langle \Phi(A)x, x \rangle - \tau(A) = 0$$

and so $\Phi(A) = \tau(A)I$. If Φ is not unital, then we have $\Phi(I) \leq \|\Phi\| I$. Then $\Psi : \mathscr{A} \to \mathbb{B}(\mathscr{H})$ defined by $\Psi(A) = [\Phi(A) + \tau(A)(\|\Phi\| I - \Phi(I))]/\|\Phi\|$ is a unital tracial positive map. Therefore, by the argument above, $\Psi(A) = \tau(A)I$. Thus, $\Phi(A) = \tau(A)\Phi(I)$ and this proves the claim.

(ii) Let $\mathscr{A}$ be a properly infinite von Neumann algebra. Then the only tracial positive linear map $\Phi : \mathscr{A} \to \mathbb{B}(\mathscr{H})$ is the zero map. To demonstrate this, we consider isometries $S_1, S_2 \in \mathscr{A}$ such that $S_1 S_1^* + S_2 S_2^* \leq I$. Such isometries exist by [61, Corollary 2, p. 298]. If Φ is a tracial positive linear map on $\mathscr{A}$, then

$$\Phi(I) \geq \Phi(S_1 S_1^*) + \Phi(S_2 S_2^*) = \Phi(S_1^* S_1) + \Phi(S_2^* S_2) = 2\Phi(I).$$

Hence, $\Phi(I) = 0$, and so Φ is the zero map.

In [57], it is assumed that $\Phi : \mathscr{A} \to \mathscr{B}$ is a tracial positive linear map and ρ is a Φ-density operator. The *generalized covariance* and the *generalized variance* of A and B are defined as

$$\text{Cov}_{\rho,\Phi}(A, B) := \Phi(\rho A^* B) - \Phi(\rho A^*)\Phi(\rho B) \text{ and } V_{\rho,\Phi}(A) := \text{Cov}_{\rho,\Phi}(A, A),$$

respectively. Furthermore, the *generalized correlation* and the *generalized Wigner–Yanase–Dyson skew information* of operators A and B are defined by

$$\text{Corr}_{\rho,\Phi}^{\alpha}(A, B) := \Phi(\rho A^* B) - \Phi(\rho^{1-\alpha} A^* \rho^{\alpha} B) \text{ and } I_{\rho,\Phi}^{\alpha}(A) := \text{Corr}_{\rho,\Phi}^{\alpha}(A, A),$$

respectively.

Then, one can establish several related inequalities; see [57] and references therein. For example, a *generalized variance–covariance inequality* can be shown for every tracial positive linear map as

$$\begin{bmatrix} V_{\rho,\Phi}(A) & \text{Cov}_{\rho,\Phi}(B,A) \\ \text{Cov}_{\rho,\Phi}(A,B) & V_{\rho,\Phi}(B) \end{bmatrix} \geq 0.$$

4.4 Exercises and Problems

Exercise 4.4.1 If $A, B \in \mathbb{M}_n$ with $n \geq 4$, then prove that

$$\|(n^2 - n - 1)\operatorname{tr}(AB)I - nAB - (n-1)\operatorname{tr}(A)\operatorname{tr}(B)I + \operatorname{tr}(B)A + \operatorname{tr}(A)B\|$$

$$\leq \frac{(n^2 - n - 1)^2}{n-1} \Delta(A, \|\cdot\|) \Delta(B, \|\cdot\|).$$

Hint: See [174].

Exercise 4.4.2 Let $\mathscr{A}$ be a unital C^*-algebra and $\Phi : \mathscr{A} \to \mathbb{B}(\mathscr{H})$ be a unital completely positive map. If $A, B \in \mathscr{A}$ belong to the balls of diameter $[m_1 I, M_1 I]$ and $[m_2 I, M_2 I]$ for some complex numbers $m_1, M_1, m_2,$ and M_2, respectively, then

$$|\operatorname{cov}(A,B)| \leq \frac{1}{4}|M_1 - m_1||M_2 - m_2|I.$$

Hint: Refer to Corollary 4.2.3.

Exercise 4.4.3 The quantity $U_\rho(A)$ as a measure of uncertainty is defined in [156] as

$$U_\rho(A) = \sqrt{V_\rho(A)^2 - (V_\rho(A) - I_\rho(A))^2}.$$

Prove that the following *Heisenberg-type uncertainty relation*

$$U_\rho(A)U_\rho(B) \geq \frac{1}{4}|\operatorname{tr}(\rho[A,B])|^2.$$

Hint: See [57, 156].

Exercise 4.4.4 Let $x \in \mathscr{H}$ be a unit vector. Show that the determinant of the positive semidefinite Gram matrix

$$\begin{bmatrix} \langle Sx, Sx \rangle & \langle Sx, Tx \rangle & \langle Sx, x \rangle \\ \langle Tx, Sx \rangle & \langle Tx, Tx \rangle & \langle Tx, x \rangle \\ \langle x, Sx \rangle & \langle x, Tx \rangle & \langle x, x \rangle \end{bmatrix}$$

is $\operatorname{var}_x(S)\operatorname{var}_x(T) - |\operatorname{cov}_x(S,T)|^2$ and it is nonnegative.
Hint: See [90].

4.4 Exercises and Problems

Exercise 4.4.5 Let $A, B \in \mathbb{B}(\mathscr{H})$ be positive. Show that

$$\langle A^2 x, x \rangle \langle B^2 x, x \rangle - |\langle ABx, x \rangle|^2 \geq \langle A^2 x, x \rangle \langle Bx, x \rangle^2 - 2|\langle ABx, x \rangle||\langle Ax, x \rangle\langle Bx, x \rangle$$
$$+ \langle B^2 x, x \rangle \langle Ax, x \rangle^2 \geq 0$$

for all unite vectors $x \in \mathscr{H}$.

Hnt: See [90].

Exercise 4.4.6 Let $A, B \in \mathbb{B}(\mathscr{H})$. We call $\{A, B\} = AB + BA$ the *Jordan product* of A and B. Show that for any self-adjoint operators A and B, and any unit vector $x \in \mathscr{H}$,

(i) $|\text{cov}_x(A, B)|^2 = \left(\frac{1}{2}\langle\{A, B\}x, x\rangle - \langle Ax, x\rangle\langle Bx, x\rangle\right)^2 + \left(\frac{1}{2i}\langle[A, B]x, x\rangle\right)^2$

(ii) $(\text{var}_x(A)\,\text{var}_x(B))^{1/2} \geq \frac{1}{2}|\langle[A, B]x, x\rangle|$.

Hint: See [91].

Exercise 4.4.7 Let $A, B \in \mathbb{B}(\mathscr{H})$ and let $x \in \mathscr{H}$ be such that $\langle A^*Bx, x \rangle \neq 0$. If

$$s\,\text{cov}_x(A, B) := \frac{\|x\|^2 \langle A^*Bx, x \rangle - \langle A^*x, x \rangle \langle Bx, x \rangle}{\langle A^*Bx, x \rangle},$$

then show that $s\,\text{cov}_x(A, B) = \|Px\|^2 + s\,\text{cov}_{(I-P)x}(A, B)$ for any projection $P \in \mathbb{B}(\mathscr{H})$ with $PA = AP = BP = PB = 0$.

Hint: See [91].

Exercise 4.4.8 Let Φ be a unital completely positive linear map on $\mathbb{B}(\mathscr{H})$ and let $A, B \in \mathbb{B}(\mathscr{H})$. Show that

(i) $|\text{cov}(A, B)| \leq U^*\text{var}(A)U\#\text{var}(B)$, where $\text{cov}(A, B) = U|\text{cov}(A, B)|$ is the polar decomposition of $\text{cov}(A, B)$.
(ii) If $\lambda, \mu \in \mathbb{C}$ and the operator $(A - \lambda I)^*(\mu I - A)$ is accretive, then

$$\text{var}(A) \leq \frac{1}{4}|\lambda - \mu|^2 - \left|\Phi(A) - \frac{\lambda + \mu}{2}\right|^2.$$

Hint: See Eq. (4.1.8).

Problem 4.4.9 Let Φ be a unital completely positive linear map on $\mathbb{B}(\mathscr{H})$, let $A, B \in \mathbb{B}_{\text{sa}}(\mathscr{H})$, and let U be the partial isometry in the polar decomposition of $\text{cov}(A, B)$. Prove that

(i) There exists an isometry $V \in \mathbb{B}(\mathcal{H})$ such that

$$U^* \operatorname{var}(A) U \# \operatorname{var}(B) \geq V^* \left(\frac{\Phi([A, B]) - [\Phi(A), \Phi(B)]}{2i} \right)_+ V,$$

where X_+ is the positive part of a self-adjoint operator $X \in \mathbb{B}(\mathcal{H})$.

(ii) If $\Phi(AB) - \Phi(A)\Phi(B)$ is normal, then

$$U^* \operatorname{var}(A) U \# \operatorname{var}(B)$$

$$= \left(\frac{1}{4} (\Phi(\{A, B\}) - \{\Phi(A), \Phi(B)\})^2 + \left(\frac{\Phi([A, B]) - [\Phi(A), \Phi(B)]}{2i} \right)^2 \right)^{1/2}$$

$$\geq \frac{1}{2} |\Phi([A, B]) - [\Phi(A), \Phi(B)]|;$$

Hint: See [92].

Problem 4.4.10 Let $2 < \eta < 12$ be integer and $\Phi : \mathbb{M}_m \to \mathbb{M}_m$ be a unital η-positive linear map. Then, is it true that

$$\|\operatorname{cov}(A, B)\| \leq \|I_n\| \, \|I_{m^2 n}\| \, \Delta(A) \, \Delta(B)?$$

for all $A, B \in \mathbb{M}_m$.

Hint: Refer to Theorem 4.2.4 and Remark 4.2.5.

Nonlinear Positive Maps

5

We can categorize maps as either linear or nonlinear. Linear maps exhibit orderly behavior while nonlinear maps are more intricate. The examination of both linear and nonlinear positive maps within the realm of C^*-algebras presents an intriguing challenge in operator theory and matrix analysis, finding applications in diverse fields such as statistics, quantum information, and mechanics.

The study of linear completely positive maps has garnered considerable attention from mathematicians, dating back to the renowned Stinespring theorem [210], as well as related work in [212]. Meanwhile, Ando and Choi [8] explored nonlinear completely positive maps between C^*-algebras. Their investigation allows us to decompose any completely positive map into completely positive maps characterized by mixed homogeneity and represent each of which through $*$-representations.

Lieb [149] introduced a class of nonlinear maps from $\mathbb{M}_n$ to $\mathbb{C}$, closely related to 2-positive maps. Hiai [115] established connections between n-positivity, n-convexity, and the differentiability of real-valued functions. He indeed introduced the notion of S-monotonicity, which states that if $A \leq B$, then $f[A] \leq f[B]$. Here, $f[A]$ is used to denote the matrix obtained by applying f entrywise to $A = [a_{ij}]$, that is, $f[A] = [f(a_{ij})]$ provided that the entries of A are in the domain of f.

Guillot et al. [99, 100] presented characterizations of power functions that preserve positivity, monotonicity, and convexity. Günther and Klotz [102] investigated the class of n-positive norms on $\mathbb{M}_n$. Meanwhile, Nagisa and Watatani [177] investigated the class of nonlinear positive maps, which consist of compositions of $*$-multiplicative maps and positive linear maps. They also introduced specific classes of general nonlinear positive maps defined over positive cones and studied their monotonicity and concavity.

Ando and Choi [8] provided a characterization of completely positive nonlinear maps between C^*-algebras. They demonstrate that such maps can be expressed as the sum of completely positive maps, which are mixed homogeneous. Moreover, they established that a nonlinear completely positive map $\Phi : \mathscr{A} \to \mathscr{B}$ is additive if and only if $\Phi(2A) = 2\Phi(A)$ for each $A \in \mathscr{A}$.

In this chapter, we aim to explore some properties of linear positive maps in comparison to general (nonlinear) positive maps. We introduce the concept of a Lieb map as an extension of the Lieb function. Subsequently, we examine the fundamental properties of these maps and establish a connection between Lieb maps and 2-positive maps. Furthermore, we prove that for any unital 3-positive map $\Phi : \mathscr{A} \to \mathscr{B}$ (which may not be linear) between C^*-algebras, the condition $\Phi(A^*A) = \Phi(A)^*\Phi(A)$ holds for some $A \in \mathscr{A}$ implies that $\Phi(XA) = \Phi(X)\Phi(A)$ for all $X \in \mathscr{A}$. Furthermore, we demonstrate that this result may not apply to arbitrary 2-positive maps. These findings are mainly derived from the works presented in [56, 58].

5.1 Lieb Maps and Their Fundamental Properties

In this section, we aim to extend the concept of a Lieb function, as originally introduced by Lieb [149]. A function $\varphi : \mathscr{A} \to \mathbb{C}$ is known as a *Lieb function* if it possesses the following properties:

(i) $\varphi(A) \geq \varphi(B) \geq 0$ whenever $A \geq B \geq 0$,
(ii) $\varphi(A^*A)\varphi(B^*B) \geq |\varphi(A^*B)|^2$ for all $A, B \in \mathscr{A}$.

Definition 5.1.1 Consider two C^*-algebras $\mathscr{A}$ and $\mathscr{B}$. A map $\Phi : \mathscr{A} \to \mathscr{B}$, which may not necessarily be continuous, is termed a *Lieb map* when it satisfies the following criteria:

(i) (Monotonicity) $\Phi(A) \geq \Phi(B)$ if $A \geq B \in \mathscr{A}_+$,
(ii) (Cauchy–Schwarz) $\begin{bmatrix} \Phi(A^*A) & \Phi(A^*B) \\ \Phi(A^*B)^* & \Phi(B^*B) \end{bmatrix} \geq 0$ for every $A, B \in \mathscr{A}$.

In [58, Remark 4.3], the authors argued that by adding a continuity assumption in the definitions of Lieb maps and Lieb functions, the monotonicity condition can be eliminated.

Example 5.1.2 (i) Every 2-positive map $\Phi : \mathscr{A} \to \mathscr{B}$ is Lieb; see Corollary 5.1.4 (1).
(ii) The determinant and the usual trace on $\mathbb{M}_n$ are Lieb since they are completely positive.

The first result of this section reads as follows.

Proposition 5.1.3 *Let $\mathscr{A}$ be a von Neumann algebra and let $\mathscr{B}$ be a unital C^*-algebra. Then a map $\Phi : \mathscr{A} \to \mathscr{B}$ is Lieb if and only if for all $A, B, C \in \mathscr{A}$, the positivity of the matrix*

5.1 Lieb Maps and Their Fundamental Properties

$$\begin{bmatrix} A & C \\ C^* & B \end{bmatrix} \quad (5.1.1)$$

implies the positivity of the matrix $\begin{bmatrix} \Phi(A) & \Phi(C) \\ \Phi(C)^* & \Phi(B) \end{bmatrix}$.

Proof ($\Longrightarrow$) Let $\begin{bmatrix} A & C \\ C^* & B \end{bmatrix} \geq 0$. There exists a contraction $K \in \mathscr{A}$ (see Remark 2.4.13) such that $C = A^{\frac{1}{2}} K B^{\frac{1}{2}}$. Since K is a contraction, we get $K^* K \leq I$, which implies $B^{\frac{1}{2}} K^* K B^{\frac{1}{2}} \leq B$. The monotonicity of Φ ensures $\Phi(B^{\frac{1}{2}} K^* K B^{\frac{1}{2}}) \leq \Phi(B)$. Hence,

$$\begin{bmatrix} \Phi(A) & \Phi(C) \\ \Phi(C)^* & \Phi(B) \end{bmatrix} \geq \begin{bmatrix} \Phi(A) & \Phi(A^{\frac{1}{2}} K B^{\frac{1}{2}}) \\ \Phi(A^{\frac{1}{2}} K B^{\frac{1}{2}})^* & \Phi(B^{\frac{1}{2}} K^* K B^{\frac{1}{2}}). \end{bmatrix}$$

The last matrix is positive by the Cauchy–Schwarz part of the definition of a Lieb map.

($\Longleftarrow$) First, let us note that Φ is a positive map. To demonstrate this, take $B = C = 0$ in (5.1.1). Let $A \geq B \geq 0$. Utilizing (2.4.4), we get $\begin{bmatrix} A & B \\ B & B \end{bmatrix} \geq 0$. Applying the assumption, we get $\begin{bmatrix} \Phi(A) & \Phi(B) \\ \Phi(B) & \Phi(B) \end{bmatrix} \geq 0$. Using (2.4.4) again, we conclude that $\Phi(B) \leq \Phi(A)$, which proves the monotonicity condition for Φ. The Cauchy–Schwarz inequality immediately follows from the positivity of the matrix $\begin{bmatrix} A^*A & A^*B \\ B^*A & B^*B \end{bmatrix}$. $\square$

As a consequence, we show that a 2-positive map is Lieb.

Corollary 5.1.4 *If Φ is a 2-positive map between unital C^*-algebras, then Φ is a Lieb map.*

Proof According to the proof of direction ($\Longleftarrow$) in Proposition 5.1.3, we only need to prove that every 2-positive map is self-adjoint. Let $A \in \mathscr{A}$. By Proposition 2.4.21, the matrix $\begin{bmatrix} |A| & A^* \\ A & |A^*| \end{bmatrix}$ is positive. Hence, the 2-positivity of Φ ensures that the matrix

$$\begin{bmatrix} \Phi(|A|) & \Phi(A^*) \\ \Phi(A) & \Phi(|A^*|) \end{bmatrix}$$

is positive, and hence self-adjoint. This implies that $\Phi(A^*) = \Phi(A)^*$. $\square$

There is a significant connection between Lieb maps and 2-positive maps. However, it is important to note that Lieb maps can sometimes be nonself-adjoint. As an illustration, consider a 2-positive map Φ in which there is a nonself-adjoint element A such that $\Phi(A) \neq 0$. In this case, we can define the map Ψ as follows:

$$\Psi(X) = \begin{cases} \Phi(X) & X \neq A \\ 0 & X = A. \end{cases}$$

We create a Lieb map (as demonstrated through a straightforward application of Proposition 5.1.3). However, it is important to note that Ψ is not a $*$-map, as deduced from the proof of Corollary 5.1.4, which establishes that every 2-positive map is indeed a $*$-map.

Now, we are ready to prove some Choi inequalities for Lieb maps.

Corollary 5.1.5 (Extended Choi inequalities) *Let $\mathscr{A}$ and $\mathscr{B}$ be unital C^*-algebras. If $\Phi : \mathscr{A} \to \mathscr{B}$ is a Lieb map such that $\Phi(I) > 0$, then the following statements hold:*

(i) $\Phi(|A|^2) \geq \Phi(A)^* \Phi(I)^{-1} \Phi(A)$ *for every $A \in \mathscr{A}$. In particular, if $0 < \Phi(I) \leq I$, then $\Phi(|A|^2) \geq |\Phi(A)|^2$.*
(ii) *If Φ is unital, then $\Phi(A^{-1}) \geq \Phi(A)^{-1}$ for every $A \in \mathscr{A}_{++}$.*

Furthermore, if $\Phi : \mathscr{A} \to \mathscr{B}$ is a 2-positive map between unital C^*-algebras, then (i) *and* (ii) *hold.*

Proof (i) The positivity and the Cauchy–Schwarz property of Φ implies that

$$\begin{bmatrix} \Phi(I) & \Phi(A) \\ \Phi(A)^* & \Phi(A^*A) \end{bmatrix} = \begin{bmatrix} \Phi(I^*I) & \Phi(I^*A) \\ \Phi(I^*A)^* & \Phi(A^*A) \end{bmatrix} \geq 0,$$

which implies that

$$\Phi(A^*A) \geq \Phi(A)^* \Phi(I)^{-1} \Phi(A).$$

The second inequality follows from the fact that $\Phi(I)^{-1} \geq I$ whenever $0 < \Phi(I) \leq I$.
(ii) According to the positivity and the Cauchy–Schwarz property of Φ, we get

$$\begin{bmatrix} \Phi(A) & \Phi(I) \\ \Phi(I)^* & \Phi(A^{-1}) \end{bmatrix} = \begin{bmatrix} \Phi(|A^{\frac{1}{2}}|^2) & \Phi(A^{\frac{1}{2}} A^{-\frac{1}{2}}) \\ \Phi(A^{\frac{1}{2}} A^{-\frac{1}{2}})^* & \Phi(|A^{-\frac{1}{2}}|^2) \end{bmatrix} \geq 0$$

for every $A > 0$. Hence,

$$\Phi(A^{-1}) \geq \Phi(I) \Phi(A)^{-1} \Phi(I) = \Phi(A)^{-1}.$$

The statement about 2-positive maps is derived directly from the statement about Lieb maps and Corollary 5.1.4. □

The following corollary is another consequence of Proposition 5.1.3.

Corollary 5.1.6 *Let $\mathscr{A}$ be a von Neumann algebra. Then an algebraic norm $\|\cdot\|$ on $\mathscr{A}$ is Lieb if and only if it is the C^*-norm on $\mathscr{A}$.*

Proof Suppose that $\|\cdot\|$ is a Lieb map. According to the positivity of the matrix $\begin{bmatrix} I & A \\ A^* & A^*A \end{bmatrix}$, we have

$$\begin{bmatrix} \|I\| & \|A\| \\ \|A\|^* & \|A^*A\| \end{bmatrix} \geq 0.$$

We therefore get

$$\|A^*A\| \geq \|A\|^2,$$

which implies that $\|\cdot\|$ is a C^*-norm.

The converse is deduced from Corollary 2.4.4. □

5.2 3-Positivity of Nonlinear Maps

In this section, we study various properties of 3-positive maps. We start by establishing a property concerning 2-positive linear maps in the context of nonlinear 3-positive maps.

Theorem 3.3.14 shows that if $\Phi : \mathscr{A} \to \mathscr{B}$ is a unital 2-positive linear map between unital C^*-algebras, and if there exists an $A \in \mathscr{A}$ such that $\Phi(A^*A) = \Phi(A^*)\Phi(A)$, then it follows that $\Phi(XA) = \Phi(X)\Phi(A)$ for all $X \in \mathscr{A}$.

It is notable that this result cannot be applied to arbitrary unital 2-positive maps. As an example, consider the C^*-norm of a C^*-algebra, which serves as a unital 2-positive map (as shown in Corollary 5.1.6). For any $A \in \mathscr{A}$, we have $\|A^*A\| = \|A^*\|\|A\|$. Nevertheless, the equality $\|AX\| = \|A\|\|X\|$ does not hold in general.

The following result gives a broader perspective by extending the previously mentioned fact to include nonlinear 3-positive maps.

Theorem 5.2.1 *Suppose that $\Phi : \mathscr{A} \to \mathscr{B}$ is a unital 3-positive map. If for some $A \in \mathscr{A}$ equality $\Phi(A^*A) = \Phi(A)^*\Phi(A)$ holds, then*

$$\Phi(A^*X) = \Phi(A^*)\Phi(X) \text{ and } \Phi(XA) = \Phi(X)\Phi(A)$$

for every $X \in \mathscr{A}$.

Proof Suppose that $A \in \mathscr{A}$ satisfies $\Phi(A^*A) = \Phi(A)^*\Phi(A)$. Let $X \in \mathscr{A}$. We have

$$\begin{bmatrix} A^*A & A^*X & A^* \\ X^*A & X^*X & X^* \\ A & X & I \end{bmatrix} = \begin{bmatrix} A^* & 0 & 0 \\ X^* & 0 & 0 \\ I & 0 & 0 \end{bmatrix} \begin{bmatrix} A & X & I \\ 0 & 0 & 0 \\ 0 & 0 & 0 \end{bmatrix} \geq 0.$$

Since Φ is 3-positive, we have

$$\begin{bmatrix} \Phi(A^*A) & \Phi(A^*X) & \Phi(A^*) & 0 \\ \Phi(X^*A) & \Phi(X^*X) & \Phi(X^*) & 0 \\ \Phi(A) & \Phi(X) & \Phi(I) & 0 \\ 0 & 0 & 0 & \Phi(I) \end{bmatrix} \geq 0. \tag{5.2.1}$$

Using the positivity of matrix (5.2.1), we deduce that

$$\begin{bmatrix} \Phi(A^*A) & \Phi(A^*X) \\ \Phi(X^*A) & \Phi(X^*X) \end{bmatrix} \geq \begin{bmatrix} \Phi(A^*) & 0 \\ \Phi(X^*) & 0 \end{bmatrix} \begin{bmatrix} \Phi(I) & 0 \\ 0 & \Phi(I) \end{bmatrix}^{-1} \begin{bmatrix} \Phi(A) & \Phi(X) \\ 0 & 0 \end{bmatrix}$$

$$= \begin{bmatrix} \Phi(A^*)\Phi(A) & \Phi(A^*)\Phi(X) \\ \Phi(X^*)\Phi(A) & \Phi(X^*)\Phi(X) \end{bmatrix}.$$

Since $\Phi(A^*A) = \Phi(A)^*\Phi(A)$ and Φ is self-adjoint, we conclude that

$$0 \leq \begin{bmatrix} \Phi(A^*A) & \Phi(A^*X) \\ \Phi(X^*A) & \Phi(X^*X) \end{bmatrix} - \begin{bmatrix} \Phi(A^*)\Phi(A) & \Phi(A^*)\Phi(X) \\ \Phi(X^*)\Phi(A) & \Phi(X^*)\Phi(X) \end{bmatrix}$$

$$= \begin{bmatrix} 0 & \Phi(A^*X) - \Phi(A^*)\Phi(X) \\ \Phi(X^*A) - \Phi(X^*)\Phi(A) & \Phi(X^*X) - \Phi(X^*)\Phi(X) \end{bmatrix}.$$

Since $\Phi(X^*X) - \Phi(X^*)\Phi(X) \geq 0$, by using Exercise 2.9.17, we get $\Phi(A^*X) = \Phi(A^*)\Phi(X)$. Due to Φ is a $*$-map, $\Phi(XA) = \Phi(X)\Phi(A)$. □

Corollary 5.2.2 *Let $\|\cdot\|$ be an algebraic norm on a unital C^*-algebra $\mathscr{A}$. If $\|\cdot\|$ is 3-positive, then*

$$\|XY\| = \|X\|\|Y\|$$

for every $X, Y \in \mathscr{A}$.

The following theorem, inspired by [8, Theorem 3], addresses the superadditivity of 3-positive maps. A map $\Phi : \mathscr{X} \subseteq \mathscr{A} \to \mathscr{B}$ is called *superadditive* on a subset $\mathscr{X}$ of $\mathscr{A}$ that is closed under addition if it satisfies the inequality:

$$\Phi(A+B) \geq \Phi(A) + \Phi(B)$$

for every $A, B \in \mathscr{X}$.

5.2 3-Positivity of Nonlinear Maps

Theorem 5.2.3 *Let $\Phi : \mathscr{A} \to \mathscr{B}$ be a 3-positive map between unital C^*-algebras. Then*

$$2\Phi(0) + \Phi(A+B) \geq \Phi(A) + \Phi(B) \tag{5.2.2}$$

for every positive elements $A, B \in \mathscr{A}$. In particular,

$$\Phi(0) = 0 \iff \Phi(A+B) \geq \Phi(A) + \Phi(B).$$

Proof Let A and B be positive. It follows from $\begin{bmatrix} A & A & 0 \\ A & A & 0 \\ 0 & 0 & 0 \end{bmatrix} \geq 0$ and $\begin{bmatrix} 0 & 0 & 0 \\ 0 & B & B \\ 0 & B & B \end{bmatrix} \geq 0$ that

$$\begin{bmatrix} A & A & 0 \\ A & A+B & B \\ 0 & B & B \end{bmatrix} \geq 0.$$

Since Φ is 3-positive, we have

$$\begin{bmatrix} \Phi(A) & \Phi(A) & \Phi(0) \\ \Phi(A) & \Phi(A+B) & \Phi(B) \\ \Phi(0) & \Phi(B) & \Phi(B) \end{bmatrix} \geq 0.$$

Therefore,

$$0 \leq \begin{bmatrix} -1 & 1 & -1 \\ 0 & 0 & 0 \\ 0 & 0 & 0 \end{bmatrix} \begin{bmatrix} \Phi(A) & \Phi(A) & \Phi(0) \\ \Phi(A) & \Phi(A+B) & \Phi(B) \\ \Phi(0) & \Phi(B) & \Phi(B) \end{bmatrix} \begin{bmatrix} -1 & 0 & 0 \\ 1 & 0 & 0 \\ -1 & 0 & 0 \end{bmatrix}$$
$$= 2\Phi(0) + \Phi(A+B) - \Phi(A) - \Phi(B),$$

which ensures that $2\Phi(0) + \Phi(A+B) \geq \Phi(A) + \Phi(B)$.
The rest is evident. $\square$

Theorems 5.2.1 and 5.2.3 underscore the profound implications of the assumption of 3-positivity on the norm of a C^*-algebra. It is worth recalling that a norm $\|\cdot\|$ on a normed space $\mathcal{X}$ is called *strictly convex* if for all nonzero elements $X, Y \in \mathcal{X}$, the equality $\|X+Y\| = \|X\| + \|Y\|$ implies $X = \lambda Y$ for some $\lambda \in \mathbb{C}$.

In the following corollary, we present certain conditions under which a C^*-algebra is $*$-isomorphic to the complex plane $\mathbb{C}$.

Corollary 5.2.4 *If $(\mathscr{A}, \|\cdot\|)$ is a unital C^*-algebra and $\|\cdot\|$ is a strictly convex 3-positive map, then $\mathscr{A} \cong \mathbb{C}$.*

Proof Theorem 5.2.3 implies that for every positive element A and B, it holds that $\|A + B\| \geq \|A\| + \|B\|$. Hence $\|A + B\| = \|A\| + \|B\|$. Since $\|\cdot\|$ is strictly convex, for any positive element A, we have $A = \lambda I$ for some $\lambda \in \mathbb{C}$. Since for any $A \in \mathscr{A}$, there exist four positive elements A_1, A_2, A_3, and A_4 such that $A = A_1 - A_2 + i(A_3 - A_4) \in \mathbb{C}I$, we infer that $\mathscr{A} \cong \mathbb{C}$. □

We end this section with the following interesting result.

Proposition 5.2.5 *Let* $\Phi : \mathscr{A} \to \mathscr{B}$ *be a 3-positive map between unital C^*-algebras. If* $P \geq Q \geq 0$ *such that* $\Phi(P) = \Phi(Q)$, *then* $\Phi(P - Q) = \Phi(0)$.

Proof If $P \geq Q \geq 0$, then we can write

$$\begin{bmatrix} P-Q & P-Q & 0 \\ P-Q & P-Q & 0 \\ 0 & 0 & 0 \end{bmatrix} \geq 0, \quad \begin{bmatrix} 0 & 0 & 0 \\ 0 & Q & Q \\ 0 & Q & Q \end{bmatrix} \geq 0.$$

Summing the above matrices and then utilizing 3-positivity of Φ give

$$\begin{bmatrix} \Phi(P-Q) & \Phi(P-Q) & \Phi(0) \\ \Phi(P-Q) & \Phi(P) & \Phi(Q) \\ \Phi(0) & \Phi(Q) & \Phi(Q) \end{bmatrix} \geq 0.$$

Given $\varepsilon > 0$, it follows from the positivity of above matrix that

$$\begin{bmatrix} \Phi(P-Q)+\varepsilon I & \Phi(P-Q)+\varepsilon I & \Phi(0)+\varepsilon I & 0 \\ \Phi(P-Q)+\varepsilon I & \Phi(P)+\varepsilon I & \Phi(Q)+\varepsilon I & 0 \\ \Phi(0)+\varepsilon I & \Phi(Q)+\varepsilon I & \Phi(Q)+\varepsilon I & 0 \\ 0 & 0 & 0 & I \end{bmatrix} \geq 0.$$

Hence, by using Remark 2.4.2, we deduce

$$\begin{bmatrix} \Phi(P-Q)+\varepsilon I & \Phi(P-Q)+\varepsilon I \\ \Phi(P-Q)+\varepsilon I & \Phi(P)+\varepsilon I \end{bmatrix}$$
$$\geq \begin{bmatrix} \Phi(0)+\varepsilon I & 0 \\ \Phi(Q)+\varepsilon I & 0 \end{bmatrix} \begin{bmatrix} (\Phi(Q)+\varepsilon I)^{-1} & 0 \\ 0 & I^{-1} \end{bmatrix} \begin{bmatrix} \Phi(0)+\varepsilon I & \Phi(Q)+\varepsilon I \\ 0 & 0 \end{bmatrix}.$$

Therefore,

$$\begin{bmatrix} \Phi(P-Q)+\varepsilon I & \Phi(P-Q)+\varepsilon I \\ \Phi(P-Q)+\varepsilon I & \Phi(P)+\varepsilon I \end{bmatrix} \geq \begin{bmatrix} (\Phi(0)+\varepsilon I)(\Phi(Q)+\varepsilon I)^{-1}(\Phi(0)+\varepsilon I) & \Phi(0)+\varepsilon I \\ \Phi(0)+\varepsilon I & \Phi(Q)+\varepsilon I \end{bmatrix}.$$

Consequently, the equality $\Phi(Q) = \Phi(P)$ implies that

$$\begin{bmatrix} * & \Phi(P-Q) - \Phi(0) \\ \Phi(P-Q) - \Phi(0) & 0 \end{bmatrix} \geq 0,$$

in which $*$ is a positive operator. Now, the positivity of the above matrix implies that $\Phi(P-Q) = \Phi(0)$ by using Exercise 2.9.17. $\square$

5.3 Continuity of 3-Positive Maps

In this section, we explore the continuity of 3-positive maps between C^*-algebras, which do not necessarily need to be linear. Furthermore, we establish that 3-positive maps are norm-SOT-continuous. In addition, we show that a 3-positive map is norm-continuous if it is norm-continuous at some positive invertible operator.

Proposition 5.3.1 *Let $\Phi : \mathscr{A} \to \mathscr{B}$ be a map between unital C^*-algebras. Then the following properties hold:*

(1) *If Φ is 2-positive and $\Phi(Z) = 0$ for some $Z > 0$, then $\Phi(X) = 0$ for all $X \in \mathscr{A}$.*
(2) *If Φ is 3-positive and $\Phi(Z) = \Phi(0)$ for some $Z > 0$, then $\Phi(X) = \Phi(0)$ for all $X \in \mathscr{A}$.*
(3) *Let Φ be 3-positive, and let $W, Z \in \mathscr{A}_{++}$ such that $W > Z$. If $\Phi(Z) = \Phi(W)$, then $\Phi(X) = \Phi(0)$ for all $X \in \mathscr{A}$.*

Proof (1) Suppose that $Z > 0$ and $\Phi(Z) = 0$. From the strict positivity of Z, we can find some $\alpha > 0$ such that $Z \geq \alpha I$. If $X \in \mathscr{A}$, then for a sufficiently large value of $n \in \mathbb{N}$, we have

$$Z \geq \alpha I \geq \frac{\|X^*X\|}{n} I \geq \frac{1}{n} X^*X.$$

Accordingly,

$$\begin{bmatrix} Z & X^* \\ X & nI \end{bmatrix} \geq 0.$$

It follows from the 2-positivity of Φ that

$$\begin{bmatrix} 0 & \Phi(X^*) \\ \Phi(X) & \Phi(nI) \end{bmatrix} = \begin{bmatrix} \Phi(Z) & \Phi(X^*) \\ \Phi(X) & \Phi(nI) \end{bmatrix} \geq 0,$$

whence $\Phi(X) = 0$.

(2) Suppose that $Z > 0$ such that $\Phi(Z) = \Phi(0)$. Let $X \in \mathscr{A}$. Applying a similar argument as in the proof of (1), there is some $n \in \mathbb{N}$ such that $\begin{bmatrix} Z & X^* \\ X & nI \end{bmatrix} \geq 0$. Hence,

$$\begin{bmatrix} Z & X^* & 0 \\ X & Z+nI & Z \\ 0 & Z & Z \end{bmatrix} \geq 0.$$

Since Φ is 3-positive, we have

$$\begin{bmatrix} \Phi(Z) & \Phi(X^*) & \Phi(0) \\ \Phi(X) & \Phi(Z+nI) & \Phi(Z) \\ \Phi(0) & \Phi(Z) & \Phi(Z) \end{bmatrix} \geq 0.$$

For a strictly positive operator $T \in \mathscr{B}_{++}$, the positivity of the above matrix gives

$$\begin{bmatrix} \Phi(Z)+T & \Phi(X^*)+T & \Phi(0)+T & 0 \\ \Phi(X)+T & \Phi(Z+nI)+T & \Phi(Z)+T & 0 \\ \Phi(0)+T & \Phi(Z)+T & \Phi(Z)+T & 0 \\ 0 & 0 & 0 & I \end{bmatrix} \geq 0.$$

This inequality, together with Theorem 2.4.1, entails that

$$\begin{bmatrix} \Phi(Z)+T & \Phi(X^*)+T \\ \Phi(X)+T & \Phi(Z+nI)+T \end{bmatrix} \geq \begin{bmatrix} (\Phi(0)+T)(\Phi(Z)+T)^{-1}(\Phi(0)+T) & \Phi(0)+T \\ \Phi(0)+T & \Phi(Z)+T \end{bmatrix},$$

whence, by the assumption $\Phi(Z) = \Phi(0) \geq 0$, we get

$$\begin{bmatrix} 0 & \Phi(X^*) - \Phi(0) \\ \Phi(X) - \Phi(0) & \Phi(Z+nI) - \Phi(Z) \end{bmatrix} \geq 0.$$

Thus, $\Phi(X) = \Phi(0)$.

(3) First, note that the matrix

$$\begin{bmatrix} W-Z & W-Z & 0 \\ W-Z & W & Z \\ 0 & Z & Z \end{bmatrix}$$

as the sum of two matrices

$$\begin{bmatrix} W-Z & W-Z \\ W-Z & W-Z \end{bmatrix} \oplus 0 \text{ and } 0 \oplus \begin{bmatrix} Z & Z \\ Z & Z \end{bmatrix}$$

is positive. Throughout the book, 0 denotes a block matrix of appropriate size with zero operators as its entries.

Suppose that $T \in \mathscr{B}_{++}$. From the 3-positivity of Φ, we can deduce that

$$\begin{bmatrix} \Phi(W-Z)+T & \Phi(W-Z)+T & \Phi(0)+T \\ \Phi(W-Z)+T & \Phi(W)+T & \Phi(Z)+T \\ \Phi(0)+T & \Phi(Z)+T & \Phi(Z)+T \end{bmatrix} \geq 0.$$

5.3 Continuity of 3-Positive Maps

Therefore,

$$\begin{bmatrix} \Phi(W-Z)+T & \Phi(W-Z)+T \\ \Phi(W-Z)+T & \Phi(W)+T \end{bmatrix} - \begin{bmatrix} (\Phi(0)+T)(\Phi(Z)+T)^{-1}(\Phi(0)+T) & \Phi(0)+T \\ \Phi(0)+T & \Phi(Z)+T \end{bmatrix} \geq 0.$$

Hence, the assumption $\Phi(Z) = \Phi(W)$ implies that $\Phi(W-Z) = \Phi(0)$. Now, the proof is complete by utilizing part (2). □

The next result reads as follows.

Theorem 5.3.2 *Let $\mathscr{A}$ and $\mathscr{B}$ be unital C^*-algebras. If $\Phi : \mathscr{A} \to \mathscr{B}$ is a 3-positive map, then Φ is norm-SOT-continuous.*

Proof We can assume that $\Phi(I) \neq \Phi(0)$. Indeed, if $\Phi(I) = \Phi(0)$, then Proposition 5.3.1 (2) ensures that Φ is a constant map and thus continuous. Furthermore, we may suppose that Φ is strictly positive (by considering the map $\Psi : \mathscr{A} \to \mathscr{B}$ defined as $\Psi(X) = \Phi(X) + T$ for some $T \in \mathscr{B}_{++}$).

Suppose that $\mathscr{B}$ acts on a Hilbert space $\mathscr{H}$. The proof is divided into three steps.

Step 1. We show that

$$\operatorname*{s-lim}_{n} \Phi\left(\frac{1}{n}I\right) = \Phi(0).$$

Let $x \in \mathscr{H}$ and $\varepsilon > 0$ be given. Since Φ is monotone on $\mathscr{A}_+$, the sequence $\left(\Phi(\frac{1}{n}I)\right)_{n=1}^{\infty} \subseteq \mathscr{B}_+$ is decreasing and bounded below. Therefore, it converges in the strong operator topology to an operator $L \in \mathbb{B}(\mathscr{H})$.

For every $n \in \mathbb{N}$, we can write

$$X_n = \begin{bmatrix} \frac{1}{n}I & \frac{1}{2n}I & 0 \\ \frac{1}{2n}I & \frac{1}{n}I & \frac{1}{2n}I \\ 0 & \frac{1}{2n}I & \frac{1}{n}I \end{bmatrix} = \left(\begin{bmatrix} \frac{1}{n}I & \frac{1}{2n}I \\ \frac{1}{2n}I & \frac{1}{2n}I \end{bmatrix} \oplus 0\right) + \left(0 \oplus \begin{bmatrix} \frac{1}{2n}I & \frac{1}{2n}I \\ \frac{1}{2n}I & \frac{1}{n}I \end{bmatrix}\right) \geq 0.$$

Accordingly, by the 3-positivity of Φ, we reach

$$\Phi_3(X_n) = \begin{bmatrix} \Phi(\frac{1}{n}I) & \Phi(\frac{1}{2n}I) & \Phi(0) \\ \Phi(\frac{1}{2n}I) & \Phi(\frac{1}{n}I) & \Phi(\frac{1}{2n}I) \\ \Phi(0) & \Phi(\frac{1}{2n}I) & \Phi(\frac{1}{n}I) \end{bmatrix} \geq 0.$$

Therefore,

$$\operatorname*{w-lim}_{n} \Phi_3(X_n) = \begin{bmatrix} L & L & \Phi(0) \\ L & L & L \\ \Phi(0) & L & L \end{bmatrix} \geq 0.$$

Assume that $Z \in \mathscr{B}_{++}$. Then, by applying the positivity of the above matrix, we can write

$$\begin{bmatrix} L & L & \Phi(0) \\ L & L & L \\ \Phi(0) & L & L \end{bmatrix} + \begin{bmatrix} Z & Z & Z \\ Z & Z & Z \\ Z & Z & Z \end{bmatrix} = \begin{bmatrix} L+Z & L+Z & \Phi(0)+Z \\ L+Z & L+Z & L+Z \\ \Phi(0)+Z & L+Z & L+Z \end{bmatrix} \geq 0.$$

Using the fact that $L + Z > 0$ and applying the positivity of the last matrix, we get

$$\begin{bmatrix} L+Z & L+Z \\ L+Z & L+Z \end{bmatrix} \geq \begin{bmatrix} (\Phi(0)+Z)(L+Z)^{-1}(\Phi(0)+Z) & \Phi(0)+Z \\ \Phi(0)+Z & L+Z \end{bmatrix}.$$

Consequently, $L + Z = \Phi(0) + Z$, which entails that $L = \Phi(0)$.

Step 2. We show that $\Phi\left(T + \frac{1}{n}I\right) \xrightarrow{\text{SOT}} \Phi(T)$ for every $T \in \mathscr{A}_+$. To this end, it is enough to define the map $\Psi : \mathscr{A} \to \mathscr{B}$ by $\Psi(X) = \Phi(T + X)$ and employ Step 1.

Step 3. We are in a position to show that if (A_m) is a sequence in $\mathscr{A}$ such that $A_m \xrightarrow{\|\cdot\|} A$, then $\Phi(A_m) \xrightarrow{\text{SOT}} \Phi(A)$. Let $x \in \mathscr{H}$ and $\varepsilon > 0$ be arbitrary. Since (A_m) is bounded, there exists a positive real number t_0 such that $A_m^* A_m \leq \|A_m\|^2 I \leq t_0 I$ for all m. Take $t := \max\{t_0, \|A\|^2, 1\}$. It follows from Step 2 that the sequence $\Phi\left(tI + \frac{1}{n}I\right)$ strongly converges to $\Phi(tI)$. Therefore, there exists some $N > 0$ with the property that

$$\left\langle \Phi\left(tI + \frac{1}{n}I\right)x, x\right\rangle - \langle\Phi(tI)x, x\rangle < \frac{\varepsilon^2}{\|\Phi(2I) - \Phi(A)\Phi(tI)^{-1}\Phi(A)^*\|} \tag{5.3.1}$$

for every $n > N$. Let us show that the denominator at (5.3.1) is nonzero.

Note that the assumption $\Phi(I) \neq \Phi(0)$ and Proposition 5.3.1 (3) ensure $\|\Phi(2I) - \Phi(I)\| > 0$. Moreover, since $t \geq \|A\|^2$, we have

$$A(tI)^{-1}A^* \leq A\|A\|^{-2}A^* \leq I. \tag{5.3.2}$$

Therefore, $\Phi(A(tI)^{-1}A^*) \leq \Phi(I)$. Furthermore, in virtue of the 2-positivity of Φ and $\begin{bmatrix} A(tI)^{-1}A^* & A \\ A^* & tI \end{bmatrix} \geq 0$, we arrive at $\Phi(A)\Phi(tI)^{-1}\Phi(A)^* \leq \Phi(A(tI)^{-1}A^*)$. Hence,

$$\Phi(2I) - \Phi(A)\Phi(tI)^{-1}\Phi(A)^* \geq \Phi(2I) - \Phi(I) \geq 0, \tag{5.3.3}$$

whence

$$\|\Phi(2I) - \Phi(A)\Phi(tI)^{-1}\Phi(A)^*\| \geq \|\Phi(2I) - \Phi(I)\| > 0,$$

as required.

Next, fix $n_0 > N$. Since $A_m \xrightarrow{\|\cdot\|} A$, there is some $M > 0$ such that $\|A_m - A\|^2 \leq \frac{1}{n_0}$ for every $m > M$. This infers that

$$(A_m - A)^*(A_m - A) \leq \frac{1}{n_0} \quad (m > M). \tag{5.3.4}$$

5.3 Continuity of 3-Positive Maps

We claim that for every $m > M$, the matrix

$$X_m = \begin{bmatrix} tI + \dfrac{1}{n_0}I & A_m^* & tI \\ A_m & 2I & A \\ tI & A^* & tI \end{bmatrix}$$

is positive. Indeed,

$$\begin{bmatrix} tI \\ A \end{bmatrix} t^{-1}I \begin{bmatrix} tI & A^* \end{bmatrix} = \begin{bmatrix} tI & A^* \\ A & A(tI)^{-1}A^* \end{bmatrix} \geq 0. \quad (5.3.5)$$

Moreover, we have from (5.3.2) that

$$2I - A(tI)^{-1}A^* \geq I, \quad (5.3.6)$$

and so

$$\begin{bmatrix} tI + \dfrac{1}{n_0}I & A_m^* \\ A_m & 2I \end{bmatrix} - \begin{bmatrix} tI \\ A \end{bmatrix} t^{-1}I \begin{bmatrix} tI & A^* \end{bmatrix} = \begin{bmatrix} tI + \dfrac{1}{n_0}I & A_m^* \\ A_m & 2I \end{bmatrix} - \begin{bmatrix} tI & A^* \\ A & A(tI)^{-1}A^* \end{bmatrix}$$

$$\geq \begin{bmatrix} \dfrac{1}{n_0}I & A_m^* - A^* \\ A_m - A & I \end{bmatrix} \quad \text{by (5.3.6)}.$$

By noting (5.3.4), the last matrix is positive for every $m > M$. Therefore, $X_m \geq 0$ for each $m > M$.

Since Φ is 3-positive, we get

$$\begin{bmatrix} \Phi\left(tI + \dfrac{1}{n_0}\right) & \Phi(A_m)^* & \Phi(tI) \\ \Phi(A_m) & \Phi(2I) & \Phi(A) \\ \Phi(tI) & \Phi(A)^* & \Phi(tI) \end{bmatrix} \geq 0 \quad (m > M).$$

Hence,

$$\begin{bmatrix} \Phi\left(tI + \dfrac{1}{n_0}I\right) & \Phi(A_m^*) \\ \Phi(A_m) & \Phi(2I) \end{bmatrix} \geq \begin{bmatrix} \Phi(tI) & \Phi(A^*) \\ \Phi(A) & \Phi(A)\Phi(tI)^{-1}\Phi(A)^* \end{bmatrix} \quad (m > M),$$

which entails that

$$\begin{bmatrix} \Phi\left(tI + \dfrac{1}{n_0}I\right) - \Phi(tI) & \Phi\left(A_m^*\right) - \Phi(A^*) \\ \Phi(A_m) - \Phi(A) & \Phi(2I) - \Phi(A)\Phi(tI)^{-1}\Phi(A)^* \end{bmatrix} \geq 0 \quad (m > M).$$

In particular,

$$\begin{bmatrix} \Phi\left(tI + \frac{1}{n_0}I\right) - \Phi(tI) & \Phi(A_m^*) - \Phi(A^*) \\ \Phi(A_m) - \Phi(A) & \|\Phi(2I) - \Phi(A)\Phi(tI)^{-1}\Phi(A)^*\| \end{bmatrix} \geq 0 \quad (m > M),$$

which deduces that

$$|\Phi(A_m) - \Phi(A)|^2 \leq \|\Phi(2I) - \Phi(A)\Phi(tI)^{-1}\Phi(A)^*\| \left(\Phi\left(t + \frac{1}{n_0}I\right) - \Phi(t)\right)$$

for every $m > M$. Thus,

$$\langle |\Phi(A_m) - \Phi(A)|^2 x, x \rangle$$
$$\leq \|\Phi(2I) - \Phi(A)\Phi(tI)^{-1}\Phi(A)^*\| \left\langle \left(\Phi\left(tI + \frac{1}{n_0}I\right) - \Phi(tI)\right) x, x \right\rangle$$
$$< \varepsilon^2$$

for every $m > M$, where the last inequality follows from (5.3.1). Hence,

$$\|\Phi(A_m)x - \Phi(A)x\|^2 < \varepsilon^2,$$

which is the desired conclusion. □

In the case where $\mathscr{B}$ is finite dimensional, the theorem stated above shows the norm-continuity of Φ. This implies that every 3-positive map $\Phi : \mathbb{M}_n \to \mathbb{M}_k$ is indeed norm-continuous.

Subsequently, we will demonstrate that a 3-positive map remains norm-continuous as long as it is norm-continuous at a strictly positive operator.

Proposition 5.3.3 *If $\Phi : \mathscr{A} \to \mathscr{B}$ is a 3-positive map between C^*-algebras, then Φ is norm-continuous if and only if there exists $Z \in \mathscr{A}_{++}$ such that $\Phi\left(Z + \frac{1}{n}I\right) \xrightarrow{\|\cdot\|} \Phi(Z)$.*

Proof We only need to prove the direction ($\Longleftarrow$). Let $Z > 0$ satisfy the hypothesis of the proposition. Without loss of generality, we can assume that $\Phi(Z) \neq \Phi(0)$; since in the case when $\Phi(Z) = \Phi(0)$, Proposition 5.3.1 entails Φ to be a constant map, and therefore, it is norm-continuous.

Let $t > 0$. Since $Z^{-1} > 0$, there exists some $\alpha > 0$ such that $Z^{-1} \geq \alpha I$. We claim that if $n \geq \frac{1}{\alpha}$, then the matrix

$$\mathfrak{X}_n = \begin{bmatrix} Z + \frac{1}{n}I & tI + \frac{1}{n}I & Z \\ tI + \frac{1}{n}I & (t+1)^2 Z^{-1} & tI \\ Z & tI & Z \end{bmatrix}$$

5.3 Continuity of 3-Positive Maps

is positive. To this end, firstly, it is easy to observe that $\begin{bmatrix} Z+\frac{1}{n}I & tI+\frac{1}{n}I \\ tI+\frac{1}{n}I & (t+1)^2Z^{-1} \end{bmatrix} \geq 0$. Second,

$$\begin{bmatrix} Z+\frac{1}{n}I & tI+\frac{1}{n}I \\ tI+\frac{1}{n}I & (t+1)^2Z^{-1} \end{bmatrix} - \begin{bmatrix} Z \\ tI \end{bmatrix} Z^{-1} \begin{bmatrix} Z & tI \end{bmatrix} = \begin{bmatrix} \frac{1}{n}I & \frac{1}{n}I \\ \frac{1}{n}I & (2t+1)Z^{-1} \end{bmatrix}. \quad (5.3.7)$$

If $n \geq \frac{1}{\alpha}$, then $Z^{-1} \geq \frac{1}{n}I$, which implies that $(2t+1)Z^{-1} \geq \frac{1}{n}I$. Hence, the matrix on the right side of (5.3.7) is positive for every $n \geq \frac{1}{\alpha}$. Therefore, $\mathfrak{X}_n \geq 0$ for all $n \geq \frac{1}{\alpha}$.

Now, we show that $\Phi(tI + \frac{1}{n}I) \xrightarrow{\|\cdot\|} \Phi(tI)$ as $n \to \infty$.

Let $\varepsilon > 0$ be given. Since $\Phi\left(Z + \frac{1}{n}I\right) \xrightarrow{\|\cdot\|} \Phi(Z)$, there is some $N > 0$ such that

$$\left\| \Phi\left(Z + \frac{1}{n}I\right) - \Phi(Z) \right\| < \frac{\varepsilon^2}{\left\| \Phi((t+1)^2 Z^{-1}) - \Phi(tI)\Phi(Z)^{-1}\Phi(tI) \right\|} \quad (5.3.8)$$

for every $n \geq N$.

Note that the assumption $\Phi(Z) \neq \Phi(0)$ and part (3) of Lemma 5.3.1 ensure that $\left\| \Phi((t+1)^2 Z^{-1}) - \Phi(t^2 Z^{-1}) \right\| > 0$. Moreover, since $\begin{bmatrix} t^2 Z^{-1} & tI \\ tI & Z \end{bmatrix} \geq 0$ and Φ is 2-positive, we have $\Phi(t^2 Z^{-1}) \geq \Phi(tI)\Phi(Z)^{-1}\Phi(tI)$, which implies that

$$\left\| \Phi\left((t+1)^2 Z^{-1}\right) - \Phi(tI)\Phi(Z)^{-1}\Phi(tI) \right\| \geq \left\| \Phi\left((t+1)^2 Z^{-1}\right) - \Phi\left(t^2 Z^{-1}\right) \right\| > 0.$$

Set $M = \max\left\{\frac{1}{\alpha}, N\right\}$. Then, the 3-positivity of Φ and the positivity of $\mathfrak{X}_n$ for $n \geq \frac{1}{\alpha}$ ensure that

$$\Phi_3(\mathfrak{X}_n) = \begin{bmatrix} \Phi(Z+\frac{1}{n}I) & \Phi(tI+\frac{1}{n}I) & \Phi(Z) \\ \Phi(tI+\frac{1}{n}I) & \Phi((t+1)^2 Z^{-1}) & \Phi(tI) \\ \Phi(Z) & \Phi(tI) & \Phi(Z) \end{bmatrix} \geq 0 \quad (n > M).$$

Accordingly,

$$\begin{bmatrix} \Phi(Z+\frac{1}{n}I) & \Phi\left(tI+\frac{1}{n}I\right) \\ \Phi(tI+\frac{1}{n}I) & \Phi((t+1)^2 Z^{-1}) \end{bmatrix} \geq \begin{bmatrix} \Phi(Z) & \Phi(tI) \\ \Phi(tI) & \Phi(tI)\Phi(Z)^{-1}\Phi(tI) \end{bmatrix} \quad (n > M).$$

Therefore,

$$\begin{bmatrix} \Phi(Z+\frac{1}{n}I) - \Phi(Z) & \Phi\left(tI+\frac{1}{n}I\right) - \Phi(tI) \\ \Phi(tI+\frac{1}{n}I) - \Phi(tI) & \Phi((t+1)^2 Z^{-1}) - \Phi(tI)\Phi(Z)^{-1}\Phi(tI) \end{bmatrix} \geq 0 \quad (n > M).$$

Thanks to the 2-positivity of the operator norm, we reach

$$\begin{bmatrix} \left\| \Phi(Z+\frac{1}{n}I) - \Phi(Z) \right\| & \left\| \Phi\left(tI+\frac{1}{n}I\right) - \Phi(tI) \right\| \\ \left\| \Phi(tI+\frac{1}{n}I) - \Phi(tI) \right\| & \left\| \Phi((t+1)^2 Z^{-1}) - \Phi(tI)\Phi(Z)^{-1}\Phi(tI) \right\| \end{bmatrix} \geq 0 \quad (n > M).$$

Thus, if $n > M$, then

$$\left\|\Phi\left(tI + \frac{1}{n}I\right) - \Phi(tI)\right\|^2 \le \left\|\Phi(Z + \frac{1}{n}I) - \Phi(Z)\right\|$$
$$\times \left\|\Phi((t+1)^2 Z^{-1}) - \Phi(tI)\Phi(Z)^{-1}\Phi(tI)\right\|$$
$$< \varepsilon^2 \qquad \text{by (5.3.8)}.$$

Thus, $\Phi\left(tI + \frac{1}{n}I\right) \xrightarrow{\|\cdot\|} \Phi(tI)$.

Next, we assume that $A_m \xrightarrow{\|\cdot\|} A$ in $\mathscr{A}$. Following, we show that $\Phi(A_m) \xrightarrow{\|\cdot\|} \Phi(A)$:

Since the sequence (A_m) is bounded, there exists a positive real number t_0 such that $A_m^* A_m \le \|A_m\|^2 I \le t_0 I$ for all m. Take $t := \max\{t_0, \|A\|^2, 1\}$. Due to the fact that the sequence $\left(\Phi\left(tI + \frac{1}{n}I\right)\right)$ is norm convergent to $\Phi(tI)$, there is $N > 0$ such that

$$\left\|\Phi\left(tI + \frac{1}{n}I\right) - \Phi(tI)\right\| < \frac{\varepsilon^2}{\|\Phi(2I) - \Phi(A)\Phi(tI)^{-1}\Phi(A)^*\|} \qquad (5.3.9)$$

for every $n > N$. Fix $n_0 > N$. By utilizing a similar argument as in *Step 3* of the proof of Theorem 5.3.2, we see that $\|\Phi(2I) - \Phi(A)\Phi(tI)^{-1}\Phi(A)^*\| > 0$ and there exists some $M > 0$ such that for every $m > M$, the matrix

$$X_m = \begin{bmatrix} tI + \frac{1}{n_0}I & A_m^* & tI \\ A_m & 2I & A \\ tI & A^* & tI \end{bmatrix}$$

is positive. It follows from the 3-positivity of Φ that

$$\begin{bmatrix} \Phi\left(tI + \frac{1}{n_0}\right) & \Phi(A_m)^* & \Phi(tI) \\ \Phi(A_m) & \Phi(2I) & \Phi(A) \\ \Phi(tI) & \Phi(A)^* & \Phi(tI) \end{bmatrix} \ge 0 \qquad (m > M).$$

Hence,

$$\begin{bmatrix} \Phi\left(tI + \frac{1}{n_0}I\right) & \Phi(A_m^*) \\ \Phi(A_m) & \Phi(2I) \end{bmatrix} \ge \begin{bmatrix} \Phi(tI) & \Phi(A^*) \\ \Phi(A) & \Phi(A)\Phi(tI)^{-1}\Phi(A)^* \end{bmatrix} \qquad (m > M).$$

Using the 2-positivity of the operator norm, we reach

$$\begin{bmatrix} \left\|\Phi\left(tI + \frac{1}{n_0}I\right) - \Phi(t)\right\| & \left\|\Phi(A_m^*) - \Phi(A^*)\right\| \\ \|\Phi(A_m) - \Phi(A)\| & \|\Phi(2I) - \Phi(A)\Phi(tI)^{-1}\Phi(A)^*\| \end{bmatrix} \ge 0 \qquad (m > M),$$

which implies that

$$\|\Phi(A_m) - \Phi(A)\|^2 \leq \|\Phi(2I) - \Phi(A)\Phi(tI)^{-1}\Phi(A)^*\|$$
$$\times \left\|\Phi\left(t + \frac{1}{n_0}I\right) - \Phi(t)\right\|$$
$$< \varepsilon^2 \text{ by inequality (5.3.9)}$$

for every $m > M$. Hence, $\|\Phi(A_m) - \Phi(A)\| < \varepsilon$. $\qquad\square$

5.4 Exercises and Problems

Exercise 5.4.1 A map $\Phi : \mathscr{A} \to \mathscr{B}$ is called an *n-monotone map* if Φ_n is monotone on $\mathbb{M}_n(\mathscr{A})_+$ in the sense that

$$[A_{ij}] \geq [B_{ij}] \geq 0 \text{ implies } \Phi_n([A_{ij}]) \geq \Phi_n([B_{ij}]).$$

Show that if $\Phi : \mathscr{A} \to \mathscr{B}$ is a 2-monotone map between C^*-algebras and if $\Phi(C) = \Phi(0)$ for some $C > 0$, then $\Phi(A) = \Phi(0)$ for each $A \in \mathscr{A}$.

Hint: See [58].

Exercise 5.4.2 Prove or disprove that n-monotonicity of a map Φ between C^*-algebras implies $(n+1)$-positivity.

Problem 5.4.3 Show that if $\alpha \geq n - 2$, then the power function $f : \mathbb{R} \to \mathbb{R}$ given by $f(x) = |x|^\alpha$ is an n-positive map in the sense that if $[a_{ij}]_{i,j=1}^n$ is a positive matrix, then the matrix $[|a_{ij}|^\alpha]_{i,j=1}^n$ is positive. Find other nontrivial functions f that have this property.

Hint: See [183].

Problem 5.4.4 Let $\Phi : \mathscr{A} \to \mathscr{B}$ be a 2-monotone map between unital C^*-algebras. Prove that $\Phi : \mathscr{A} \to \mathscr{B}$ is norm-continuous. Does this statement hold for n-positive maps for some specific values $n \geq 2$?

Hint: See [58].

Problem 5.4.5 Are 3-positive maps acting on C^*-algebras norm-continuous?

Hint: Refer to Proposition 5.3.3.

General Notation

Notation	Description	Section
$\mathbb{N}$	Set of natural numbers	1.1
$\mathbb{R}$	Set of real numbers	1.1
$\mathbb{C}$	Set of complex numbers	1.1
$(\mathcal{X}, \|\cdot\|)$	Normed space	1.1
$\mathcal{X}'$	Dual of a normed space $\mathcal{X}$	1.1
$(\mathcal{H}, \langle\cdot,\cdot\rangle)$	Inner product space	1.1
$\mathcal{M}^\perp$	Orthogonal complement of $\mathcal{M} \subseteq \mathcal{H}$	1.1
$\mathscr{A}, \mathscr{B}, \ldots$	C^*-algebra	1.1
(Ω, μ)	Measure space	1.1
$\mathbb{B}(\mathcal{H})$	Algebra of bounded linear operators on a Hilbert Space	1.1
$\mathbb{K}(\mathcal{H})$	Algebra of compact operators on a Hilbert space	1.3
$\mathcal{C}_p$	p-Schatten class	1.3
$\mathbb{M}_n$	Algebra of $n \times n$ complex matrices	1.1
$\simeq$	$*$-isomorphism between C^*-algebras	1.1

$\|\cdot\|$	Operator norm	1.1
$\|\cdot\|_\infty$	Supremum norm	1.1
$\|\|\cdot\|\|$	Unitarily invariant norm	1.3
$\|\cdot\|_{(k)}$	Ky Fan k-norm	1.3
$\|\cdot\|_p$	p-Schatten norm	1.3
$r, s, t, \ldots$	Real number	1.1
$z, w, \ldots, \alpha, \beta, \ldots, \lambda, \ldots$	Complex number	1.1
$x, y, \ldots$	Element of a vector or normed space	1.1
$x, y, \ldots$	Element of a Hilbert space	1.1
$A, B, \ldots, T, S, \ldots$	Matrix or operator	1.1
(e_n)	Orthonormal basis	1.1
$A = [A_{ij}]$	Matrix A with the entries A_{ij}	1.1
A^*	Adjoint of a matrix or operator A	1.1
A^t	Transpose of a matrix A	1.1
$A^{1/2}$	Positive square root	1.1
$A^\dagger$	Moore–Ponrose inverse of A	5.1
$|A|$	Absolute value of A	1.1
$f(A)$	Functional calculus applied to a function f	1.1
$A \geq 0$	Positive operator or positive semidefinite matrix A	1.1
$A > 0$	Strictly positive operator or positive definite matrix A	1.1
$\leq$	Löwner order	1.1
$\{E_\lambda\}_{\lambda \in \mathbb{R}}$	Spectral family	1.1
$(E_{ij})_{1 \leq i,j \leq n}$	System of matrix units	1.1

5.4 Exercises and Problems

ran(A)	Range of A	1.1
ker(A)	Kernel of A	1.1
tr(A)	Trace of A	1.1
ReA (ImA)	Real (imaginary) part of A	1.1
det(A)	Determinant of a matrix A	2.1
$W(A)$	Numerical range of A	1.1
$w(A)$	Numerical radius of A	1.1
$r(A)$	Spectral radius of A	1.1
sp(A)	Spectrum of A	1.1
$\lambda_j(A)$	The jth eigenvalue of A	1.1
$s_j(A)$	The jth singular value of A	3.1

diag($\lambda_1, \ldots, \lambda_n$)	Diagonal matrix with the entries $\lambda_1, \ldots, \lambda_n$	1.1
I_n	Identity matrix of order n	1.1
I	Identity operator (matrix)	1.1
$f, \phi, \tau, \ldots$	Bounded linear functional	1.1
Φ	Map between C^*-algebras	1.1
π	Homomorphism	1.1
$[A, B]$	Commutator of operators A and B	2.7
$\{A, B\}$	Jordan product of operators A and B	4.4
$A \circ B$	Hadamard product of matrices A and B	1.1
cov(A, B)	Covariance of A and B	4.1
var(A)	Variance of A	4.1
$\Delta(A)$	$\|\cdot\|$-distance of A from the scalar operators	4.1

$A\sigma B = \Phi_{\tilde{f}}(A, B)$	Operator mean of positive operators A and B	3.2
OM_+	Set of positive operator monotone functions on $(0, \infty)$	3.2
∇_α	Weighted arithmetic mean	3.2
$\#_\alpha$	Weighted geometric mean	3.2
$!_\alpha$	Weighted harmonic mean	3.2
Corr$^\alpha_\rho(A, B)$	One-parameter correlation	4.3
$I^\alpha_\rho(A)$	One-parameter Wigner–Yanase skew information	4.3

Bibliography

1. Abu-omar, A., Kittanneh, F.: Numerical radius inequalities for $n \times n$ operator matrices. Linear Algebra Appl. **468**, 18–26 (2015)
2. Amyari, M., Chakoshi, M., Moslehian, M.S.: Quasi-representations of Finsler modules over C^*-algebras. J. Operator Theory **70**(1), 181–190 (2013)
3. W. N. Jr. Anderson and R. J. Duffin, *Series and parallel addition of matrices*, J. Math. Anal. Appl. **26** (1969), 576–594
4. W. N., Jr. Anderson and G. E. Trapp, *Shorted operators. II.*, SIAM J. Appl. Math. **28** (1975), 60–71
5. Ando, T.: On a pair of commutative contractions. Acta Sci. Math. (Szeged) **24**, 88–90 (1963)
6. Ando, T.: Topics on operator inequalities. In: Division of Applied Mathematics. Research Institute of Applied Electricity, Hokkaido University, Sapporo (1978)
7. Ando, T.: Majorizations and inequalities in matrix theory. Linear Algebra Appl. **199**, 17–67 (1994)
8. Ando, T., Choi, M.D.: nonlinear completely positive maps. In: Aspects of Positivity in Functional Analysis, vol. 122, pp. 3–13. North Holland Mathematical Studies (1986)
9. Ando, T., Li, C.-K., Mathias, R.: Geometric means. Linear Algebra Appl. **385**, 305–334 (2004)
10. Ando, T., Hiai, H.: Operator log-convex functions and operator means. Math. Ann. **350**, 611–630 (2011)
11. Arambašić, L., Bakić, D., Moslehian, M.S.: A treatment of the Cauchy-Schwarz inequality in C^*-modules. J. Math. Anal. Appl. **381**(2), 546–556 (2011)
12. Arveson, W.B.: Subalgebras of C^*-algebras. Acta Math. **123**, 141–224 (1969)
13. Arveson, W.B.: Interpolation problems in nest algebras. J. Funct. Anal. **20**(3), 208–233 (1975)
14. Audenaert, K.M.R.: On a norm compression inequality for $2 \times N$ partitioned block matrices. Linear Algebra Appl. **428**(4), 781–795 (2008)
15. Berberian, S.K.: Note on a theorem of Fuglede and Putnam. Proc. Am. Math. Soc. **10**, 175–182 (1959)
16. Bernal-González, L., Moslehian, M.S., Seoane Sepúlveda, J.B.: Dichotomy between operators acting on finite and infinite dimensional Hilbert spaces. J. Iran. Math. Soc. **3**(1), 33–41 (2022)
17. Bhat, B.V.R., Elliott, G.A., Fillmore, P.A. (eds.): Lectures on Operator Theory. Fields Inst. Monographs, Vol. 13 (1999)

18. Bhat, R., Osaka, H.: A factorization property of positive maps on C^*-algebras. Int. J. Quantum Inf. **18**(5), 2050019 (2020)
19. Bhatia, R.: Matrix Analysis. Springer, New York (1997)
20. Bhatia, R.: Positive Definite Matrices. Princeton University Press, Princeton (2007)
21. Bhatia, R., Choi, M.D.: Corners of normal matrices. Proc. Indian Acad. Sci. Math. Sci. **116**(4), 393–399 (2006)
22. Bhatia, R., Choi, M.D., Davis, C.: Comparing a matrix to its off-diagonal part. The Gohberg anniversary collection, Vol. I (Calgary, AB, 1988), pp. 151–164, Oper. Theory Adv. Appl., 40, Birkhäuser, Basel (1989)
23. Bhatia, R., Jain, T.: The numerical radius and positivity of block matrices. Linear Algebra Appl. **656**, 463–482 (2023)
24. Bhatia, R., Kittaneh, F.: On the singular values of a product of operators. SIAM J. Matrix Anal. Appl. **11**, 272–277 (1990)
25. Bhatia, R., Kittaneh, F.: Norm inequalities for positive operators. Lett. Math. Phys. **43**, 225–231 (1998)
26. Bhatia, R., Kittaneh, F.: Notes on matrix arithmetic-geometric mean inequalities. Linear Algebra Appl. **308**, 203–211 (2000)
27. Bhatia, R., Kittaneh, F.: The matrix arithmetic-geometric mean inequality revisited. Linear Algebra Appl. **428**(8–9), 2177–2191 (2008)
28. Bhatia, R., Davis, C.: A better bound on the variance. Am. Math. Mon. **107**(4), 353–357 (2000)
29. Bhatia, R., Sharma, R.: Some inequalities for positive linear maps. Linear Algebra Appl. **436**, 1562–1571 (2012)
30. Bhatia, R., Šemrl, P.: Orthogonality of matrices and some distance problems, Special issue celebrating the 60th birthday of Ludwig Elsner. Linear Algebra Appl. **287**(1–3), 77–85 (1999)
31. Bhunia, P., Dragomir, S.S., Moslehian, M.S., Paul, K.: Lectures on Numerical Radius Inequalities, Infosys Science Foundation Series, Infosys Science Foundation Series in Mathematical Sciences. Springer, Cham (2022)
32. Bikchentaev, A.M., Kittaneh, F., Moslehian, M.S., Seo, Y.: Trace Inequalities: For Matrices and Hilbert Space Operators, in Press
33. Böttcher, A., Spitkovsky, I.M.: A gentle guide to the basics of two projections theory. Linear Algebra Appl. **432**(6), 1412–1459 (2010)
34. Bourin, J.-C., Lee, E.-Y.: Unitary orbits of Hermitian operators with convex and concave functions. Bull. Lond. Math. Soc. **44**(6), 1085–1102 (2012)
35. Bourin, J.-C., Lee, E.-Y.: Decomposition and partial trace of positive matrices with Hermitian blocks. Internat. J. Math. **24**(1), 1350010 (2013)
36. Bourin, J.-C., Lee, E.-Y.: Eigenvalue inequalities for positive block matrices with the inradius of the numerical range. Int. J. Math. **33**(1), Paper No. 2250009, 10 pp (2022)
37. Bourin, J.-C., Lee, E.-Y., Lin, M.: On a decomposition lemma for positive semi-definite block-matrices. Linear Algebra Appl. **437**, 1906–1912 (2012)
38. Bunce, J.W.: Models for n-tuples of noncommuting operators. J. Funct. Anal. **57**(1), 21–30 (1984)
39. Cao, X.H., Guo, M.Z., Meng, B.: Semi-Fredholm spectrum and Weyl's theorem for operator matrices. Acta Math. Sin. (Engl. Ser.) **22**(1), 169–178 (2006)
40. Choi, M.-D.: Positive linear maps on C^*-algebras. Canad. J. Math. **24**, 520–529 (1972)
41. Choi, M.-D.: A Schwarz inequality for positive linear maps on C^*-algebras. Illinois J. Math. **18**, 565–574 (1974)
42. Choi, M.-D.: Completely positive linear maps on complex matrices. Linear Algebra Appl. **10**, 285–290 (1975)
43. Choi, M.-D.: Positive semidefinite biquadratic forms. Linear Algebra Appl. **12**, 95–100 (1975)

44. Choi, M.-D.: Some assorted inequalities for positive linear maps on C^*-algebras. J. Oper. Theory **4**(2), 271–285 (1980)
45. Choi, M.-D.: Almost commuting matrices need not be nearly commuting. Proc. Am. Math. Soc. **102**(3), 529–533 (1988)
46. Choi, M.-D., Hadwin, D., Nordgren, E., Radjavi, H., Rosenthal, P.: On positive linear maps preserving invertibility. J. Funct. Anal. **59**(3), 462–469 (1984)
47. Choi, M.-D., Li, C.-K.: Constrained unitary dilations and numerical ranges. J. Oper. Theory **46**(2), 435–447 (2001)
48. Choi, M.-D., Tsui, S.K.: Tracial positive linear maps of C^*-algebras. Proc. Am. Math. Soc. **87**(1), 57–61 (1983)
49. Choi, H., Choi, H., Kim, S., Lee, H.: Extension of block matrix representation of the geometric mean. J. Korean Math. Soc. **57**(3), 641–653 (2020)
50. Conway, J.B.: A Course in Operator Theory. Springer, New York (1990)
51. Cvetković-Ilić, D.S.: A note on the representation for the Drazin inverse of 2×2 block matrices. Linear Algebra Appl. **429**(1), 242–248 (2008)
52. Cvetković-Ilić, D.S., Wang, Q.-W., Xu, Q.: Douglas' + Sebestyén's lemmas = a tool for solving an operator equation problem. J. Math. Anal. Appl. **482**(2), 123599, 10 pp (2020)
53. Cvetković-Ilić, D.S.: Completion problems on operator matrices. In: Mathematical Surveys and Monographs, vol. 267. American Mathematical Society, Providence, RI (2022)
54. Cvetković-Ilić, D.S., Wei, Y.: Algebraic Properties of Generalized Inverses, Developments in Mathematics, vol. 52. Springer, Singapore (2017)
55. Dadkhah, A., Moslehian, M.S.: Grüss type inequalities for positive linear maps on C^*-algebras. Linear Multilinear Algebra **65**(7), 1386–1401 (2017)
56. Dadkhah, A., Moslehian, M.S.: Non-linear positive maps between C^*-algebras. Linear Multilinear Algebra **68**(8), 1501–1517 (2020)
57. Dadkhah, A., Moslehian, M.S.: Quantum information inequalities via tracial positive linear maps. J. Math. Anal. Appl. **447**(1), 666–680 (2017)
58. Dadkhah, A., Moslehian, M.S., Kian, M.: Continuity of positive nonlinear maps between C^*-algebras. Studia Math. **263**(3), 241–266 (2022)
59. Davis, C.: A Schwarz inequality for convex operator functions. Proc. Am. Math. Soc. **8**, 42–44 (1957)
60. Deng, C.Y., Du, H.K.: Representations of the Moore-Penrose inverse of 2×2 block operator valued matrices. J. Korean Math. Soc. **46**(6), 1139–1150 (2009)
61. Dixmier, J.: Les algébres d'opérateurs dans l'espace hilbertien, (French) Cahiers Scientifiques. Fasc. XXV Gauthier-Villars, Paris (1957)
62. Djordjević, D.S.: Perturbations of spectra of operator matrices. J. Operator Theory **48**(3), suppl., 467–486 (2002)
63. Djordjević, D.S., Kolundžija, M.Z., Radosavljević, S., Mosić, D.: A survey on 2×2 operator matrices and their applications. Zb. Rad. (Beogr.) **20**(28) (2022). Topics in Operator Theory, 7–81
64. Douglas, R.G.: On majorization, factorization, and range inclusion of operators on Hilbert space. Proc. Am. Math. Soc. **17**, 413–415 (1966)
65. Douglas, R.G.: Banach Algebra Techniques in Operator Theory, 2nd edn. Springer, New York (2003)
66. Donoghue, W.F., Jr.: Monotone Matrix Functions on Analytic Continuation. Springer, New York (1974)
67. Drury, S.W.: On a question of Bhatia and Kittaneh. Linear Algebra Appl. **437**, 1955–1960 (2012)
68. Drury, S.W.: Positive semidefiniteness of a 3×3 matrix related to partitioning. Linear Algebra Appl. **446**, 369–376 (2014)

69. Drury, S.: Principal powers of matrices with positive definite real part. Linear Multilinear Algebra **63**(2), 296–301 (2015)
70. Du, H.K., Jin, P.: Perturbation of spectrums of 2 × 2 operator matrices. Proc. Am. Math. Soc. **121**(3), 761–766 (1994)
71. Dym, H., Gohberg, I.: Extensions of band matrices with band inverses. Linear Algebra Appl. **36**, 1–24 (1981)
72. Effros, E.G., Ruan, Z.-J.: Operator Spaces, London Mathematical Society Monographs, New Series, VOL. 23. The Clarendon Press, Oxford University Press, New York (2000)
73. Enomoto, M.: Commutative relations and related topics. Surikaisekikenkyusho Kokyuroku. Kyoto Univ. **1039**, 135–140 (1998)
74. Eom, M.-H., Kye, S.-H.: Duality for positive linear maps in matrix algebras. Math. Scand. **86**(1), 130–142 (2000)
75. Eskandari, R., Fang, X., Moslehian, M.S., Xu, Q.: Pedersen-Takesaki operator equation and operator equation $AX = B$ in Hilbert C^*-modules. J. Math. Anal. Appl. **521**(1), Paper No. 126878, 14 pp (2023)
76. Fillmore, P.A.: Sums of operators with square zero. Acta Sci. Math. (Szeged) **28**, 285–288 (1967)
77. Fialkow, L.A., Salas, H.: Majorization, factorization and systems of linear operator equations. Math. Balkanica (N. S.) **4**(1), 22–34 (1990)
78. Fiedler, M.: Über eine Ungleichung für positiv definite Matrizen, (German) Math. Nachr. **23**, 197–199 (1961)
79. Fong, C.K., Holbrook, J.A.R.: Unitarily invariant operators norms. Canad. J. Math. **35**, 274–299 (1983)
80. Fong, C.K., Wu, P.Y.: Diagonal operators: dilation, sum and product. Acta Sci. Math. (Szeged) **57**(1–4), 125–138 (1993)
81. Froelich, J.: Compact operators, invariant subspaces, and spectral synthesis. J. Funct. Anal. **81**(1), 1–37 (1988)
82. Fu, X., Gumus, M.: Trace inequalities involving positive semi-definite block matrices. Linear Multilinear Algebra **70**(20), 5987–5994 (2022)
83. Fuglede, B.: A commutativity theorem for normal operators. Proc. Nat. Acad. Sci. U.S.A. **36**, 35–40 (1950)
84. Fujii, J.I.: Operator-valued inner product and operator inequalities. Banach J. Math. Anal. **2**(2), 59–67 (2008)
85. Fujii, J.I.: Moore-Penrose inverse and operator mean. Sci. Math. Jpn. **82**(2), 125–129 (On-line 2017) (2019)
86. Fujii, M.: Cauchy-Schwarz inequality and Riccati equation for positive semidefinite matrices. In: Andrica, D. et al., (ed.) Differential and Integral Inequalities. Springer, Cham. Springer Optim. Appl. **151**, 341–350 (2019)
87. Fujii, J.I., Fujii, M.: Jensen's Inequalities on Any Interval for Operators, Nonlinear Analysis and Convex Analysis, PP. 29–39. Yokohama Publ, Yokohama (2004)
88. Fujii, J.I., Fujii, M., Nakamoto, R.: Riccati equation and positivity of operator matrices. Kyungpook Math. J. **49**(4), 595–603 (2009)
89. Fujii, M., Furuta, T., Nakamoto, R., Takahasi, S.-E.: Operator inequalities and covariance in noncommutative probability. Math. Jpn. **46**(2), 317–320 (1997)
90. Fujii, M., Izumino, S., Nakamoto, R., Seo, Y.: Operator inequalities related to Cauchy-Schwarz and Hölder-McCarthy inequalities. Nihonkai Math. J. **8**(2), 117–122 (1997)
91. Fujii, M., Nakamoto, R., Seo, Y.: Covariance in Bernstein's inequality for operators. Nihonkai Math. J. **8**(1), 1–6 (1997)
92. Fujimoto, M., Seo, Y.: The Schwarz inequality via operator-valued inner product and the geometric operator mean. Linear Algebra Appl. **561**, 141–160 (2019)

93. Furuichi, S.: Inequalities for Tsallis entropy and generalized skew information. Linear Multilinear Algebras **59**(10), 1143–1158 (2011)
94. Furuta, T., Mićić Hot, J., Pečarić, J., Seo, Y.: Recent Developments of Mond-Pečarić method in operator inequalities. In: Monographs in Inequalities, vol. 4, Element, Zagreb (2012)
95. Furuta, T., Nakamoto, R.: On the numerical range of an operator. Proc. Jpn. Acad. **47**, 279–284 (1971)
96. Glimm, J.: On a certain class of operator algebras. Trans. Am. Math. Soc. **95**, 318–340 (1960)
97. Gohberg, I.C., Krein, M.G.: Introduction to the theory of linear nonselfadjoint operators. Transl. Math. Monogr, 18, Providence, R.I: Amer. Math. Soc. (1969)
98. Groetsch, C.W.: Inclusions and identities for the Moore-Penrose inverse of a closed linear operator. Math. Nachr. **171**, 157–164 (1995)
99. Guillot, D., Khare, A., Rajaratnam, B.: Complete characterization of Hadamard powers preserving Löwner positivity, monotonicity, and convexity. J. Math. Anal. Appl. **425**, 489–507 (2015)
100. Guillot, D., Rajaratnam, B.: Functions preserving positive definiteness for sparse matrices. Trans. Am. Math. Soc. **367**, 627–649 (2015)
101. Gumus, M., Liu, J., Raouafi, S., Tam, T.-Y.: Positive semi-definite 2×2 block matrices and norm inequalities. Linear Algebra Appl. **551**, 83–91 (2018)
102. Günther, M., Klotz, L.: Lieb functions and m-positivity of norms. Linear Algebra Appl. **456**, 54–63 (2014)
103. Gurvits, L.: Classical complexity and quantum entanglement. J. Comput. Syst. Sci. **69**, 448–484 (2004)
104. Ha, K.-C.: Atomic positive linear maps in matrix algebras. RIMS **34**, 591–599 (1998)
105. Hai, G., Chen, A.: The semi-Fredholmness of 2×2 operator matrices. J. Math. Anal. Appl. **352**(2), 733–738 (2009)
106. Halmos, P.R.: Two subspaces. Trans. Am. Math. Soc. **144**, 381–389 (1969)
107. Halmos, P.R.: A Hilbert Space Problem Book, 2nd edn., Graduate Texts in Mathematics, vol. 19. Encyclopedia of Mathematics and its Applications, p. 17. Springer, New York (1982)
108. Hansen, F.: An operator inequality. Math. Ann. **246**(3), 249–250 (1979/1980)
109. Hansen, F., Pedersen, K.G.: Jensen's inequality for operators and Löwner's theorem. Math. Ann. **258**, 229–241 (1982)
110. Hansen, F.: Functions of matrices with nonnegative entries. Linear Algebra Appl. **166**, 29–43 (1992)
111. Hansen, F., Pedersen, K.G.: Jensen's operator inequality. Bull. Lond. Math. Soc. **35**(4), 553–564 (2003)
112. Hansen, F., Ji, G., Tomiyama, J.: Gaps between classes of matrix monotone functions. Bull. Lond. Math. Soc. **36**, 53–58 (2004)
113. Hansen, F., Tomiyama, J.: Differential analysis of matrix convex functions. Linear Algebra Appl. **420**, 102–116 (2007)
114. Hawkins, J.B., Kammerer, W.J.: A class of linear transformations which can be written as the product of projections. Proc. Am. Math. Soc. **19**, 739–745 (1968)
115. Hiai, F.: Monotonicity for entrywise functions of matrices. Linear Algebra Appl. **431**(8), 1125–1146 (2009)
116. Hiai, F., Petz, D.: Introduction to Matrix Analysis and Applications. Universitext Springer, ChamHindustan Book Agency, New Delhi (2014)
117. Hiai, F., Yanagi, K.: Hilbert Spaces and Linear Operators. Makino Pub., Ltd. (1995)
118. Hiroshima, T.: Majorization criterion for distillability of abipartite quantum state. Phys. Rev. Lett. **91**(5), 057902, 4pp (2003)
119. Hirzallah, O., Kittaneh, F., Shebrawi, K.: Numerical radius inequalities for certain 2×2 operator matrices. Integr. Eqn. Oper. Theory **71**(1), 129–147 (2011)

120. Hochwald, S.K.: Linear algebra by analogy. Am. Math. Mon. **98**(10), 918–926 (1991)
121. Horn, R.A., Johnson, C.R.: Matrix Analysis, 2nd edn. Cambridge University Press, Cambridge (2013)
122. Hoa, D.T.: On characterization of operator monotone functions. Linear Algebra Appl. **487**, 260–267 (2015)
123. Hoa, D.T., Khue, V.T.B., Osaka, H.: A generalized reverse Cauchy inequality for matrixes. Linear Multilinear Algebra **64**(7), 1415–1423 (2016)
124. Horodecki, M., Horodecki, P., Horodecki, R.: Separability of mixed states: necessary and sufficient conditions. Phys. Lett. A **223**, 1–8 (1996)
125. Hou, J.C., Du, H.K.: Norm inequalities of positive operator matrices. Integr. Eqn. Oper. Theory **22**(3), 281–294 (1995)
126. Hua, L.-K.: Inequalities involving determinants (in Chinese). Acta Math. Sin. **5**, 463–470 (1955)
127. Jarosz, K.: When is a linear functional multiplicative? Contemp. Math. **22**, 201–210 (1999)
128. Jeribi, A.: Spectral Theory and Applications of Linear Operators and Block Operator Matrices. Springer, Cham (2015)
129. Johnson, C.R.: Partitioned and Hadamard product matrix inequalities. J. Res. Nat. Bur. Standards **83**(6), 585–591 (1978)
130. Johnson, C.R., Spitkovsky, I., Gottlieb, S.: Inequalities involving the numerical radius. Linear Multilinear Algebra **37**, 13–24 (1994)
131. Kadison, R.V.: A generalized Schwarz inequality and algebraic invariants for operator algebras. Ann. Math. **56**(2), 494–503 (1952)
132. Kadison, R.V.: On the orthogonalization of operator representations. Am. J. Math. **77**, 600–620 (1955)
133. Kadison, R.V., Pedersen, G.K.: Means and convex combinations of unitary operators. Math. Scand. **57**(2), 249–266 (1985)
134. Kadison, R.V., Ringrose, J.R.: Fundamentals of the Theory of Operator Algebras, I. Acad Press (1983)
135. King, C., Nathanson, M.: New trace norm inequalities for 2×2 blocks of diagonal matrices. Linear Algebra Appl. **389**, 77–93 (2004)
136. Kirchberg, E.: The derivation and the similarity problem are equivalent. J. Oper. Theory **36**, 59–62 (1996)
137. Kittaneh, F.: A note on the arithmetic-geometric-mean inequality for matrices. Linear Algebra Appl. **171**, 1–8 (1992)
138. Kittaneh, F.: Inequalities for commutators of positive operators. J. Funct. Anal. **250**, 132–143 (2007)
139. Kittaneh, F., Lin, M.: Trace inequalities for positive semidefinite block matrices. Linear Algebra Appl. **524**, 153–158 (2017)
140. Klotz, L., Mädler, C.: Some functions preserving positive semidefiniteness of 2×2 block matrices. Linear Algebra Appl. **507**, 68–76 (2016)
141. Kraus, K.: General state changes in quantum theory. Ann. Phys. **64**, 311–335 (1971)
142. Kubo, K., Ando, T.: means of positive linear operators. Math. Ann. **246**, 205–224 (1980)
143. Kubo, F., Nakamura, N., Ohno, K., Wada, S.: Barbour path of operator monotone functions. Far East J. Math. Sci. (FJMS) **57**(2), 181–192 (2011)
144. Kye, S.-H.: Facial structures for various notions of positivity and applications to the theory of entanglement. Rev. Math. Phys. **25**(2), 1330002, 52 pp (2013)
145. Levy, E., Shalit, O.M.: Dilation theory in finite dimensions: the possible, the impossible and the unknown. Rocky Mountain J. Math. **44**(1), 203–221 (2014)
146. Li, Y.: Extensions of some matrix inequalities related to trace and partial traces. Linear Algebra Appl. **639**, 205–224 (2022)

147. Li, C.-K., Zhang, F.: Positivity of partitioned Hermitian matrices with unitarily invariant norms. Positivity **19**(3), 439–444 (2015)
148. Lieb, E.H.: Convex trace functions and the Wigner-Yanas-Dyson conjecture. Adv. Math. **11**, 267–288 (1973)
149. Lieb, E.: Inequalities for some operator and matrix functions. Adv. Math. **20**, 174–178 (1976)
150. Lin, C.S.: On variance and covariance for bounded linear operators. Acta Math. Sin. (Engl. Ser.) **17**, 657–668 (2001)
151. Lin, M.: A completely PPT map. Linear Algebra Appl. **459**, 404–410 (2014)
152. Lin, M.: Inequalities related to 2×2 block PPT matrices. Oper. Matrices **9**(4), 917–924 (2015)
153. Lin, M.: The Hua matrix and inequalities related to contractive matrices. Linear Algebra Appl. **511**, 22–30 (2016)
154. Lin, M.: On Drury's solution of Bhatia & Kittaneh's question. Linear Algebra Appl. **528**, 33–39 (2017)
155. Lin, M., Wolkowicz, H.: Hiroshima's theorem and matrix norm inequalities. Acta Sci. Math. (Szeged) **81**(1–2), 45–53 (2015)
156. Luo, S.: Heisenberg uncertainty relation for mixed states. Phys. Rev. A **72**, 042110, pp. 3 (2005)
157. Luo, W., Moslehian, M.S., Xu, Q.: Halmos' two projections theorem for Hilbert C^*-module operators and the Friedrichs angle of two closed submodules. Linear Algebra Appl. **577**, 134–158 (2019)
158. Manuilov, V., Moslehian, M.S., Xu, Q.: Douglas factorization theorem revisited. Proc. Am. Math. Soc. **148**(3), 1139–1151 (2020)
159. Marcoux, L.W., Radjavi, H., Zhang, Y.: Around the closure of the set of commutators of idempotents in $\mathscr{B}(\mathscr{H})$: biquasitriangularity and factorisation. J. Funct. Anal. **284**(8), Paper No. 109854, 45 pp (2023)
160. Marcus, M., Watkins, W.: Partitioned hermitian matrices. Duke Math. J. **38**, 237–249 (1971)
161. Matharu, J.S., Moslehian, M.S.: Grüss inequality for some types of positive linear maps. J. Oper. Theory **73**(1), 265–278 (2015)
162. Mathias, R.: A note on: "More operator versions of the Schwarz inequality". Positivity **8**(1), 85–87 (2004)
163. Mathieu, M.: Derivations and completely bounded maps on C^*-algebras. Irish Math. Soc. Bull. **26**, 17–41 (1991)
164. McCarthy, C.A.: c_p. Isr. J. Math. **5**, 249–271 (1967)
165. Miao, J.M.: General expressions for the Moore-Penrose inverse of a 2×2 block matrix. Linear Algebra Appl. **151**, 1–15 (1991)
166. Moslehian, M.S.: On 2×2 matrices over C^*-algebras. Acta Math. Acad. Paedagog. Nyházi. (N.S.) **19**(1), 51–53 (2003)
167. Moslehian, M.S.: Trick with 2×2 matrices over C^*-algebras. Austral. Math. Soc. Gaz. **30**(3), 150–157 (2003)
168. Moslehian, M.S.: On product of projections. Arch. Math. **40**(4), 355–357 (2004)
169. Moslehian, M.S., Fujii, J.I.: Operator inequalities related to weak 2-poitivity. J. Math. Inequal. **7**(2), 175–182 (2013)
170. Moslehian, M.S., Joiţa, M., Ji, U.C.: KSGNS type constructions for α-completely positive maps on Krein C^*-modules. Complex Anal. Oper. Theory **10**(3), 617–638 (2016)
171. Moslehian, M.S., Kian, M., Xu, Q.: Positivity of 2×2 block matrices of operators. Banach J. Math. Anal. **13**, 726–743 (2019)
172. Moslehian, M.S., Kusraev, A., Pliev, M.: Matrix KSGNS construction and a Radon-Nikodym type theorem. Indag. Math. (N.S.) **28**(5), 938–952 (2017)
173. Moslehian, M.S., Muñoz-Fernández, G.A., Seoane-Sepulveda, J.B., Peralta, A.M.: Similarities and differences between real and complex Banach spaces, An overview and recent developments. Rev. R. Acad. Cienc. Exactas Fís. Nat. Ser. A Mat. RACSAM **116**(2), 88, 80 pp (2022)

174. Moslehian, M.S., Rajić, R.: A Grüss inequality for n-positive linear maps. Linear Algebra Appl. **433**(8–10), 1555–1560 (2010)
175. Moslehian, M.S., Sharifi, K., Forough, M., Chakoshi, M.: Moore-Penrose inverses of Gram operators on Hilbert C^*-modules. Studia Math. **210**(2), 189–196 (2012)
176. Murphy, G.J.: C^*-Algebras and Operator Theory. Academic Press Inc., Boston, MA (1990)
177. Nagisa, M., Watatani, Y.: Nonlinear monotone positive maps. J. Oper. Theory **87**(1), 203–228 (2022)
178. Najafi, H.: Operator means and positivity of block operators. Math. Z. **289**(1–2), 445–454 (2018)
179. Nakayama, R., Seo, Y., Tojo, R.: Matrix Hölder-McCarthy inequality via matrix geometric. Adv. Op. Theory **5**, 744–767
180. Osaka, H.: A series of absolutely indecomposable positive maps in matrix algebras. Linear Algebra Appl. **153**, 73–83 (1991)
181. Osaka, H.: Indecomposable positive maps in low dimensional matrix algebras. Linear Algebra Appl. **186**, 45–53 (1993)
182. Osaka, H., Silvestrov, S., Tomiyama, J.: Monotone operator functions on C^*-algebras. Int. J. Math. **16**(2), 181–196 (2005)
183. Osaka, H., Silvestrov, S., Tomiyama, J.: Monotone operator functions, gaps and power moment problem. Math. Scand. **100**(1), 161–183 (2007)
184. Osaka, H., Tomiyama, J.: Double piling structure of matrix monotone functions and of matrix convex functions. Linear Algebra Appl. **431**(10), 1825–1832 (2009)
185. Osaka, H., Wada, S.: Unexpected relations which characterize operator means. Proc. Am. Math. Soc. Ser. B **3**, 9–17 (2016)
186. Osaka, H., Tsurumi, Y., Wada, S.: Generalized reverse Cauchy inequality and applications to operator means. J. Math. Inequal. **12**(4), 1029–1039 (2018)
187. Osaka, H., Wada, S.: Perspectives, Means and Their Inequalities, pp. 131–178. Cham, Trends Math. Birkhäuser/Springer (2022)
188. Parrott, S.: Unitary dilations for commuting contractions. Paci c J. Math. **34**, 81–490 (1970); Pusz, W., Woronowicz, S.L.: Functional calculus for sesquilinear forms and the purification map. Rep. Math. Phys. 8(2), 159–170 (1975)
189. Paulsen, V.I.: Completely bounded homomorphisms of operator algebras. Proc. Am. Math. Soc. **92**(2), 225–228 (1984)
190. Paulsen, V.I.: Completely Bounded Maps and Dilations. Pitman Research Notes in Math, vol. 146. Longmans (1986)
191. Paulsen, V.I.: Completely Bounded Maps and Operator Algebras, Cambridge Studies in Advanced Mathematics, vol. 78. Cambridge University Press, Cambridge (2002)
192. Pedersen, G.K.: Analysis Now Graduate Texts in Mathematics, vol. 118. Springer, New York (1989)
193. Pedersen, G.K., Takesaki, M.: The operator equation $THT = K$. Proc. Am. Math. Soc. **36**, 311–312 (1972)
194. Peres, A.: Separability criterion for density matrices. Phys. Rev. Lett. **77**, 1413–1415 (1996)
195. Plastiras, J.: C^*-algeras isomorphic after tensoring. Proc. Am. Math. Soc. **66**(2), 276–278 (1977)
196. Plastiras, J.: Compact perturbations of certain von Neumann algebras. Trans. Am. Math. Soc. **234**(2), 561–577 (1977)
197. Behncke, H., Leptin, H.: C^*-algebras with a two point dual. J. Funct. Anal. **10**(3), 330–335 (1972)
198. Piani, M., Mora, C.E.: Class of positive-partial-transpose bound entangled states associated with almost any set of pure entangled states. Phys. Rev. A **75**, 012305 (2007)

199. Pisier, G.: Similarity problems and completely bounded maps. In: Springer Lecture Notes, vol. 1618 (1995)
200. Pusz, W., Woronowicz, S.L.: Functional calculus for sesquilinear forms and the purification map. Rep. Math. Phys. **8**(2), 159–170 (1975)
201. Raeburn, I., Williams, D.R.: Morita Equivalence and Continuous-Trace C^*-*Algebras*, vol. 60. Math. Survey and Monographs (1998)
202. Raïssouli, M., Moslehian, M.S., Furuichi, S.: Relative entropy and Tsallis entropy of two accretive operators. C R Acad Sci Paris Ser **I**(355), 687–693 (2017)
203. Renaud, P.F.: A matrix formulation of Grüss inequality. Linear Algebra Appl. **335**, 95–100 (2001)
204. Ruan, Z.-J.: Real operator spaces. In: International workshop on operator algebra and operator theory. Linfen (2001); Acta Math. Sin. (Engl. Ser.) **19**(3), 485–496 (2003)
205. Russo, B., Dye, H.A.: A note on unitary operators in C^*-algebras. Duke Math. J. **33**, 413–416 (1966)
206. Schrödinger, E.: About Heisenberg uncertainty relation. Proc. Russian Acad. Sci. Phys. Math. Sect. **XIX**, 293 (1930)
207. Shapiro, J.H.: Notes on the numerical range. AMS Open Notes, availabel at https://www.ams.org/open-math-n
208. Simon, B.: Loewner's theorem on monotone matrix functions. Grundlehren Math. Wiss. **354** [Fundamental Principles of Mathematical Sciences] xi+459 pp. Springer, Cham (2019)
209. Stampfli, J.G.: The norm of a derivation. Pac. J. Math. **33**, 737–747 (1970)
210. Stinespring, W.F.: Positive functions on C^*-algebras. Proc. Am. Math. Soc. **6**, 211–216 (1955)
211. Størmer, E.: Positive linear maps of operator algebras. Acta Math. **110**, 233–278 (1963)
212. Størmer, E.: Positive Linear Maps of Operator Algebras. Springer Monographs in Mathematics, Heidelberg, Springer (2013)
213. Størmer, E.: Positive linear maps on C^*-algebras. Acta Math. **110**, 233–278 (1963)
214. Størmer, E.: Decomposable positive maps on C^*-algebras. Proc. Am. Math. Soc. **86**(3), 401–404 (1982)
215. Nagy, B. Sz., Foiaş, C.: Harmonic Analysis of Operators in Hilbert Space. Akadémiai Kiadó, Budapest; North-Holland Publishing Company, Amsterdam-London
216. Takesaki, M.: Theory of Operator Algebras. I, Reprint of the first (1979) edition. Encyclopaedia of Mathematical Sciences, vol. 124. Operator Algebras and Non-commutative Geometry, p. 5. Springer, Berlin (2002)
217. Takasaki, T., Tomiyama, J.: On the geometry of positive maps in matrix algebras. Math. Z. **184**(1), 101–108 (1983)
218. Tao, Y.: More results on singular value inequalities of matrices. Linear Algebra Appl. **416**(2–3), 724–729 (2006)
219. Tarcsay, Zs.: Operator extensions with closed range. Acta Math. Hungar. **135**(4), 325–341 (2012)
220. Thompson, R.C.: A determinantal inequality for positive definite matrices. Canad. Math. Bull. **4**, 57–62 (1961)
221. Tomiyama, J.: On the difference of n-positivity and complete positivity in C*-algebras. J. Funct. Anal. **49**(1), 1–9 (1982)
222. Tretter, C.: Spectral Theory of Block Operator Matrices and Applications. Imperial College Press, London (2008)
223. Wang, G., Wei, Y., Qiao, S.: Generalized Inverses: Theory and Computations, 2nd edn., Developments in Mathematics, vol. 53. Science Press, Beijing, Springer (2018)
224. Weiss, G.: Commutators of Hilbert-Schmidt operators. II. Integr. Eqn. Oper. Theory **3**(4), 574–600 (1980)

225. Woerdeman, H.J.: Strictly contractive and positive completions for block matrices. Linear Algebra Appl. **136**, 63–105 (1990)
226. Wegge-Olsen, N.E.: K-Theory and C^*-Algebras. Oxford University Press (1993)
227. Wittstock, G.: On matrix order and convexity. In: Functional analysis: surveys and recent results, III (Paderborn, 1983), pp. 175–188. North-Holland Math. Stud., 90, Notas Mat., 94, North-Holland, Amsterdam (1984)
228. Woronowicz, S.L.: Positive maps of low dimensional matrix algebras. Rep. Math. Phys. **10**(2), 165–183 (1976)
229. Wu, P.Y.: Products of normal operators. Can. J. Math. **XL**(6), 1322–1330 (1988)
230. Wu, P.Y.: The operator factorization problems. Linear Algebra Appl. **117**, 35–63 (1989)
231. Wu, X., Huang, J., Chen, A.: Weylness of 2×2 operator matrices. Math. Nachr. **291**(1), 187–203 (2018)
232. Yamagami, S.: Cyclic inequalities. Proc. Am. Math. Soc. **118**, 521–527 (1993)
233. Yamazaki, T.: On upper and lower bounds for the numerical radius and an equality condition. Studia Math. **178**(1), 83–89 (2007)
234. Yanagi, K., Furuichi, S., Kuriyama, K.: A generalized skew information and uncertainty relation. IEEE Trans. Inf. Theo. **51**, 4401–4404 (2005)
235. Yang, Y., Leung, D.H., Tang, W.-S.: All 2-positive linear maps from $\mathbb{M}_3(\mathbb{C})$ to $\mathbb{M}_3(\mathbb{C})$ are decomposable. Linear Algebra Appl. **503**, 233–247 (2016)
236. Zhan, X.: Singular values of differences of positive semidefinite matrices. SIAM J. Matrix Anal. Appl. **22**(3), 819–823 (2000)
237. Zhan, X.: Matrix Inequalities. Lecture Notes in Mathematics, vol. 1790. Springer, Berlin (2002)
238. Zhan, X.: On some matrix inequalities. Linear Algebra Appl. **376**, 299–303 (2004)
239. Zhang, F.: Matrix Theory, Basic Results and Techniques, 2nd edn. Universitext. Springer, New York (2011)
240. Zhang, F.: Positivity of matrices with generalized matrix functions. Acta Math. Sin. (Engl. Ser.) **28**(9), 1779–1786 (2012)

Index

Symbols
∗-algebra, 5
∗-homomorphism, 5
∗-map, 130
C^*-algebra, 5
Φ-density, 181
π-derivation, 54
π-inner derivation, 54
k-copositive, 151
k-simple vector, 160
n-convex, 108
n-dimensional Euclidean space, 2
n-monotone, 108
n-monotone map, 201
n-positive, 130
r-fold tensor power, 77
(orthogonal) projection, 6

A
Absolute value, 11
Accretive, 72
Adjoint operation, 5
Ando's characterization, 117
Approximate identity, 145
Approximate point spectrum, 99
Arithmetic mean, 86
Arveson's extension theorem, 165
Atomic map, 157

B
Banach space, 1
Barbour transform, 124
Bounded below, 3
Bounded operator, 3

C
Cartesian decomposition, 15
Cauchy–Schwarz inequality, 2
Choi–Davis–Jensen theorem, 142
Choi indecomposable map, 155
Choi Inequality, 138
Choi Inequality for subnormal operators, 139
Choi inequality for unital 2-positive maps, 138
Choi–Kraus representation, 151
Classical covariance, 180
Classical variance, 180
Coisometry, 6
Commutant of a set, 9
Commutator, 80
Compact operator, 23
Completely bounded, 131
Completely copositive, 151
Completely positive, 130
Contraction, 6
Covariance–variance inequality, 167
C^*-algebra, 5

D

Decomposable, 151
Decomposition of the identity, 140
Derivavtion, 54
Diagonalizable, 45
Dilation, 43
Douglas majorization theorem, 19
Douglas Solution, 29
Drazin inverse, 30
Dual space, 1

E

Entangled, 162
Entanglement, 162
Entanglement witness, 162
Essential spectrum, 98, 99
Extended Choi inequalities, 188

F

Fiedler's inequality, 61
Final space, 6
Finite-rank operator, 23
Fredholm, 47, 96
Fuglede theorem, 53

G

Gelfand–Beurling formula, 6
Gelfand–Naimark–Segal representation, 5
Generalized CauchyâŁ"Schwarz inequality, 64
Generalized correlation, 181
Generalized covariance, 181
Generalized variance, 181
Generalized variance–covariance inequality, 181
Generalized Wigner–Yanase–Dyson skew information, 181
Geometric mean, 86
Gram matrix, 38

H

Hadamard product, 9
Halving, 52
Hansen–Pedersen–Jensen inequality, 105
Hansen's inequality, 105
Heinz means, 126
Heisenberg-type uncertainty relation, 182
Heisenberg uncertainty relation, 180
Hermitian matrix, 6
Hilbert space, 2
Hua matrix, 90
Hua's determinantal inequality, 90

I

Idempotent, 6
Imaginary part, 6
Indecomposable, 151
Initial space, 6
Inner product space, 2
Invariant subspace, 5
Irreducible representation, 5
Isometry, 6

J

Jamiolkowski–Choi isomorphism, 151
Jordan decomposition, 9
Jordan homomorphism, 146
Jordan product, 183

K

Kadison inequality, 140
Kadison's similarity problem, 54
Ky Fan dominance theorem, 24
Ky Fan norms, 24

L

Löwner–Heinz theorem, 102
Löwner order, 9
Lehmar operator mean, 126
Lieb function, 186
Lieb map, 186
Logarithmic mean, 116

M

Matrix concave, 108
Matrix convex, 108
Matrix monotone of order n, 108
Maximal entangled state, 162
McCarthy's inequalities, 78
Moore–Penrose inverse, 26
Murray–von Neumann equivalent, 10

Index

N
Negative part, 9
Nilpotent, 6
Noncommutative covariance–variance inequality, 168
Nonnegative matrix, 38
Norm, 1
Normal, 6, 173
Normal linear functional, 10
Normed space, 1
Numerical radius, 7
Numerical range, 7

O
Off-diagonal part, 14
One-parameter correlation, 180
One-parameter Wigner–Yanase skew information, 180
Operator Cauchy–Schwarz inequality (I), 122
Operator Cauchy–Schwarz inequality (II), 122
Operator Cauchy–Schwarz inequality (III), 123
Operator concave, 101
Operator connection, 112
Operator convex, 101
Operator decreasing, 101
Operator mean, 115
Operator monotone, 101
Operator norm, 3
Operator space, 129, 130
Operator system, 129
Orthogonal, 2
Orthogonal complement, 2
Orthonormal, 2

P
Parseval identity, 3
Partial isometry, 6
Partial transpose, 160
Partition of unity, 137, 143, 144
Paulsen's 2×2 matrix trick, 130
Pinching, 14
Point spectrum, 98, 99
Polar decomposition, 20
Polarization identity, 2
Positive, 8
Positive map, 130
Positive part, 9
Positive Partial Transpose (PPT), 76
Positive square root, 11
Power dilation, 43
Principal submatrix, 4
Properly infinite, 10
Pusz–Woronowicz functional calculus, 115
Putnam–Fuglede theorem, 53

Q
quantum state, 162

R
Rank-one operator, 23
Real part, 6
Reflexive, 53
Representation, 5
Reverse Cauchy–Schwarz inequality, 126
Reverse order law, 27
Riesz representation theorem, 4
Russo–Dye theorem, 22, 139

S
Schatten p-norms, 24
Schmidt number, 160
Schur complement, 58
Schur product, 9
Schur product theorem, 9
Self-adjoint, 6
Self-adjoint map, 130
Self-adjoint operator mean, 126
Semi-inner product, 2
Semi-inner product space, 2
Seminorm, 1
Separable, 162
Shorted operator, 30
Singular value decomposition, 23
Space of square integrable functions, 2
Spectral family, 10
Spectral radius, 6
Spectral representation, 10
Spectral theorem, 9
Størmer's criterion, 155
Stinespring theorem, 134
Strictly convex, 191
Strictly positive, 130
Strong operator topology, 3

Submatrix, 3
Subnormal, 139
Superadditive, 190
Symmetric operator mean, 125
System of matrix units, 3
Sz.-Nagy's dilation theorem, 45

T
Tensor product of matrices, 4
Tracial map, 180
Transpose, 5

U
Unitarily invariant norm, 24
Unitary, 6
Unitary congruence, 65
Unitary orbit, 168

V
von Neumann algebra, 9
von Neumann inequality, 45

W
Weak operator topology, 3
Weakly 2-positive, 147
Weakly invariant norm, 41
Weight of operator mean, 125
Weighted arithmetic mean, 116
Weighted geometric mean, 116
Weighted harmonic mean, 116
Weyl, 96
Weyl monotonicity principle, 26, 87
Weyl spectrum, 98, 99
Weyl's theorem, 70
Wittstock's extension theorem, 166

SPRINGER NATURE

GPSR Compliance

The European Union's (EU) General Product Safety Regulation (GPSR) is a set of rules that requires consumer products to be safe and our obligations to ensure this.

If you have any concerns about our products, you can contact us on ProductSafety@springernature.com

In case Publisher is established outside the EU, the EU authorized representative is:

Springer Nature Customer Service Center GmbH
Europaplatz 3
69115 Heidelberg, Germany

The manufacturer's authorised representative in the EU is Springer Nature Customer Service Centre GmbH, Europaplatz 3, 69115 Heidelberg, Germany. If you have any concerns regarding our products, please contact ProductSafety@springernature.com

Printed and bound by CPI Group (UK) Ltd, Croydon, CR0 4YY

25/03/2026

02078188-0019